Nanostructured Materials for the Detection of CBRN

NATO Science for Peace and Security Series

This Series presents the results of scientific meetings supported under the NATO Programme: Science for Peace and Security (SPS).

The NATO SPS Programme supports meetings in the following Key Priority areas: (1) Defence Against Terrorism; (2) Countering other Threats to Security and (3) NATO, Partner and Mediterranean Dialogue Country Priorities. The types of meeting supported are generally "Advanced Study Institutes" and "Advanced Research Workshops". The NATO SPS Series collects together the results of these meetings. The meetings are co-organized by scientists from NATO countries and scientists from NATO's "Partner" or "Mediterranean Dialogue" countries. The observations and recommendations made at the meetings, as well as the contents of the volumes in the Series, reflect those of participants and contributors only; they should not necessarily be regarded as reflecting NATO views or policy.

Advanced Study Institutes (ASI) are high-level tutorial courses to convey the latest developments in a subject to an advanced-level audience

Advanced Research Workshops (ARW) are expert meetings where an intense but informal exchange of views at the frontiers of a subject aims at identifying directions for future action.

Following a transformation of the programme in 2006 the Series has been re-named and re-organised. Recent volumes on topics not related to security, which result from meetings supported under the programme earlier, may be found in the NATO Science Series.

The Series is published by IOS Press, Amsterdam, and Springer, Dordrecht, in conjunction with the NATO Emerging Security Challenges Division.

Sub-Series

A.	Chemistry and Biology	Springer
B.	Physics and Biophysics	Springer
C.	Environmental Security	Springer
D.	Information and Communication Security	IOS Press
E.	Human and Societal Dynamics	IOS Press

http://www.nato.int/science
http://www.springer.com
http://www.iospress.nl

Series A: Chemistry and Biology

Nanostructured Materials for the Detection of CBRN

edited by

Janez Bonča
J. Stefan Institute
Ljubljana, Slovenia

and

Sergei Kruchinin
Bogolyubov Institute for Theoretical Physics, NASU
Kiev, Ukraine

Springer
Published in Cooperation with NATO Emerging Security Challenges Division

Proceedings of the NATO Advanced Research Workshop
on Nanostructured Materials for the Detection of CBRN
Kiev, Ukraine
14–17 August 2017

Library of Congress Control Number: 2018944258

ISBN 978-94-024-1306-9 (PB)
ISBN 978-94-024-1303-8 (HB)
ISBN 978-94-024-1304-5 (e-book)
DOI 10.1007/978-94-024-1304-5

Published by Springer,
Van Godewijckstraat 30, 3311 GX Dordrecht, The Netherlands

www.springer.com

Printed on acid-free paper

Preface

The proceedings of the NATO ARW "Nanostructured Materials for the Detection of CBRN" emerged as a result of many presentations and discussions between participants of the workshop held in the "Rus" Hotel, Kiev, Ukraine, in August 14–17, 2017.

The program of the workshop allowed presentations and opened discussions on several modern research topics such as new nanomaterials and sensors. There was intense discussion in the field of nanotechnologies and safety systems that include nanosensors, nanocomposite multifunctional coatings for safety systems, bionanosensors, and nanoanalyzers. On the session devoted to nanostructured materials and spintronics, the physical properties of graphene, carbon nanotubes, and new composite materials were analyzed. The latest developments in nanotechnology and measurement techniques facilitate the detection of explosives. The most promising new materials and methods for the detection of hazardous materials including explosives are carbon nanotubes, graphene, and NMR techniques.

Participants benefited from the presentations of new methods for the detection of chemical, biological, radiological, and nuclear (CBRN) agents with the use of chemical and biochemical sensors. The identification, protection, and decontamination are the main scientific and technological responses to the modern challenges of CBRN agents. The contemporary open problems of the physics of sensors include the determination of sizes of nanoparticles, identification of particles, and determination of concentrations and mobilities of nanoparticles.

We are grateful to members of the International Advisory Committee, especially to F. Peeters and B. Vlahovic, for their consistent help and suggestions.

We thank the NATO Science for Peace and Security Programme for the essential financial support, without which the meeting could not have taken place. We also acknowledge the generous support by the National Academy of Science of Ukraine, J. Stefan Institute (Ljubljana, Slovenia), and Faculty of Mathematics and Physics at the University of Ljubljana, Slovenia.

Ljubljana, Slovenia
Kiev, Ukraine
January 2018

Janez Bonča
Sergei Kruchinin

Contents

List of Contributors

F. Accattato Laboratory of Biomedical Applications, Technologies and Sensors – BATS, Department of Health Sciences, University Magna Græcia of Catanzaro, Catanzaro, Italy

O. Adiguzel Firat University, Department of Physics, Elazig, Turkey

L. Alfonta Avram and Stella Goldstein-Goren Department of Biotechnology Engineering, Ben-Gurion University of the Negev, Beer-Sheva, Israel

S. Bellucci INFN-Laboratori Nazionali di Frascati, Frascati, Italy

M. G. Bianco Laboratory of Biomedical Applications, Technologies and Sensors – BATS, Department of Health Sciences, University Magna Græcia of Catanzaro, Catanzaro, Italy

V. Boev Institute of Electrochemistry and Energy Systems, Bulgarian Academy of Sciences, Sofia, Bulgaria

M. Bondarenko Frantsevich Institute for Problems of Materials Science of NASU, Kyiv, Ukraine

Yu. Bondaruk South-Ukrainian National Pedagogical University after K.D. Ushinskij, Odessa, Ukraine

J. Briancin Institute of Geo-technics, Slovak Academy of Sciences, Kosice, Slovakia

F. Bruno SOGIN S.p.A., Rome, Italy

M. Carestia University of Rome "Tor Vergata", Rome, Italy

L. Chkhartishvili Department of Engineering Physics, Georgian Technical University, Tbilisi, Georgia

Laboratory for Boron-Containing & Composite Materials, Ferdinand Tavadze Institute of Metallurgy & Materials Science, Tbilisi, Georgia

M. Civica SOGIN S.p.A., Rome, Italy

C. D. Critello Laboratory of Biomedical Applications, Technologies and Sensors – BATS, Department of Health Sciences, University Magna Græcia of Catanzaro, Catanzaro, Italy

I. Donchev South-Ukrainian National Pedagogical University after K.D. Ushinskij, Odessa, Ukraine

M. Figge Fraunhofer Institute for Microelectronic Circuits and Systems IMS, Duisburg, Germany

L. Filevska Odessa I.I. Mechnikov National University, Odessa, Ukraine

D. Fink Departamento de Fisica, Universidad Autónoma Metropolitana-Iztapalapa, México, D.F., México

Nuclear Physics Institute, Řež, Czech Republic

A. S. Fiorillo Laboratory of Biomedical Applications, Technologies and Sensors – BATS, Department of Health Sciences, University Magna Græcia of Catanzaro, Catanzaro, Italy

H. García-Arrellano Departamento de Ciencias Ambientales, Division de Ciencias Biológicas y de la Salud, Universidad Autónoma Metropolitana-Lerma, Lerma de Villada, Municipio de Lerma, Estado de México

P. Gaudio University of Rome "Tor Vergata", Rome, Italy

S. Glenis Department of Solid State Physics, Faculty of Physics, University of Athens, Athens, Greece

A. Goehlich Fraunhofer Institute for Microelectronic Circuits and Systems IMS, Duisburg, Germany

M. Goertz Fraunhofer Institute for Microelectronic Circuits and Systems IMS, Duisburg, Germany

V. Gopeynko ISMA University, Riga, Latvia

Ventspils University College, Ventspils, Latvia

V. Grinevych Odessa I.I. Mechnikov National University, Odessa, Ukraine

H. V. Grushevskaya Physics Department, Belarusian State University, Minsk, Belarus

N. Gubareni Frantsevich Institute for Problems of Materials Science of NASU, Kyiv, Ukraine

N. Guskos Institute of Physics, Faculty of Mechanical Engineering and Mechatronics, West Pomeranian University of Technology, Al. Piastow 48, 70-311 Szczecin, Poland

V. Hnatowicz Nuclear Physics Institute, Řež, Czech Republic

G. Janusas Faculty of Mechanical Engineering and Design, Kaunas University of Technology, Kaunas, Lithuania

A. Jupe Fraunhofer Institute for Microelectronic Circuits and Systems IMS, Duisburg, Germany

S. Kahnert Fraunhofer Institute for Microelectronic Circuits and Systems IMS, Duisburg, Germany

H. Kappert Fraunhofer Institute for Microelectronic Circuits and Systems IMS, Duisburg, Germany

M. Kandyla NHRF-National Hellenic Research Foundation, Theoretical and Physical Chemistry Institute-TPCI, Athens, Greece

T. S. Kavetskyy Drohobych Ivan Franko State Pedagogical University, Drohobych, Ukraine

The John Paul II Catholic University of Lublin, Lublin, Poland

A. Kiv Department of Materials Engineering, Ben-Gurion University of the Negev, Beer-Sheva, Israel

South-Ukrainian National Pedagogical University after K.D. Ushinskij, Odessa, Ukraine

O. Kharlamov Frantsevich Institute for Problems of Materials Science of NASU, Kyiv, Ukraine

G. Kharlamova Taras Shevchenko National University of Kyiv, Kyiv, Ukraine

O. Khyzhun Frantsevich Institute for Problems of Materials Science of NASU, Kyiv, Ukraine

S. Kondratiuk Department of Theoretical Cybernetics, Taras Shevchenko National University of Kyiv, Kyiv, Ukraine

E. Kovacova Institute of Electrical Engineering Slovak Academy of Sciences, Bratislava, Slovakia

Iu. Krak Department of Theoretical Cybernetics, Taras Shevchenko National University of Kyiv, Kyiv, Ukraine

R. Krause-Rehberg Department of Physics, University Halle, Halle, Germany

S. P. Kruchinin Bogolyubov Institute for Theoretical Physics, Kiev, Ukraine

K. Kruchynin Department of Theoretical Cybernetics, Taras Shevchenko National University of Kyiv, Kyiv, Ukraine

G. G. Krylov Physics Department, Belarusian State University, Minsk, Belarus

S. B. Kryvyi V. Lashkaryov Institute of Semiconductor Physics, National Academy of Sciences of Ukraine, Kyiv, Ukraine

M. Ksibi Laboratory of Environment Engineering and Ecotechnology, ENIS, University of Sfax, Sfax, Tunisia

D. Kupka Institute of Geo-technics, Slovak Academy of Sciences, Kosice, Slovakia

M. O. Liedke Institute of Radiation Physics, Helmholtz-Zentrum Dresden-Rossendorf, Dresden, Germany

P. Livshits Fraunhofer Institute for Microelectronic Circuits and Systems IMS, Duisburg, Germany

T. Lobanova-Shunina Faculty of Mechanical Engineering, Transport and Aeronautics, Riga Technical University, Riga, Latvia

L. S. Maksimenko V. Lashkaryov Institute of Semiconductor Physics, National Academy of Sciences of Ukraine, Kyiv, Ukraine

A. Malizia University of Rome "Tor Vergata", Rome, Italy

A. Mansharipova Kazakh-Russian Medical University, Almaty, Kazakhstan

S. P. Milovanović University of Antwerp, Antwerp, Belgium

V. I. Morozov National Aviation University, Kiev, Ukraine

I. V. Morozova National Aviation University, Kiev, Ukraine

C. Moslah NHRF-National Hellenic Research Foundation, Theoretical and Physical Chemistry Institute-TPCI, Athens, Greece

Laboratory of Environment Engineering and Ecotechnology, ENIS, University of Sfax, Sfax, Tunisia

G. A. Mousdis NHRF-National Hellenic Research Foundation, Theoretical and Physical Chemistry Institute-TPCI, Athens, Greece

S. Mross Fraunhofer Institute for Microelectronic Circuits and Systems IMS, Duisburg, Germany

R. Mukhamediyev Kazakh-Russian Medical University, Almaty, Kazakhstan

H. G. Muñoz Departamento de Fisica, Universidad Autónoma Metropolitana-Iztapalapa, México, DF, México

U. Narkiewicz Department of Chemical and Environment Engineering, West Pomeranian University of Technology, Szczecin, Poland

W. Nawrocki Faculty of Electronics and Telecommunications, Poznan University of Technology, Poznan, Poland

A. Palevicius Faculty of Mechanical Engineering and Design, Kaunas University of Technology, Kaunas, Lithuania

P. Palevicius Faculty of Mathematics and Natural Sciences, Kaunas University of Technology, Kaunas, Lithuania

F. M. Peeters University of Antwerp, Antwerp, Belgium

T. Petkova Institute of Electrochemistry and Energy Systems, Bulgarian Academy of Sciences, Sofia, Bulgaria

G. Petropoulou NHRF-National Hellenic Research Foundation, Theoretical and Physical Chemistry Institute-TPCI, Athens, Greece

S. A. Pullano Laboratory of Biomedical Applications, Technologies and Sensors – BATS, Department Health Sciences, University Magna Græcia of Catanzaro, Catanzaro, Italy

M. Ragulskis Faculty of Mathematics and Natural Sciences, Kaunas University of Technology, Kaunas, Lithuania

S. P. Rudenko V. Lashkaryov Institute of Semiconductor Physics, National Academy of Sciences of Ukraine, Kyiv, Ukraine

O. Šauša Institute of Physics, Slovak Academy of Sciences, Bratislava, Slovakia

R. Sciacqua SOGIN S.p.A., Rome, Italy

B. K. Serdega V. Lashkaryov Institute of Semiconductor Physics, National Academy of Sciences of Ukraine, Kyiv, Ukraine

Yu. M. Shukrinov Join Institute for Nuclear Research, BLTP, Dubna, Russia

Yu. Shunin Institute of Solid State Physics of the University of Latvia, Riga, Latvia

Ventspils University College, Ventspils, Latvia

D. Sibera Department of Chemical and Environment Engineering, West Pomeranian University of Technology, Szczecin, Poland

P. Silenko Frantsevich Institute for Problems of Materials Science of NASU, Kyiv, Ukraine

V. Skobeeva Research Institute of Physics, I.I. Mechnikov National University, Odessa, Ukraine

V. Smatko Institute of Electrical Engineering Slovak Academy of Sciences, Bratislava, Slovakia

V. Smyntyna Odessa I.I. Mechnikov National University, Odessa, Ukraine

B. A. Snopok V.E. Lashkaryov Institute of Semiconductor Physics, National Academy of Sciences of Ukraine, Kyiv, Ukraine

O. B. Snopok V.E. Lashkaryov Institute of Semiconductor Physics, National Academy of Sciences of Ukraine, Kyiv, Ukraine

A. Sodah Faculty of Mechanical Engineering and Design, Kaunas University of Technology, Kaunas, Lithuania

U. Spezia SOGIN S.p.A., Rome, Italy

N. Srinivasan Institute of Radiation Physics, Helmholtz-Zentrum Dresden-Rossendorf, Dresden, Germany

A. L. Stepanov Kazan Physical-Technical Institute, Russian Academy of Sciences, Kazan, Russia

Kazan Federal University, Kazan, Russia

M. O. Stetsenko V. Lashkaryov Institute of Semiconductor Physics, National Academy of Sciences of Ukraine, Kyiv, Ukraine

F. Troiani SOGIN S.p.A., Rome, Italy

J. Typek Institute of Physics, Faculty of Mechanical Engineering and Mechatronics, West Pomeranian University of Technology, Szczecin, Poland

J. Vacik Nuclear Physics Institute, Řež, Czech Republic

A. Vaseashta International Clean Water Institute, Herndon, VA, USA

NJCU – State University of New Jersey, Jersey, NJ, USA

Ghitu Institute of Electronic Engineering and Nanotechnology, Academy of Sciences of Moldova, Chisinau, Moldova

J. D. Vinko ERA Physics, Velletri, Italy

B. Vlahovic North Carolina Central University, Durham, NC, USA

H. Vogt Fraunhofer Institute for Microelectronic Circuits and Systems IMS, Duisburg, Germany

A. Wagner Institute of Radiation Physics, Helmholtz-Zentrum Dresden-Rossendorf, Dresden, Germany

V. Yatsenko Space Research Institute of NASU-NSAU, Kyiv, Ukraine
Yu. Zhukovskii Institute of Solid State Physics of the University of Latvia, Riga, Latvia

G. Zolnierkiewicz Institute of Physics, Faculty of Mechanical Engineering and Mechatronics, West Pomeranian University of Technology, Al. Piastow 48, 70-311 Szczecin, Poland

Part I
Nanostructured Materials

Chapter 1
Strained Graphene Structures: From Valleytronics to Pressure Sensing

S. P. Milovanović and F. M. Peeters

Abstract Due to its strong bonds graphene can stretch up to 25% of its original size without breaking. Furthermore, mechanical deformations lead to the generation of pseudo-magnetic fields (PMF) that can exceed 300 T. The generated PMF has opposite direction for electrons originating from different valleys. We show that valley-polarized currents can be generated by local straining of multi-terminal graphene devices. The pseudo-magnetic field created by a Gaussian-like deformation allows electrons from only one valley to transmit and a current of electrons from a single valley is generated at the opposite side of the locally strained region. Furthermore, applying a pressure difference between the two sides of a graphene membrane causes it to bend/bulge resulting in a resistance change. We find that the resistance changes linearly with pressure for bubbles of small radius while the response becomes non-linear for bubbles that stretch almost to the edges of the sample. This is explained as due to the strong interference of propagating electronic modes inside the bubble. Our calculations show that high gauge factors can be obtained in this way which makes graphene a good candidate for pressure sensing.

Keywords Graphene · Pseudo-magnetic fields · Valleytronics

1.1 Introduction

Graphene is the first ever two-dimensional material to be synthesized in a laboratory [1]. Before K. Novoselov and A. Geim used mechanical exfoliation to isolate one layer of carbon atoms from a piece of graphite, it was firmly believed that two-dimensional (2D) materials can not exist in nature except as part of three dimensional structures. This belief was based on various experiments with thin films and on theoretical considerations, i.e. the Mermin–Wagner theorem [2]. These

S. P. Milovanović (✉) · F. M. Peeters
University of Antwerp, Antwerp, Belgium
e-mail: slavisa.milovanovic@uantwerpen.be; francois.peeters@uantwerpen.be

J. Bonča, S. Kruchinin (eds.), *Nanostructured Materials for the Detection of CBRN*, NATO Science for Peace and Security Series A: Chemistry and Biology,
https://doi.org/10.1007/978-94-024-1304-5_1

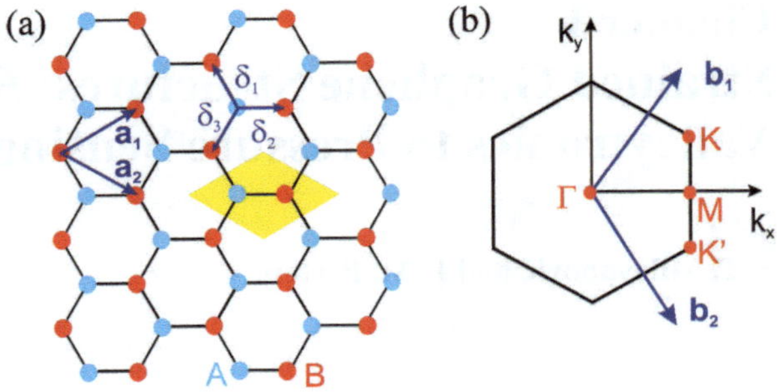

Fig. 1.1 (**a**) Graphene lattice with basis vectors $\mathbf{a}_1$ and $\mathbf{a}_2$. We also show the space vectors that connect two sublattices (A and B) $\boldsymbol{\delta}_1$, $\boldsymbol{\delta}_2$, and $\boldsymbol{\delta}_3$. Yellow rhombus represents the primitive unit cell. (**b**) First Brillouin zone with reciprocal lattice vectors $\mathbf{b}_1$ and $\mathbf{b}_2$. (Taken from Ref. [3])

experimental studies showed that the melting temperature of thin films rapidly decreases with decreasing thickness [4, 5]. Hence, it was believed that thermal fluctuations would lead to segregation and decomposition of 2D materials.

"Material that shouldn't exist" triggered a lot of attention in the scientific community. Graphene showed that the two-dimensional (2D) crystals are not only feasible but turn out to be high quality structures with unique properties and numerous potential applications. This caused an enormous interest in 2D materials and a quest for new atomic thin materials started. Dozens of them have been discovered and together with materials created by stacking different 2D materials, i.e. heterostructures, the number goes well above a hundred. It is expected that there are around 500 different two-dimensional materials [6].

Monolayer graphene consists of carbon atoms arranged in a two-dimensional honeycomb crystal structure, as shown in Fig. 1.1. The honeycomb structure consists of the triangular Bravais lattice with a basis of two atoms, labeled A and B. All the atoms from the same sublattice can be reached with primitive lattice vectors

$$\mathbf{a}_1 = \left(\frac{1}{2}, \frac{\sqrt{3}}{2}\right) a, \quad \text{and} \quad \mathbf{a}_2 = \left(\frac{1}{2}, -\frac{\sqrt{3}}{2}\right) a, \tag{1.1}$$

where a is the lattice constant and for graphene its value is $a = 2.46$ Å.

The first Brillouin zone defined by these vectors is shown in Fig. 1.1b. Of particular interest in the physics of graphene are the six corners of the first Brillouin zone which consist of three pairs of inequivalent points $\mathbf{K}$ and $\mathbf{K}'$. The position of these points in momentum space is given by

$$\mathbf{K} = \frac{4\pi}{3\sqrt{3}a_{cc}}(1, 0) \quad \text{and} \quad \mathbf{K}' = \frac{4\pi}{3\sqrt{3}a_{cc}}(-1, 0). \tag{1.2}$$

The remaining corners can be connected to one of these points via translation by a reciprocal lattice vector. Note that the $\mathbf{K}$ and $\mathbf{K}'$ points are not related to the different sublattices of the graphene lattice but are a consequence of the two-dimensional hexagonal lattice structure.

Due to its strong covalent bonds, graphene exhibits excellent mechanical properties. Enormous values of Young's modulus and intrinsic strength of a defect-free graphene sheet were reported in Ref. [7]. In this paper, Lee et al. measured the elastic properties and intrinsic breaking strength of free-standing monolayer graphene membranes through nanoindentation by an atomic force microscope (AFM). They found a Young modulus of $E \approx 1$ TPa. This value is very close to the theoretical value $E = 1.05$ TPa [8]. Experiment further showed that the brittle fracture of graphene occurs at a critical stress equal to its intrinsic strength of $\sigma_{\text{int}} = 130\,\text{GPa}$. This is the highest value ever measured for a material.

1.2 Strain Engineering

To describe graphene one can use the standard nearest-neighbor tight-binding Hamiltonian given by

$$H = \sum_i \epsilon_i c_i^\dagger c_i + \sum_{i,j} t_{ij} c_i^\dagger c_j, \tag{1.3}$$

where $c_i^\dagger (c_i)$ is the creation (annihilation) operator for an electron at site i, ϵ_i is the onsite potential at site i, and t_{ij} is the hopping energy between sites i and j. Stretching graphene results in changes of the bond length between neighboring atoms in its lattice. The new positions of the carbon atoms are given by $\mathbf{r}_i = \mathbf{r}_i^{\text{eq}} + \mathbf{u}$, where $\mathbf{r}_i^{\text{eq}}$ is the equilibrium position of atom i and $\mathbf{u}$ is the displacement field. This change results in a modification of the hopping energy given by

$$t_{ij} = t_0 e^{-\beta(d_{ij}/a_0 - 1)}, \tag{1.4}$$

where $t_0 = 2.8$ eV is the equilibrium hopping energy, $a_0 = 0.142\,\text{nm}$ is the length of the unstrained $C - C$ bond, and d_{ij} is the length of the strained bond between atoms i and j. The decay factor $\beta = \partial \log t / \partial \log a \mid_{a=a_0} \approx 3.37$ [9] is the strained hopping energy modulation factor and has been extracted from experiment. As a check, one can calculate the energy of the next-nearest neighbour in graphene which gives $t' = 0.23$ eV and agrees well with estimates obtained from other techniques [10, 11]. On the other hand, d_{ij} can be calculated from the strain tensor as

$$d_{ij} = \frac{1}{a_0} \mathbf{r}_{ij} \boldsymbol{\varepsilon} \mathbf{r}_{ij} = \frac{1}{a_0} \left(a_0^2 + \varepsilon_{xx} x_{ij}^2 + \varepsilon_{yy} y_{ij}^2 + 2\varepsilon_{xy} x_{ij} y_{ij} \right), \tag{1.5}$$

where $\boldsymbol{\varepsilon}$ is the strain tensor obtained from classical continuum mechanics and given by [12]

$$\varepsilon_{ij} = \frac{1}{2} \left(\partial_j u_i \partial_i u_j + (\partial_i u_z)(\partial_j u_z) \right), \quad i, j = x, y. \tag{1.6}$$

The spatial variation of the hopping energy is equivalent to the generation of a magnetic vector potential, $\mathbf{A} = (A_x, A_y, 0)$, which can be evaluated around the **K** point using[13]

$$A_x - \mathrm{i}A_y = -\frac{1}{ev_\mathrm{F}}\sum_j \delta t_{ij} e^{\mathrm{i}\mathbf{K}\cdot\mathbf{r}_{ij}}, \tag{1.7}$$

where the sum runs over all neighboring atoms of atom i, v_F is the Fermi velocity, $\delta t_{ij} = (t_{ij} - t_0)$, and $\mathbf{r}_{ij} = \mathbf{r}_i - \mathbf{r}_j$. We can obtain δt_{ij} by expanding Eq. (1.4) to e.g. linear order

$$t_{ij} = t_0\left(1 - \frac{\beta}{a_0}(d_{ij} - a_0)\right), \tag{1.8}$$

which using Eq. (1.5) transforms into

$$t_{ij} = t_0\left(1 - \frac{\beta}{a_0^2}\boldsymbol{\delta}_i\boldsymbol{\varepsilon}\boldsymbol{\delta}_j\right). \tag{1.9}$$

Inserting Eq. (1.9) into Eq. (1.7) one can show that the vector potential has the following form [14]

$$\mathbf{A} = -\frac{\hbar\beta}{2ea_{cc}}\begin{pmatrix}\varepsilon_{xx} - \varepsilon_{yy}\\ -2\varepsilon_{xy}\end{pmatrix}. \tag{1.10}$$

Equation (1.10) leads to an effective low-energy Hamiltonian ($\mathbf{k} = \mathbf{K} + \mathbf{q}$ and $|\mathbf{q}| \to 0$) given by

$$H_{\pm\mathbf{K}}(\mathbf{q}) = v_\mathrm{F}\boldsymbol{\sigma}(\mathbf{q} \pm e\mathbf{A}). \tag{1.11}$$

The pseudo-magnetic field (PMF) is then obtained as

$$\mathbf{B}_{ps} = \nabla \times \mathbf{A} = \left(0, 0, \partial_x A_y - \partial_y A_x\right) = \left(0, 0, B_{ps}\right). \tag{1.12}$$

Here we use the subscript "*ps*" to differentiate between the pseudo-magnetic field generated by strain from the applied external magnetic field. It is important to mention that the PMF calculated for the **K'** point has the opposite direction compared to the one in the **K** point (see Eqs. (1.11)).

It is obvious from Eq. (1.7) that different strain configurations result into different profiles of the pseudo-magnetic field. This is illustrated in Fig. 1.2 where we show a few PMF profiles (bottom panel) generated by displacement fields shown in the top panel of the same figure. Guinea et al. showed that a triaxial (triagonal) stretch results into a quasi-uniform PMF [15] (See Fig. 1.2a). A similar type of field is generated if one bends a graphene nanoribbon as in Ref. [16]. Unfortunately,

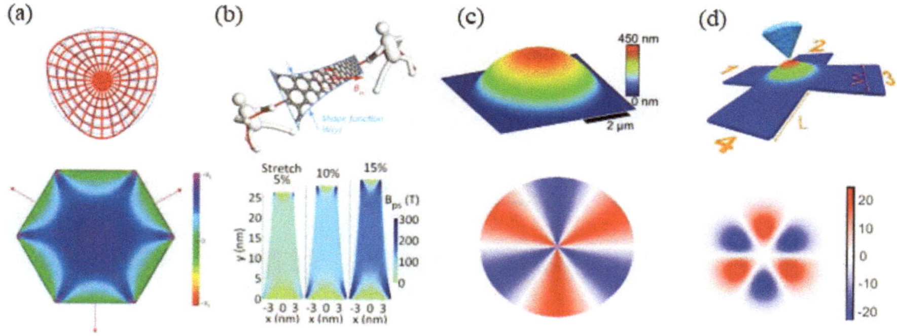

Fig. 1.2 (**a–d**) Different strain configurations (top panel) with the corresponding PMF profiles (bottom panel). (Figures taken from Refs. [15, 17, 19, 20])

uniaxial stretch does not produce a pseudo-magnetic field. However, Zhu et al. showed that in the case of graphene ribbons of a non-uniform width uniaxial stretch will result in a quasi-uniform PMF profile [17]. This is shown in Fig. 1.2b. The first experimental observation of the generation of Landau levels by a pseudo-magnetic field in graphene was reported in Ref. [18]. In their experiment, authors grew graphene on a Pt(111) substrate and found that a few triangular bubbles appear where graphene is stretched. In the strained region, Levy et al. measured dI/dV curves that exhibit well-defined Landau levels. From the Landau levels spacing, authors were able to extract a value of the magnetic field which was higher than 300 T. Obtaining a real magnetic field of such a high intensity in a laboratory is probably an impossible task.

On the other hand, most of the out-of-plane deformations, e.g. bumps and bubbles, result in a non-uniform distribution of the PMF. Numerous studies [19–25] showed that these types of strain result in alternating regions of positive and negative magnetic field, as shown in Fig. 1.2c, d. Of course, all the pseudo-magnetic field profiles shown in Fig. 1.2 are calculate for one valley (at **K** point). The PMF in the other valley (at $\mathbf{K}'$ point), is similar to the one shown in Fig. 1.2 with the difference that positive and negative regions are now switched. Hence, half of the electrons feel magnetic field in one direction and the other half feel a magnetic field of the same strength but with opposite sign. Thus, the total average magnetic field in the whole structure is zero.

In a recent paper by Jones and Pereira, a numerical recipe was given how to design a particular PMF profile by patterning the substrate on which a graphene layer is deposited [26]. When external pressure is applied, graphene stretches following the topology of the substrate. The presented method relies on a PDE (partial differential equations) – constrained optimization strategy to minimize the generic objective function which penalizes significant deviations between the induced and target PMFs [26]. In Fig. 1.3 we show the results of their calculations. Left column shows targeted PMF. Middle column shows the field profile calculated

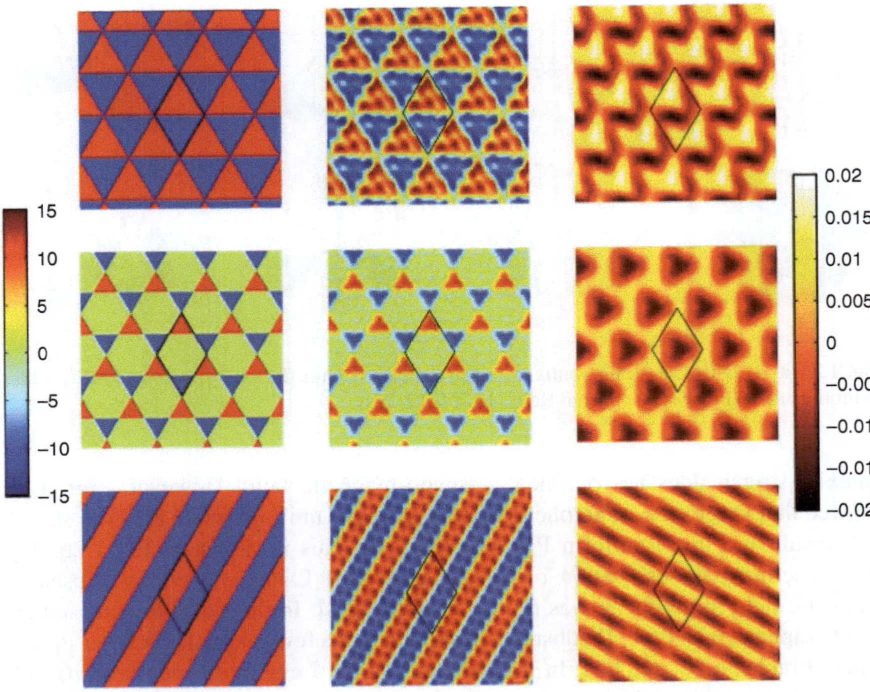

Fig. 1.3 Patterning a substrate can be used to engineer various PMF profiles. Left panel: targeted PMF; middle panel: calculated PMF profile; right panel: required substrate topology. Left color scale: PMF value in Tesla. Right color scale: out-of-plane deformation in units of 10 nm. (Figure taken from Ref. [26])

from the deformation field shown in the right column. Recently, those authors and co-workers used the same technique to optimize a graphene Corbino device for valley filltering applications [27].

1.3 Valleytronics

After the discovery of graphene, one of the major tasks was to use this material for electronic circuitry. Low dimensionality, high conductivity, and electron mobility makes this material ideal for high-frequency logic operations. Unfortunately, graphene is a semimetal. It has no band gap which severely limits its application in electronic industry. Hence, many groups tried to find a way to overcome this problem by opening a gap in the graphene spectrum. One of the most popular routes to do this is by enhancing the spin-orbit interaction in graphene [28]. Another route envolves using graphene for spintronics. Exploiting the spin degree of freedom proved to be a very challenging task. Graphene has extremely long spin diffusion length (of order of few micrometers at room temperature [29–31]) making this material suitable as a spin channel material. However, graphene is diamagnetic which means that in order to have spin-polarized current one would need either to inject it from a ferromagnetic leads or to induce spin ordering in graphene using e.g. ferromagnetic defects. Both approaches have a number of limitations [32]. In the former case, injection efficiency is strongly limited by the conduction mismatch between graphene and a ferromagnetic lead [33]. In the latter case, transport properties of graphene would be highly affected by the presence and density of adatoms and/or vacancies.

Opening a band gap in graphene and exploiting its spin degree of freedom are both very challenging tasks. Efficiency of these devices would be very low and the implentation of the above mentioned proposals require extremely complex device geometries. However, graphene has one additional degree of freedom that we didn't make use of – two valleys. Recently, there has been a lot of interest to use the valley degree of freedom to encode information. The idea is similar as for spintronics where the spin degree of freedom is used as a bit of information, however, valleytronic devices would recquire less complex device designs. Using two valleys to encode information was first proposed in Ref. [34] for the AlAs quantum point contact (QPC) structure. In this study authors were able to create valley polarized currents due to the fact that the lowest energy levels of a QPC are occupied by heavy mass states along the QPC lateral dimension. Hence, depending on the orientation of the QPC one is able to filter either electons from the Y or X valley.

The first proposal to use graphene for valleytronics was given in Ref. [35] where Rycerz et al. showed that graphene nano-constrictions can be used to create valley polarized currents. The idea relies on the fact that the lowest energy sub-band in the case of a zigzag graphene nanoribbon is valley polarized. Thus, if one tunes the Fermi level in such a way that in the region of the constriction only this sub-band is occupied electrons from different valleys would propagate in opposite direction and consequently, current becomes valley polarized (See Fig. 1.4a). By adding a gate in the constriction region one can move the Fermi region to the valence band and even completely block the current flow (valley valve). This was later on explained as a consequence of parity of the lowest mode under the switch of the sublattice, i.e. the incident and transmitted modes have opposite parity for even number of atoms across the ribbon, leading to a complete reflection, while they have the same parity for odd number of atoms, leading to a complete transmission [36].

Since then, there have been numerous different proposals to use graphene for valleytronics applications. Several of them rely on the usage of line defects in graphene. In these systems, electron transmission through the defect depends on the valley degree of freedom [37–40]. Electrons approaching the line defect are reflected or transmitted by it depending on the angle of incidence (See top part of Fig. 1.4b). Hence, for some angles transmission of electrons from one valley is completely blocked while the electrons from the opposite valley transmit without

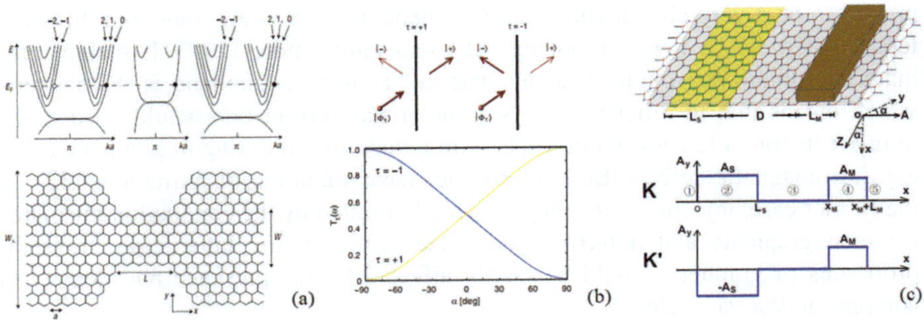

Fig. 1.4 (**a**) Energy spectrum of a graphene constriction. By tuning the Fermi level inside the constriction one is able to create valley polarized current. (Figure taken from Ref. [35]). (**b**) (Top) The sublattice symmetric $|+\rangle$ and antisymmetric $|-\rangle$ components of the incident state $|\Phi_\tau\rangle$ are transmitted and reflected, respectively. The thickness of each arrow indicates the probability the electron will follow the respective path. (Bottom) The probability that an incident electron at the Fermi level with valley index τ and angle of incidence α will transmit through the line defect. (Figure taken from Ref. [37]). (**c**) Graphene nanoribbon with a stretched region (yellow) and a deposited ferromagnetic stripe (brown region). Bottom plot shows the profile of the gauge field along the ribbon for electrons from two valleys. (Figure taken from Ref. [44])

loss, as shown in the bottom part of Fig. 1.4b. Other proposals are based on the addition of the specific mass term to the Hamiltonian that controls the valley isospin [41, 42]. Combination of strain and an external magnetic field is also a very popular choice to obtain a valley filter [43–45]. Here, the pseudo-gauge field is used to break valley degeneracy and the external magnetic field to filter electrons from one valley (See Fig. 1.4c). In the rest of this section, we discuss the most interesting valley filters that rely solely on the use of strain for the generation of valley polarized currents and the first experimental realizations of valley filters.

Controlling the edges of the sample is an extremely difficult task. Easier route to separate electrons from different valleys is by the use of strain. As mentioned in Sect. 1.2, straining the graphene lattice results in the appearance of a PMF. This field has opposite direction for electrons that originate from different valleys (See Eqs. (1.7) and (1.12)). We can use this fact to separate electrons from the two valleys. The idea is sketched in Fig. 1.5a. Current carried by both, **K** and **K**′ electrons, flows towards a bump region. Here, electrons from one valley, e.g. **K**, feel the magnetic field as plotted in Fig. 1.5a and move along the zero-PMF line towards the other end of the bump. Electrons from **K**′ valley, feel a magnetic field of opposite sign and consequently move in opposite direction, i.e. they are reflected by the bump. Hence, on the opposite side of the bump a valley-polarized current is generated. We tested this on a four-terminal structure with a bump in the middle, as shown in Fig. 1.5b. The bump has a Gaussian-like profile given by $z(\mathbf{r}) = h_0 exp(-\mathbf{r}^2/2\sigma^2)$. Current is injected at lead i and collected at the opposite side of the bump, lead c. The two perpendicular leads are added to collect any current reflected from the bump. The results of our calculations are shown in Fig. 1.5c. Here we show how the transmission of electrons from **K**, **K**′, as well as the polarization t_K

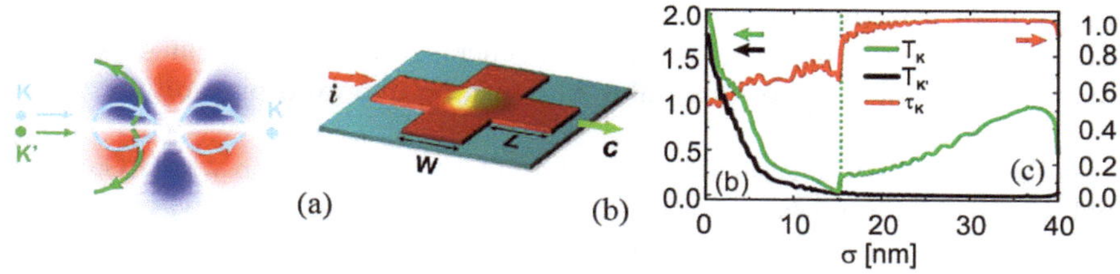

Fig. 1.5 (**a**) Sketch of flow of electrons from different valleys through the strained region. (**b**) Schematics of a cross shaped Hall bar with a bump in the center. (**c**) Transmission probabilities T_K, $T_{K'}$, τ_K versus the width of the bump, σ. Calculations are performed for $E_F = 0.1$ eV and $h_0 = 20\,\text{nm}$. (Figure taken from Ref. [46])

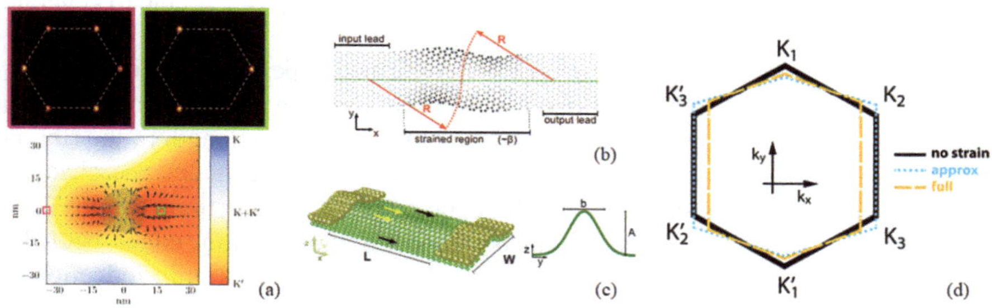

Fig. 1.6 (**a**) Top panel shows the k-space occupation of the magenta and green square from the bottom panel. Bottom panel shows local current map in the sample. (Figure taken from Ref. [47]). (**b**) Sketch of the strained graphene ribbon. (Figure taken from Ref. [48]). (**c**) Graphene nanoribbon with a Gaussian fold. (Figure taken from Ref. [49]). (**d**) Change of the first Brillouin zone of graphene with strain. The black, blue, and orange line show the unstrained case, using linear approximation of Eq. (1.7), and using Eq. (1.7), respectively. (Figure taken from Ref. [24])

($t_K = T_K/(T_K + T_{K'})$), change with the width of the bump. One notices that as σ increases the total transmission first goes down which simply means that the current is heavily scattered by the bump. However, after a certain value of σ, transmission probability T_K starts to increase while $T_{K'}$ stays low. In this regime the bump works as a valley filter. Conducting channels inside the bump open and valley polarized current is generated at the opposite side of the bump [46].

Similar observations were made in Ref. [47]. In this paper authors calculated the spectral density for each **k** value at the two opposite sides of the Gaussian bump for an infinitely extending graphene sheet. Fourier maps showed contributions from both valleys for the region before the bubble while in the region after the bubble only electrons from a single valley were detected, as shown in Fig. 1.6a. Furthermore, upon a change of the sign of the Fermi energy one can switch between the two valleys in the filtered region.

Cavalcante et al. studied snake states transport in a graphene nanoribbon with a bended region as in Ref. [15]. This type of strain, shown in Fig. 1.6b, generates a (pseudo-) magnetic barrier which allows snake state transport along the length of the ribbon. In their study (Ref. [48]), authors use the fact that electrons from opposite valleys propagate in opposite direction to generate valley dependent snake state transport. The efficiency of this valley filter can be improved by carefully tuning the Fermi energy.

In Ref. [49] Carillo–Bastos et al. used a graphene nanoribbon with a Gaussian fold, as shown in Fig. 1.6c. Enhanced LDOS around the fold region was noticed as a consequence of the sublattice symmetry breaking. LDOS plots revealed a current split in this region. Namely, part of the current along the center of the strained fold was coming from states from one valley. The region around sides of the fold, on the other hand, is filled with states from the other valley. Thus, using a fold-like structure one is able to spatially separate electrons from different valleys. If edge disorder is present in the structure, the edge states as well as states further away from the center of the ribbon would be destroyed. In this way, one is able to generate valley polarized currents at a given point [49].

The results of Rycerz et al. were recently confirmed experimentally in Ref. [50] where authors used a graphene ribbon with a non-uniform width (in)conjunction with a p–n junction. The p–n interface can be moved along the length of the ribbon by means of electrostatic gating. Authors observed that the conductance changed as the p–n interface crosses the point where the width of the junction change. They argue that this is due to the change of the isospin configuration of the edges. Namely, if the bottom and the top edge of the ribbon have the same isospin, reflection is forbidden. This will change as the p–n interface is moved along the ribbon and a drop in the conductance will occur. A set of oscillations (of order of e^2/h) appear in the conductance plot which are sensitive to the position of the p–n interface but do not change with doping or magnetic field. This suggests that these conductance oscillations are the first experimental confirmation of the effect of the local isospin configuration of the edges on transport in graphene.

Recently, Georgi at al. observed sublattice polatization in a strained graphene sample [51]. Authors used an STM tip to locally deform the graphene lattice and at the same time map changes in the LDOS. By carefully tuning the strain they were able to observe up to 50% difference in the LDOS at different sublattices. Although, sublattice polarization itself does not directly imply valley polarization, results of Georgi et al. showed that the PMF generated by the STM tip indeed shows regions with positive and negative pseudo-magnetic field. Furthermore, the strength of this field is large enough to cause valley dependent deflection of electrons. Hence, this result represents an important advance towards the experimental realization of valley filters.

1.4 Strain Sensing

Straining graphene locally results in a change of the global electronic properties of the system, e.g. resistance changes. This is due to the fact that the **K** point changes position [52] in the strained region resulting in electron scattering due to the

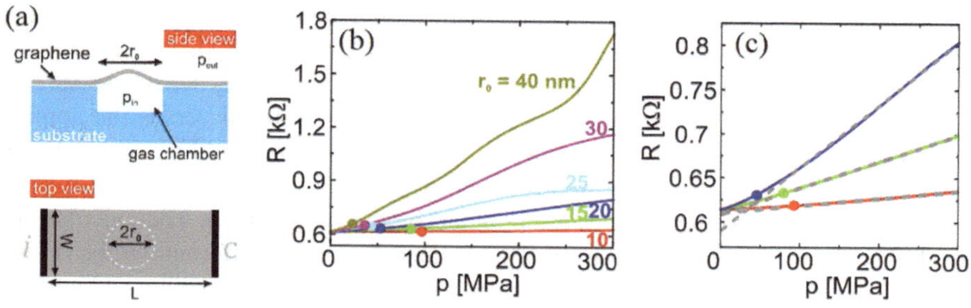

Fig. 1.7 (**a**) Schematics of a graphene pressure sensor. (**b**) Resistance versus applied pressure for different values of the bubble radius, r_0. Circles show the pressure at which the bubble height reaches 10% of its radius (membrane model is not valid below this value [57]). (**c**) The resistances for the three smallest bubbles from (**b**) together with the corresponding linear fits (gray dashed lines). (Figure taken from Ref. [58])

mismatch of the Dirac points in the strained and unstrained region (See Fig. 1.6d). The efficiency of electron scattering is proportional to the amount of applied strain. This provides the possibility to use a resistance measurement to calculate the amount of strain. Furthermore, if straining of the graphene sheet is a consequence of external pressure we can use such a resistance measurement to get information about the pressure, i.e. realize a pressure sensor.

Graphene membranes are ideal candidates for pressure sensing applications due to their large stretchability and their electrical conducting properties. Graphene is an impermeable membrane for almost all standard gases including helium [53–55]. It clamps firmly to almost all substrates achieving extremely large adhesion energies [19]. Kleshtanova et al. showed that graphene membranes with small radius (less than 10 nm) can sustain up to 1 GPa of pressure without rapture [56]. This makes graphene a suitable candidate for extremely high pressure sensing applications.

To test this we calculated the resistance of a graphene membrane shown in Fig. 1.7a. The system consists of a substrate with a small chamber of radius r_0 etched in it. Over the substrate a graphene layer is deposited. Thus, above the etched part the graphene sheet is not supported by a substrate, i.e. behaves as a free-standing membrane of radius r_0. The difference between the pressure inside and outside the chamber causes a bending/bulging of the graphene membrane. This will generate a pseudo-magnetic field and electrons will be scattered by the strained region. All relevant formulas regarding the problem of straining circular graphene membranes are given in Ref. [57]. We assume that due to the strong adhesion energy graphene clamps firmly to the substrate, hence, the bending stiffness of graphene can be neglected. The resulting PMF has a three-fold symmetry with alternating positive and negative regions similar to the case of Gaussian-like bumps (See Figs. 1.2c, d and 1.5a).

The results of our calculations are shown in Fig. 1.7b. Here we used $L = 2W = 200$ nm and $E_F = 0.2$ eV. Notice that for membranes of small radius the resistance changes linearly with pressure. This is clearly shown in Fig. 1.7c where we plot the resistance of the three smallest bubbles with their corresponding linear fits (gray dashed lines). Sensitivity, on the other hand, increases with the size of the bubble. For the largest bubble, $r_0 = 40$ nm, resistance increases 3 times when pressure of 300 MPa is applied. The curve shows a highly non-linear trend versus the applied pressure. This is explained by the fact that in the case of large bubbles, interference between different propagating modes occurs which differs from the case of small bubbles where only a few modes are affected by the bubble [58]. The calculations of the resistance were repeated for a few graphene stripes of different widths (in the range from 80 to 200 nm), and the results always showed a linear dependence on the pressure for bubbles with $r_0 < 0.15W$. This opens the possibility of scaling our results due to the fact that the ratio $r_0 = W$ is the important quantity for defining the behavior of the resistance [58]. Efficiency of the sensor is characterized by the gauge factor given by

$$GF = \frac{\Delta R/R}{\varepsilon}, \tag{1.13}$$

where $\Delta R/R$ is the relative change of the resistance with the strain and ε is the maximal strain in the structure. Our simulations reveal that for $r_0 = 20$ nm we obtain $GF \approx 8$, while for $r_0 = 30$ nm we have $GF \approx 18$. However, linearity is lost in the latter case.

In Ref. [59] authors measured the resistance of a suspended graphene nanoribbon in a nanoindentation experiment. Here, they used a wedge indentation tip to uniaxially strain a graphene ribbon and at the same time measure the change in the resistance. Measured Young's modulus was of the order of 335 N/m, which is in good agreement with previous experiments, and a gauge factor of $GF = 1.9$ was obtained. The low gauge factor can be due to the fact that uniaxial stretch does not generate a psuedo-magnetic field. Similar results were obtained in Ref. [60] with $GF = 1.6$. On the other hand, Lee et al. measured a gauge factor of 6.1 in their device made of printed graphene films [61]. They observe a resistance increase from ≈492 to ≈522 kΩ when the graphene film was strained up to 1%. Experiment showed that the resistance change is reproducible even after hundreds of repetitions. This value of gauge factor is higher than for conventional strain gauges based on metal alloys and thus shows a great potential for application of graphene in strain sensing.

Acknowledgements This work was supported by the Flemish Science Foundation (FWO-Vl).

References

1. Novoselov KS, Geim AK, Morozov SV, Jiang D, Zhang Y, Dubonos SV, Grigorieva IV, Firsov AA (2004) Electric field effect in atomically thin carbon films Science 306:666
2. Mermin ND, Wagner H (1966) Absence of Ferromagnetism or Antiferromagnetism in One- or Two-Dimensional Isotropic Heisenberg Models Phys Rev Lett 17:1133

3. Milovanovic SP (2017) Electronic transport properties in nano- and micro-engineered graphene structures (University of Antwerp), web: nano.uantwerpen.be/cmt/PhD_slavisa_milovanovic.pdf
4. Takagi M (1954) Electron-Diffraction Study of Liquid-Solid Transition of Thin Metal Films J Phys Soc Jpn 9:359
5. Venables JA, Spiller GDT, Hanbücken M (1984) Nucleation and growth of thin films Rep Prog Phys 47:399
6. Gibney E (2015) The super materials that could trump graphene Nature 522:274
7. Lee C, Wei X, Kysar JW, Hone J (2008) Measurement of the Elastic Properties and Intrinsic Strength of Monolayer Graphene Science 321:385
8. Liu F, Ming P, Li J (2007) Ab initio calculation of ideal strength and phonon instability of graphene under tension Phys Rev B 76:064120
9. Pereira VM, Castro Neto AH, Peres NMR (2009) Tight-binding approach to uniaxial strain in graphene Phys Rev B 80:045401
10. Castro Neto AH, Guinea F, Peres NMR, Novoselov KS, Geim AK (2009) The electronic properties of graphene Rev Mod Phys 81:109
11. Partoens B, Peeters FM (2006) From graphene to graphite: Electronic structure around the K point Phys Rev B 74:075404
12. Landau LD, Lifshitz EM (1986) Theory of elasticity – course of theoretical physics, Vol 7. Springer, Berlin
13. Pereira VM, Castro Neto AH (2009) Strain Engineering of Graphene's Electronic Structure Phys Rev Lett 103:046801
14. Vozmediano M, Katsnelson M, Guinea F (2010) Gauge fields in graphene Phys Rep 496:109
15. Guinea F, Katsnelson MI, Geim AK (2009) Energy gaps and a zero-field quantum Hall effect in graphene by strain engineering Nat Phys 6:30
16. Guinea F, Geim AK, Katsnelson MI, Novoselov KS (2010) Generating quantizing pseudomagnetic fields by bending graphene ribbons Phys Rev B 81:035408
17. Zhu S, Stroscio JA, Li T (2015) Programmable Extreme Pseudomagnetic Fields in Graphene by a Uniaxial Stretch Phys Rev Lett 115:245501
18. Levy N, Burke SA, Meaker KL, Panlasigui M, Zettl A, Guinea F, Castro Neto AH, Crommie MF (2010) Strain-Induced Pseudo–Magnetic Fields Greater Than 300 Tesla in Graphene Nanobubbles Science 329:544
19. Koenig SP, Wang L, Pellegrino J, Bunch JS (2012) Selective Molecular Sieving through Porous Graphene Nat Nanotech 7:728
20. Milovanovic SP, Peeters FM (2016) Strained graphene Hall bar J Phys Condens Matter 29:075601
21. Kim K-J, Blanter Ya M, Ahn K-H (2011) Interplay between real and pseudomagnetic field in graphene with strain Phys Rev B 84:081401(R)
22. de Juan F, Cortijo A, Vozmediano MAH, Cano A (2011) Aharonov-Bohm interferences from local deformations in graphene Nat Phys 7:810
23. Klimov NN, Jung S, Zhu S, Li T, Wright CA, Solares SD, Newell DB, Zhitenev NB, Stroscio JA (2012) Electromechanical Properties of Graphene Drumheads Science 336:1557
24. Masir MR, Moldovan D, Peeters FM (2013) Pseudo magnetic field in strained graphene: Revisited Solid State Commun 76:175–176
25. Settnes M, Power SR, Jauho A-P (2016) Pseudomagnetic fields and triaxial strain in graphene Phys Rev B 93:035456
26. Jones GW, Pereira VM (2014) Designing electronic properties of two-dimensional crystals through optimization of deformations New J Phys 16:093044
27. Jones GW, Bahamon DA, Castro Neto AH, Pereira VM (2017) Quantized Transport, Strain-Induced Perfectly Conducting Modes, and Valley Filtering on Shape-Optimized Graphene Corbino Devices Nano Lett 17:5304
28. Kane CL, Mele EJ (2005) Quantum Spin Hall Effect in Graphene Phys Rev Lett 95:226801
29. Tombros N, Jozsa C, Popinciuc M, Jonkman HT, van Wees BJ (2007) Electronic spin transport and spin precession in single graphene layers at room temperature Nature 448:571

30. Yang T-Y, Balakrishnan J, Volmer F, Avsar A, Jaiswal M, Samm J, Ali SR, Pachoud A, Zeng M, Popinciuc M, Güntherodt G, Beschoten B, Özyilmaz B (2011) Observation of Long Spin-Relaxation Times in Bilayer Graphene at Room Temperature Phys Rev Lett 107:047206
31. Dlubak B, Martin M-B, Deranlot C, Servet B, Xavier S, Mattana R, Sprinkle M, Berger C, De Heer WA, Petroff F, Anane A, Seneor P, Fert A (2012) Highly efficient spin transport in epitaxial graphene on SiC Nat Phys 8:557
32. Han W, Kawakami RK, Gmitra M, Fabian J (2014) Graphene spintronics Nat Nanotech 9:794
33. Schmidt G, Ferrand D, Molenkamp LW, Filip AT, van Wees BJ (2000) Fundamental obstacle for electrical spin injection from a ferromagnetic metal into a diffusive semiconductor Phys Rev B 62:R4790(R)
34. Gunawan O, Habib B, De Poortere EP, Shayegan M (2006) Quantized conductance in an AlAs two-dimensional electron system quantum point contact Phys Rev B 74:155436
35. Rycerz A, Tworzydło J, Beenakker CWJ (2007) Valley filter and valley valve in graphene Nat Phys 3:172
36. Akhmerov AR, Bardarson JH, Rycerz A, Beenakker CWJ (2008) Theory of the valley-valve effect in graphene nanoribbons Phys Rev B 77:205416
37. Gunlycke D, White CT (2011) Graphene Valley Filter Using a Line Defect Phys Rev Lett 106:136806
38. Liu Y, Song J, Li Y, Liu Y, Sun Q (2013) Controllable valley polarization using graphene multiple topological line defects Phys Rev B 87:195445
39. Chen J-H, Autés G, Alem N, Gargiulo F, Gautam A, Linck M, Kisielowski C, Yazyev OV, Louie SG, Zettl A (2014) Controlled growth of a line defect in graphene and implications for gate-tunable valley filtering Phys Rev B 89:121407(R)
40. Nguyen VH, Dollfus P, Charlier J-C (2016) Valley Filtering and Electronic Optics Using Polycrystalline Graphene Phys Rev Lett 117:247702
41. Moldovan D, Ramezani Masir M, Covaci L, Peeters FM (2012) Resonant valley filtering of massive Dirac electrons Phys Rev B 86:115431
42. Ramezani Masir M, Matulis A, Peeters FM (2011) Scattering of Dirac electrons by circular mass barriers: Valley filter and resonant scattering Phys Rev B 84:245413
43. Wu Z, Zhai F, Peeters FM, Xu HQ, Chang K (2011) Valley-Dependent Brewster Angles and Goos-Hänchen Effect in Strained Graphene Phys Rev Lett 106:176802
44. Zhai F, Zhao X, Chang K, Xu HQ (2010) Magnetic barrier on strained graphene: A possible valley filter Phys Rev B 82:115442
45. Fujita T, Jalil MBA, Tan SG (2010) Valley filter in strain engineered graphene Appl Phys Lett 97:043508
46. Milovanovic SP, Peeters FM (2016) Strain controlled valley filtering in multi-terminal graphene structures Appl Phys Lett 109:203108
47. Settnes M, Power SR, Brandbyge M, Jauho A-P (2016) Graphene Nanobubbles as Valley Filters and Beam Splitters Phys Rev Lett 117:276801
48. Cavalcante LS, Chaves A, da Costa DR, Farias GA, Peeters FM (2016) All-strain based valley filter in graphene nanoribbons using snake states Phys Rev B 94:075432
49. Carrillo-Bastos R, Leon C, Faria D, Latgé A, Andrei EY, Sandler N (2016) Strained fold-assisted transport in graphene systems Phys Rev B 94:125422
50. Handschin C, Makk P, Rickhaus P, Maurand R, Watanabe K, Taniguchi T, Richter K, Liu M-H, Schönenberger C (2017) Giant Valley-Isospin Conductance Oscillations in Ballistic Graphene Nano Lett 17:5389
51. Georgi A, Nemes-Incze P, Carrillo-Bastos R, Faria D, Viola Kusminskiy S, Zhai D, Schneider M, Subramaniam D, Mashoff T, Michael Freitag N, Liebmann M, Pratzer M, Wirtz L, Woods CR, Vladislavovich Gorbachev R, Cao Y, Novoselov KS, Sandler N, Morgenstern M (2017) Tuning the Pseudospin Polarization of Graphene by a Pseudomagnetic Field Nano Lett 17:2240
52. Pereira VM, Castro Neto AH (2009) Strain Engineering of Graphene's Electronic Structure Phys Rev Lett 103:046801
53. Leenaerts O, Partoens B, Peeters FM (2008) Graphene: A perfect nanoballoon Appl Phys Lett 93:193107

54. Nair RR, Wu HA, Jayaram PN, Grigorieva IV, Geim AK (2012) Unimpeded Permeation of Water Through Helium-Leak–Tight Graphene-Based Membranes Science 335:442
55. Bunch JS, Verbridge SS, Alden JS, van der Zande AM, Parpia JM, Craighead HG, McEuen PL (2008) Impermeable Atomic Membranes from Graphene Sheets Nano Lett 8:2458
56. Khestanova E, Guinea F, Fumagalli L, Geim AK, Grigorieva IV (2016) Nat Commun 7:12587
57. Yue K, Gao W, Huang R, Liechti KM (2012) Universal shape and pressure inside bubbles appearing in van der Waals heterostructures J Appl Phys 112:083512
58. Milovanovic SP, Tadić MŽ, Peeters FM (2017) Graphene membrane as a pressure gauge Appl Phys Lett 111:043101
59. Huang M, Pascal TA, Kim H, Goddard WA, Greer JR (2011) Electronic–Mechanical Coupling in Graphene from in situ Nanoindentation Experiments and Multiscale Atomistic Simulations Nano Lett 11:1241
60. Zhu S-E, Krishna Ghatkesar M, Zhang C, Janssen GCAM (2013) Graphene based piezoresistive pressure sensor Appl Phys Lett 102:161904
61. Lee Y, Bae S, Jang H, Jang S, Zhu S-E, Sim SH, Song YI, Hong BH, Ahn J-H (2010) Wafer-Scale Synthesis and Transfer of Graphene Films Nano Lett 10:490

Chapter 2
Development of Biosensors Using Carbon Nanotubes and Few Layered Graphene

S. Bellucci

Abstract The main task in the antiterrorist defense is the prevention and/or the early detection of terroristic attacks. In the case of attacks carried out using the biological warfare, e.g. living agents or toxins, DNA – biosensors yield one of the most effective tools for early diagnostics. The main goal of this paper is to review recent results obtained in the past few years, within the framework of a NATO Science for Peace project creation of a reliable, sensitive, selective and noise protected prototype of a DNA biosensor. Currently we have: quantitative understanding of the mechanisms of hybridization in the bulk and on the surface; the main ways to improve DNA sensors sensitivity and selectivity; the electric scheme of the DNA – sensor prototype.

Keywords Biosensors · Carbon nanotubes · Few layered graphene

There is a growing interest in using graphene for many applications [1, 2], ranging from novel electronic circuitry, to antennas design, electromagnetic shielding, radar absorbing materials, space applications, just to mention a few. In our nanotechnology group called NEXT, i.e. Nanoscience EXperiments for Technological applications, we synthetize nanomaterials by physical (arc discharge) as well as chemical (vapor deposition) methods (Figs. 2.1 and 2.2).

To the memory of the late Prof. Yuri Shunin (March 6, 1951 - April 5, 2018), a noble and gentle Soul, a kind Man and a deep Thinker, engaged constantly towards good things in Life and Science.

S. Bellucci (✉)
INFN-Laboratori Nazionali di Frascati, Frascati, Italy
e-mail: Stefano.Bellucci@lnf.infn.it

J. Bonča, S. Kruchinin (eds.), *Nanostructured Materials for the Detection of CBRN*, NATO Science for Peace and Security Series A: Chemistry and Biology, https://doi.org/10.1007/978-94-024-1304-5_2

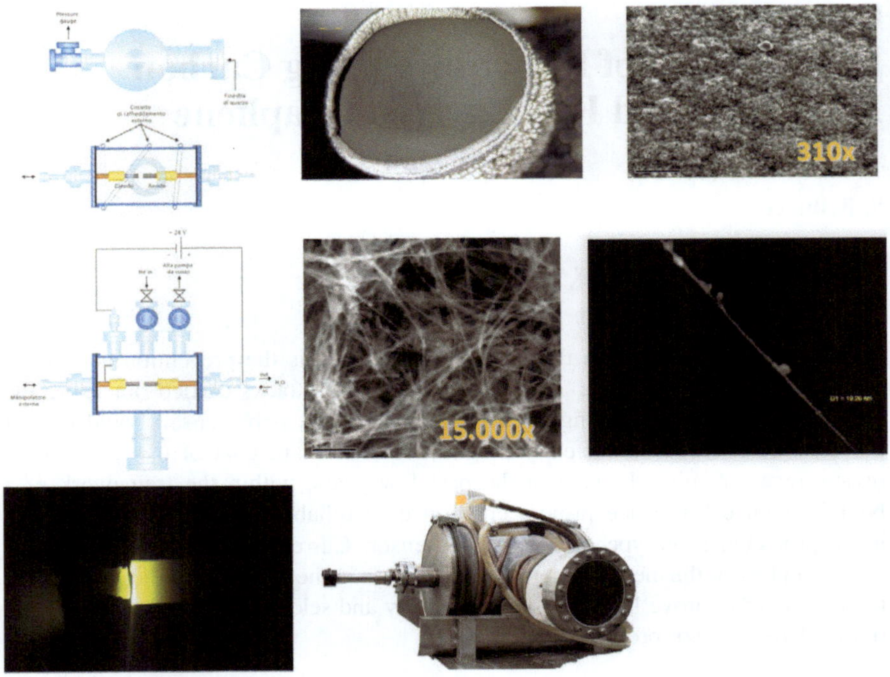

Fig. 2.1 Synthesis of nanomaterials (multiwall c arbon nanotubes) by physical (arc discharge method)

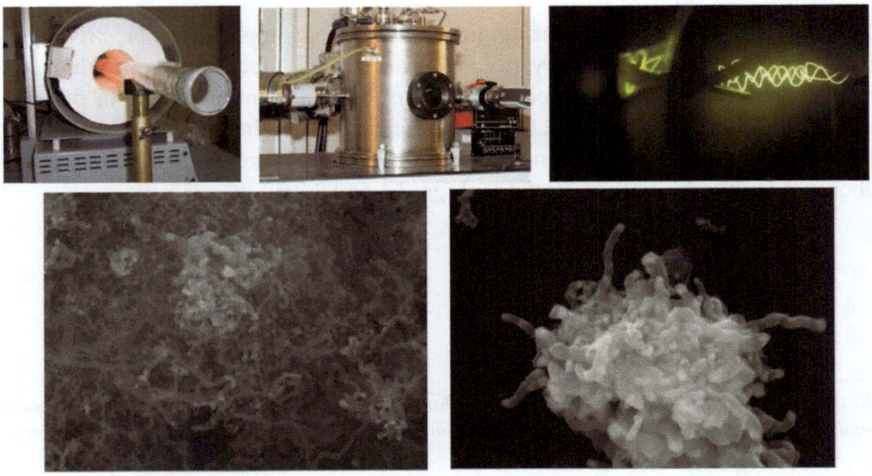

Fig. 2.2 Synthesis of nanomaterials (graphene) by (thermal and hot filament) chemical vapor deposition

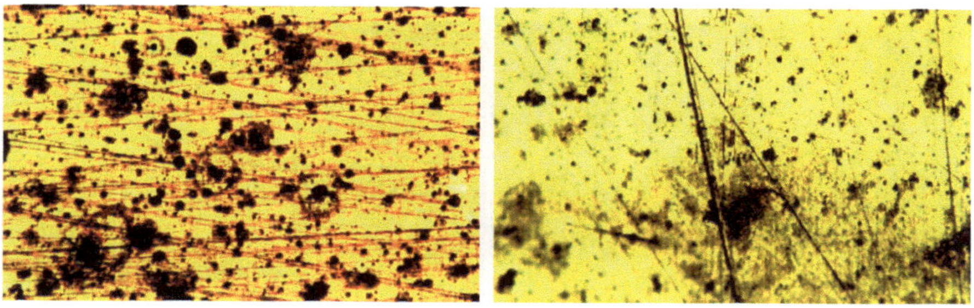

Fig. 2.3 **Graphene growth** in Chemical Vapor Deposition with acetylene and 2h annealing (before annealing shown on the left, after annealing on the right)

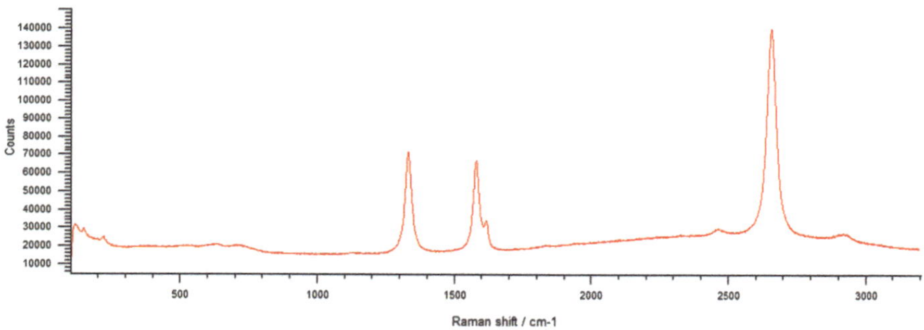

Fig. 2.4 Raman spectrum 2D FWHM $\Gamma_{2D} = 32.3\,\text{cm}^{-1}$ showing nanodomains (hundreds nm, with the presence of D and D' bands). $I_{2D}/I_G = 3.7$

The hydrocarbon gases we use are CH_4 – Methane and C_2H_2 – Acetylene. **Graphene growth** in Chemical Vapor Deposition is carried out on a common Cu slab, with 2h annealing (Fig. 2.3) with H_2.

The single layer nature of the produced graphene, in nanodomains having hundreds nm size is documented in Fig. 2.4.

An alternative method we devised to produce graphene is in the form of few-layered flakes obtained by microwaves (800 W) irradiation, starting expandable graphite generously provided by ASBURY [3–7] followed by a sonication in ultrasonic bath in water-alcohol solution for size reduction, and further enriched by centrifugation. Images (SEM and TEM) are seen in Fig. 2.5.

Raman spectra document the low density of defects (absence of a D band), seen in Fig. 2.6 (left), and the 2D band evolution (Fig. 2.6 right).

The few-layered graphene was used in electronics as a thin film to realize a tunable attenuator device made of a patch deposited in two configurations; firstly in the gap of a microstrip line (see Fig. 2.7 [9, 10]) and secondly, as a novel enhanced design, in two graphene patches located between the main microstrip line and two metal vias (see Fig. 2.8 [11]). The results show for the first configuration a wide band functionality from DC to 20 GHz, with a tunability of 7 dB and minimum insertion

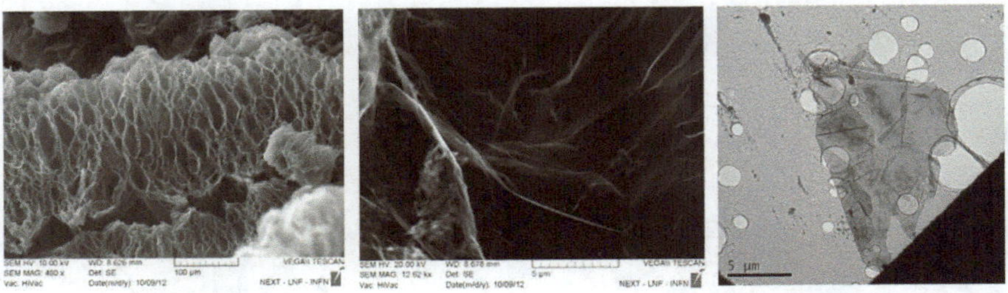

Fig. 2.5 SEM (first two panels) and TEM (last panel) of the produced flakes

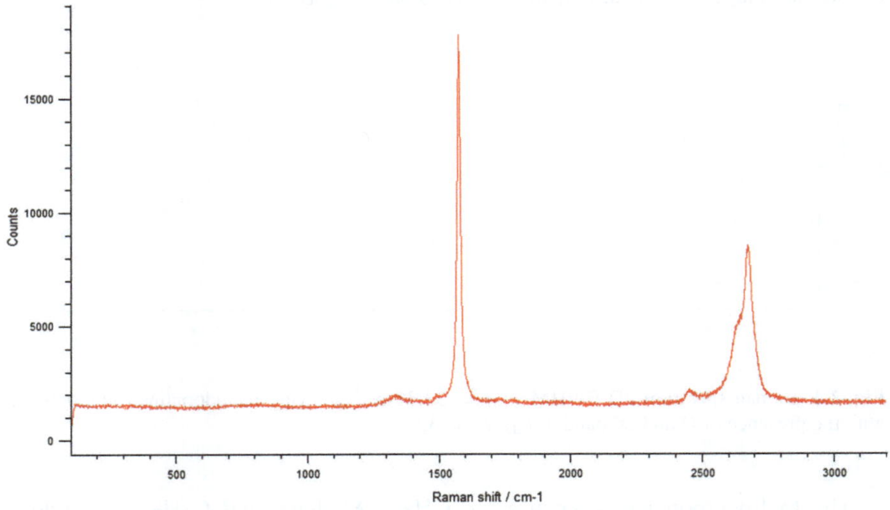

Fig. 2.6 Raman spectra with the 2D band evolution

loss of 5 dB, and for the second an operation in a frequency band of DC to 5 GHz, with 14 dB tunability and minimum insertion loss of 0.3 dB.

The Enhanced Graphene Attenuator in Fig. 2.8 shows a reduction in the graphene resistance on an increase in the bias voltage, along with lower transmission on lower resistance and vice versa, as well as a higher dissipative attenuation, in comparison with the reflective contribution.

The graphene flakes thin film was used for a tunable patch antenna [12], obtaining a change in the radiating frequency of the antenna, as well as almost 500 MHz of shift in its resonant frequency at 5 GHz, see Fig. 2.9.

One of the important directions of DNA – chips improvement is the increasing selectivity and sensitivity in expense of enhancement of electric signal and target probe hybridization stability. Efficiency of such devices as DNA-sensors and DNA-chips depends on precise prediction of experimental parameters responsible for thermal stability of nucleic acids duplexes and specific times of formation of DNA

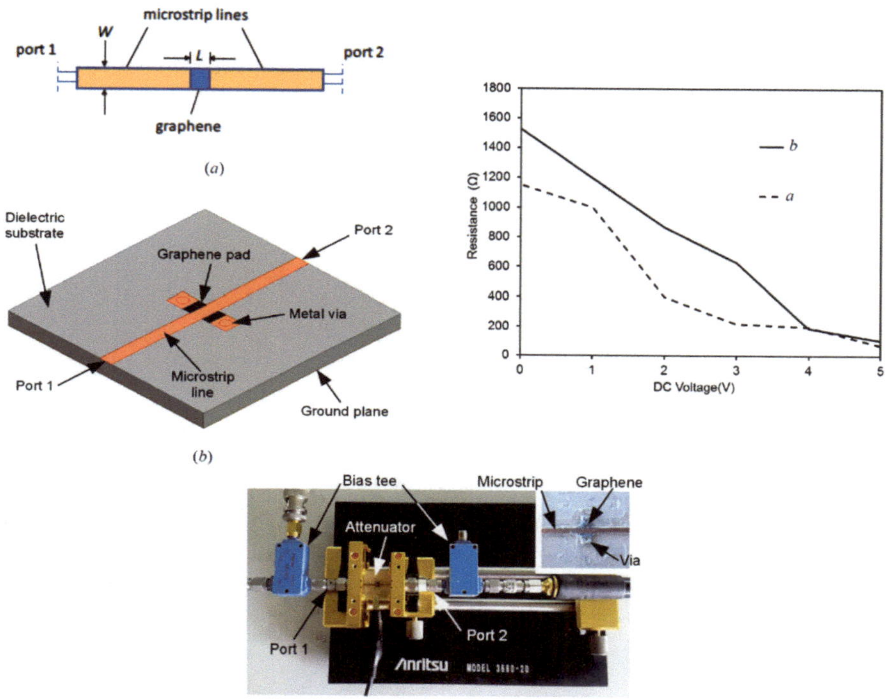

Fig. 2.7 Tunable attenuator based on graphene

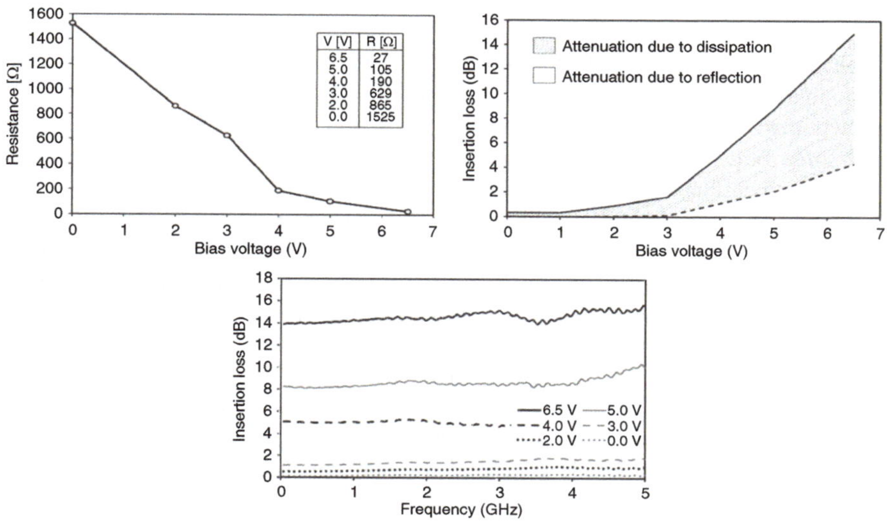

Fig. 2.8 Performance of the Enhanced Graphene Attenuator

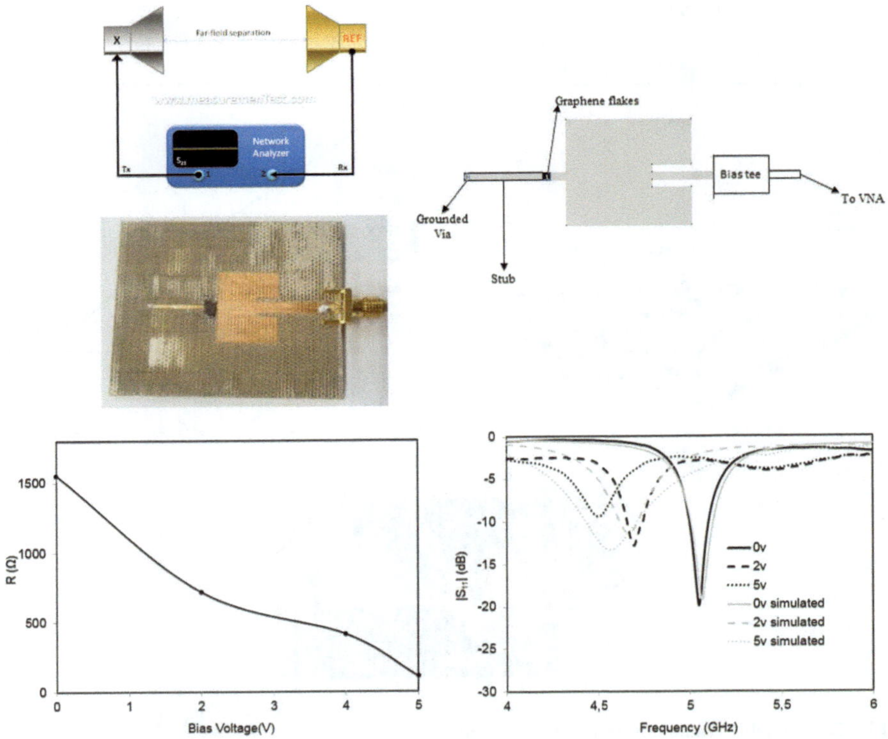

Fig. 2.9 The design and realization of a tunable patch antenna

duplexes. Some following factors influence on the thermodynamics of hybridization, in particular: the density of single-strand DNA assays (the length 25–49 nucleotides) immobilized on the surface; the presence of competing hybridization; and some other factors. Development of DNA-sensors imposes specific requirements on the effectiveness of hybridization on the interface solid-solution.

One of the main requirements to the DNA-sensors is the high sensitivity and selectivity, which in its turn, demands a maximal effectiveness of hybridization and minimal non-specific adsorption on the interface of solid and liquid phases. The nucleic acids hybridization depends substantially on the temperature, salt concentration, viscosity, GC-content and other physical-chemical features. The increase of selectivity and sensitivity of DNA-sensors can be reached by using electro-chemically active compounds with higher affinity to the dsDNA than to the ssDNA. This kind of compounds can substantially increase the dsDNA stability and at the same time, the amplitude of generated signal, which increases the DNA-sensor sensitivity.

The crystallization of polymers is a way of modification of parameters for the polymeric systems and composite materials. We plan to use it in the frameworks of the current project to control the features of composite electrodes. As a result of the crystallization only a part of macromolecules are involved in crystallite formation, while remained parts, such as folds, macromolecules pass through crystallites and the ends of chains constitute an amorphous phase of inter crystalline sheet. The formed amorphous-crystalline structure is in a metastable state, whose properties, in general, are controlled b kinetic effects. It is important to mention that the crystallization temperature of polymers in general is lower than melting temperature. Supercooling of polymer in laboratory conditions may be carried even at 100 K.

It is known that the process of crystallization begins in the so-called crystallization centers, which are located near different types of heterogeneities and structural defects. One may assume that in some cases the mixture of the nanoparticles and flexible chain polymers may manifest as this kind of defects. Our project is focused mainly on the carbon nanotubes, as nanofillers for the polymer composites. Carbon nanotubes (CNTs) can be produced in a number of different ways. One usually distinguishes two kinds of carbon nanotubes, referred to as single-walled nanotubes (SWNTs) and multi-walled nanotubes (MWNTs). This means, that the transfer of the SWCNT from the flexible chain's melt to the crystallite is thermodynamically favored and indicated the possibility of the nanohybrid shish–kebab (NHSK) structures formation around SWCNTs as centers of crystallization for polymers. The predicted NHSK structure is a unique polymer single crystal/CNT hybrid material, which was experimentally observed in Fig. A1.1, reported in [13].

Electrochemical sensing using composite coatings based on carbon nanostructures, were obtained by a modification of screen printed electrodes (SPE) with nanocomposite materials made of epoxy resin as a matrix and 10% of different fillers, from carbon nanotubes (CNT) to graphene nanoplatelet (GNP) flakes (see Figs. 2.10, 2.11 and 2.12).

Figure 2.12 It is seen that the SPE modified with Epoxy-CNT-NH_2 10% and Epoxy-CNT-NH_2 10% are far more sensitive than SPE unmodified, see Fig. 2.13.

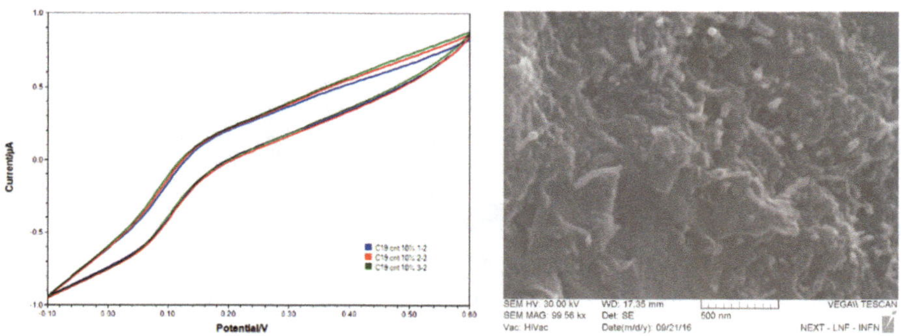

Fig. 2.10 Cyclic Voltammetry and SEM images of nanocomposite Epoxy-CNT 10%

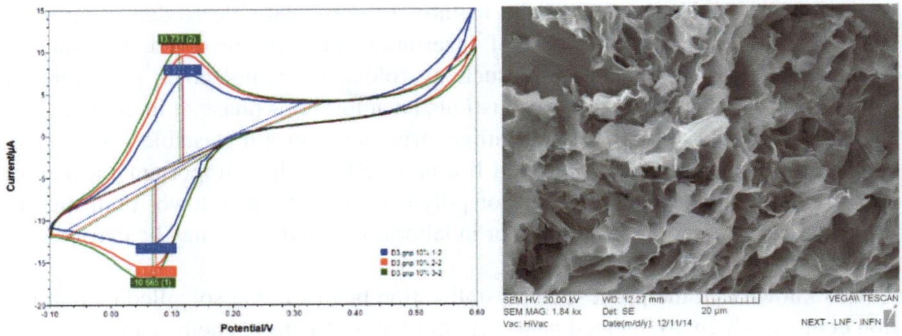

Fig. 2.11 Cyclic Voltammetry and SEM images of nanocomposite Epoxy-GNP

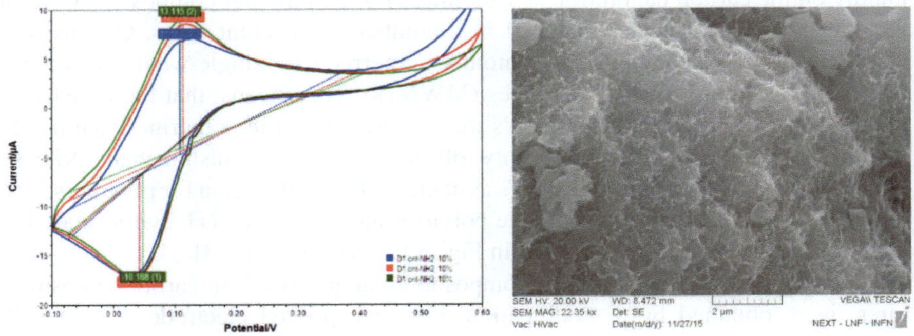

Fig. 2.12 Cyclic Voltammetry and SEM images of nanocomposite Epoxy-10% CNT-NH_2 functionalized (<5% amine groups)

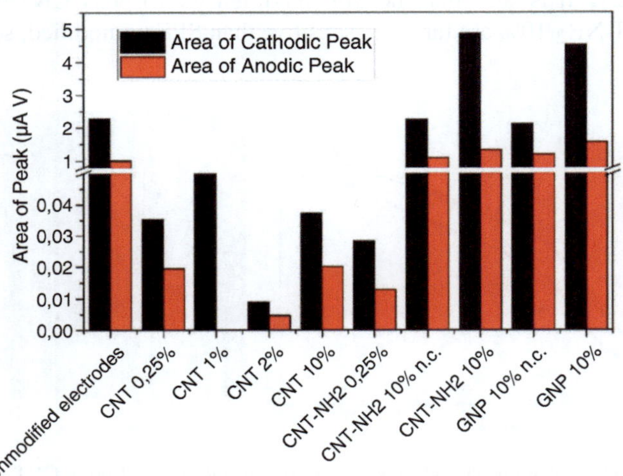

Fig. 2.13 Comparison of sensitivity of different sensor made of nanocomposites obtained varying both filler kind and concentration (percentage weight)

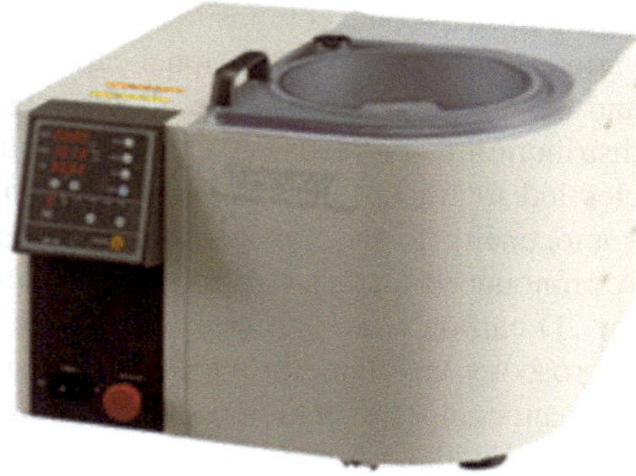

Fig. 2.14 The Thinky mixer

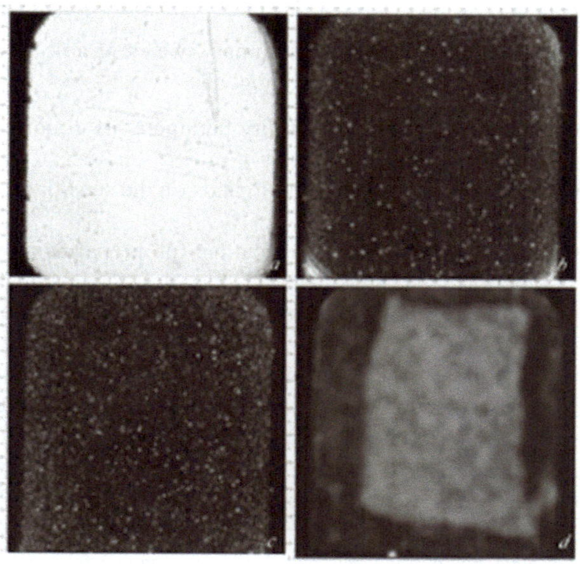

Fig. 2.15 Internal microstructure of carbon–epoxy nanocomposite with addition of graphite nanoplatelets (epoxy + 1.0 wt.% GNP). Acoustic images (60-mm thick layers) are given at different depths inside the 1.56 mm thick specimen: (**a**) upper surface; (**b**) at 260 mm deep; (**c**) at 380 mm deep; (**d**) bottom image. The dark area corresponds to the specimen fixing zone. Operation frequency is 100 MHz; scanning field is $10 \times 11\ \text{mm}^2$. Bright spots are agglomerates of carbon particles

The Thinky mixer (see Fig. 2.14) was essential in handling the high amount of filler, as the latter increases dramatically the viscosity of the resin. In these conditions, the Thinky planetary movement worked efficiently and dispersed nanoparticles in the resin. Also in this case, we deposited a droplet of nanocomposite with a stick, trying to cover completely the surface of working electrode.

The high-resolution ultrasonic technique (impulse acoustic microscopy) was applied for observing the bulk microstructure of nanocomposites with various types of carbon nanoparticles as filler (see Fig. 2.15 [14–17]). It was shown that ultrasonic methods are excellent tools for studying nanoparticle distribution over the material bulk and for revealing probable nonuniformity. The acoustic microscopy technique allows for looking through the bulk microstructure of objects by means of layer-by-layer visualization and object cross-sectional imaging. In composites based on dispersed micrometer-sized particles of exfoliated graphite, the particle distribution over the composite material bulk was observed. In nanocomposites with various kinds of low-dimensional carbon nanofillers, i.e., nanoflakes, nanoplatelets, and nanotubes, the technique allowed the cluster architecture of the nanoparticle distribution to be revealed. Contact conjugation of low-dimensionality nanoparticles led to fractal clusters despite significant technological efforts for providing homogeneity to the nanocomposite materials and uniformity to their properties. A pronounced tendency to form micrometer-sized fractal agglomerates was found for 2D carbon nano-particles: nanoflakes and nanoplatelets. The impulse acoustic microscopy technique provides visualization of the agglomerate distribution over the nanocomposite material bulk. Another kind of nanoparticle distribution was observed with carbon nanotubes (CNTs). The latter formed CNT packings having different densities. Such regions were seen in acoustical images as small-sized areas of various brightness.

In conclusion and as a summary, we were able to carry out the following tasks and achieve the corresponding goals:

- Measurement of the stability parameter of double – stranded DNAs complexes with ligands.
- The effect of intercalating ligands on the sensitivity and selectivity of the DNA – sensor.
- Implementation of spin coating for preparing polymer substrate suitable for functionalization.
- Raman observation of small percentage of functionalizing molecules put on the substrate.
- The impulse acoustic microscopy characterization of the nanocarbon cluster architecture in the composite to modify screen printed electrodes is made.
- The Raman characterization of nanocomposite modified screen printed electrodes was made.
- Development of the electric scheme for the DNA – biosensor prototype.

Acknowledgements I acknowledge partial funding from NATO SfP-984537 and from the EU H2020 project GrapheneCore1 – Grant Agreement AMD-696656-4.

References

1. Novoselov KS, Fal'ko VI, Colombo L, Gellert PR, Schwab MG, Kim K (2012) A roadmap for graphene. Nature 490:192–200
2. Yu X, Cheng H, Zhang M, Zhao Y, Qu L, Shi G (2017) Graphene-based smart materials. Nature Reviews Materials 2:17046
3. Dabrowska A, Bellucci S, Cataldo A, Micciulla F, Huczko A (2014) Nanocomposites of epoxy resin with graphene nanoplates and exfoliated graphite: synthesis and electrical properties. Phys Status Solidi B 251(12):2599–2602
4. Pierantoni L, Mencarelli D, Bozzi M, Moro R, Bellucci S (2014) Graphene-based electronically tuneable microstrip attenuator. Nanomater Nanotechnol 4:1–6

5. Bozzi M, Pierantoni L, Bellucci S (2015) Applications of graphene at microwave frequencies. Radioengineering 24:661–669
6. Pierantoni L, Mencarelli D, Bozzi M, Moro R, Moscato S, Perregrini L, Micciulla F, Cataldo A, Bellucci S (2015) Broadband microwave attenuator based on few layer graphene flakes. IEEE Trans Microwave Theory Tech 63:2491–2497
7. Maffucci A, Micciulla F, Cataldo A, Miano G, Bellucci S (2016) Bottom-up realization and electrical characterization of a graphene-based device. Nanotechnology 27:095204
8. Mencarelli D, Pierantoni L, Stocchi M, Bellucci S (2016) Efficient and versatile graphene-based multilayers for EM field absorption. Appl Phys Lett 109:093103
9. Pierantoni L, Mencarelli D, Bozzi M, Moro R, Moscato S, Perregrini L, Micciulla F, Cataldo A, Bellucci S (2015) IEEE Trans Microwave Theory Tech 63(8):2491–2497
10. Yasir M, Bozzi M, Perregrini L, Bistarelli S, Cataldo A, Bellucci S (2016) Innovative tunable microstrip attenuators based on few-layer graphene flakes. In: 2016 16th mediterranean microwave symposium (MMS), pp 1–4
11. Yasir M, Bistarelli S, Cataldo A, Bozzi M, Perregrini L, Bellucci S (2017) Enhanced tunable microstrip attenuator based on few layer graphene flakes. IEEE Microwave Wirel Compon Lett 27(4):332–334
12. Yasir M, Savi P, Bistarelli S, Cataldo A, Bozzi M, Perregrini L, Bellucci S (2017) A planar antenna with voltage-controlled frequency tuning based on few-layer graphene. IEEE Antennas Wirel Propag Lett 16:2380–2383
13. Li L, Li B, Hood MA, Li CY (2009) Carbon nanotube induced polymer crystallization: the formation of nanohybrid shish-kebabs. Polymer 50:953–965
14. Levin V, Morokov E, Petronyuk Y, Cataldo A, Bistarelli S, Micciulla F, Bellucci S (2017) Cluster microstructure and local elasticity of carbon-epoxy nanocomposites studied by impulse acoustic microscopy. Poly Eng Sci 57(7):697–702
15. Levin V, Petronyuk Y, Morokov E, Chernozatonskii L, Kuzhir P, Fierro V, Celzard A, Mastrucci M, Tabacchioni I, Bistarelli S, Bellucci S (2016) The cluster architecture of carbon in polymer nanocomposites observed by impulse acoustic microscopy. Phys Status Solidi B 253(10):1952–1959
16. Levin V, Petronyuk Y, Morokov E, Chernozatonskii L, Kuzhir P, Fierro V, Celzard A, Bellucci S, Bistarelli S, Mastrucci M, Tabacchioni I (2016) Bulk microstructure and local elastic properties of carbon nanocomposites studied by impulse acoustic microscopy technique. AIP Conf Proc 1736(1):020056
17. Levin VM, Petronyuk YS, Morokov ES, Celzard A, Bellucci S, Kuzhir PP (2015) What does see the impulse acoustic microscopy inside nanocomposites? Phys Procedia 70:703–706

Chapter 3
Loaded Electrospun Nanofibers: Chemical and Biological Defense

A. Vaseashta

Abstract In a complex, asymmetric and kinetic warfare environment, it is critical to deploy the latest technological innovations for tactical leverage over adversaries. Among others, improvements in outerwear of soldiers' uniform provides significant tactical advantage. We have studied use of loaded nanofibers, which demonstrate capabilities such as light weight, mechanically resilient, and breathable, i.e. allow perspiration to evaporate. In addition, the fibres are capable of sensing the chemical/biological environment. The nanofibers can be fabricated to include communication capabilities for enhanced situational awareness for emergency responders arising from chemical/biological environments. Due to the bioactive nature of the additive materials, the resulting textile can deliver therapeutics on-site and on-demand, thus providing temporary relief from injury. This investigation outlines an ongoing research on fabrication using loading of electrospun nanofibers, characteristics and bioactivities of selected nanomaterials to produce outerwear for soldiers for providing protection from adverse chemical and biological environment. We further discuss future pathways to integrate additional capabilities to provide tactical superiority in combat theatre.

Keywords Electrospinning · Nanofibers · Filtration · Chem.-bio agents · Sensors

A. Vaseashta (✉)
International Clean Water Institute, Herndon, VA, USA

NJCU – State University of New Jersey, Jersey City, NJ, USA

Ghitu Institute of Electronic Engineering and Nanotechnology, Academy of Sciences of Moldova, Chisinau, Moldova
e-mail: avaseash@norwich.edu

J. Bonča, S. Kruchinin (eds.), *Nanostructured Materials for the Detection of CBRN*, NATO Science for Peace and Security Series A: Chemistry and Biology,
https://doi.org/10.1007/978-94-024-1304-5_3

3.1 Introduction: Justification and Methodology

In most conventional warfare scenarios, the adversaries deploy forces of a comparable type and the outcome can be predicted by estimating the quantity of the opposing forces and/or by their quality. In the current asymmetric warfare scenario, this assumption does not hold true because the composition or strategy of the opposition forces makes it impossible for either side to close-in during encounter. Naturally, the first line of defence is self "protection". With this as the basis, the study presented here outlines a systematic methodology to develop micro/nanofibers using electrospinning, loaded with selected nanomaterials which are mixed into the solutions of high-performance polymers. The methodology involves systematic identification and selection of novel combinations of loading of nanomaterials with high-performance polymers to provide a desired set of characteristics, which include increased mechanical strength, improved sensitivity to contaminants, better filtering capabilities, improved response to electromagnetic stimuli, on-demand or time-released drug delivery, optical tunability or adaptation and conductivity thus leading to textile, termed as system-on-fibers (SoF). Additionally, the high surface area of nanofibers coupled with their potential biocompatibility and biodegradable nature, offers tremendous promise for diverse applications in sensing of chemical and biological environments, targeted drug/vaccine delivery, non-thrombogenic materials for blood contacting applications and information and communication platforms.

Parallel to this work, our objective has been aimed at use of advanced science convergence methodologies to develop automated data mining tools such as automated data analysis, mining and surveillance (ADAMS^{TM}), technology foresight and road-mapping (TechFARM^{TM}) and New and Emerging Science and Technology Trends and Surveillance (NESTTS^{TM}) to provide predictive intelligence and database. The objective of such data mining tools has been to learn technological capabilities and gaps to gain tactical advantage, as introduced earlier. Preliminary data of fibres using loaded high-performance polymers are presented here with the objective to fabricate multi-functional nano/microfibers filled with several novel and nanomaterials for military outerwear for suits and masks, and gears with embedded sensors that provide an integrated solution pathway through a revolutionary approach.

3.2 Electrospinning and Nanofibers Formation

Electrospinning is a process in which polymer fibres are produced through the action of an electric field applied to a polymer solution. The concept of electrospinning was first introduced by Formalas in his series of patents in 1934. This technique has gained scientific popularity in the last couple of decades, as several groups have developed several micro/nano fibres [1–10]. A simple electrospinning apparatus

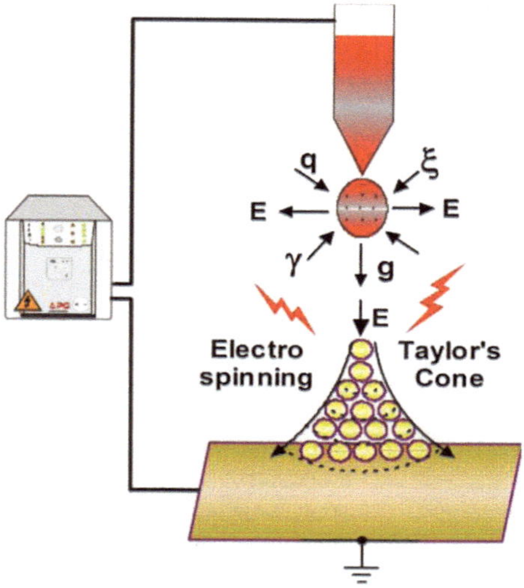

Fig. 3.1 A typical electrospinning apparatus

consists of a high voltage power supply, a capillary to deliver polymer and a collector electrode, as shown in Fig. 3.1. A typical laboratory electrospinning system consists of a microprocessor-controlled syringe pump with high metering precision at low, pulse-free pre-set rate and volume control. The internal diameter of the syringe is used by the control program to calibrate the pump and deliver a pre-determined volume and flow rate. The electric field is applied with a high voltage (typically $\sim$30 kV) power supply by connecting the metal discharge needle and a ground stationary or rotating collector electrode. The strong electric field on a polymer fluid generates an electrostatic force on the surface against the viscous stresses of the polymeric mass, thus producing elongation jets towards the other electrode. A Taylor cone is generated in the jet when hydraulic effect produced by the syringe plunger to the polymer solution is modified by the strong electrostatic field. The shape of the cone, employing parameters such as viscosity; conservation of mass, charge (q) and momentum; electric field (E); surface tension (ξ); air drag (γ); and Coulombic and Gravitational forces (g) has been modelled by several investigators [7]. Upon application of high voltage, the polymer jet is formed and undergoes an elongation process before reaching the collector electrode, the solvent evaporates, and the polymer is collected as a solidified interconnected network of nanofibers.

Electrospinning, under appropriate conditions, can produce three-dimensional porous network of nanofibrous mat with large global pore volume, predictable pore size distribution, and tunable interconnected porosity. Several synthetic and natural polymers, polymers composites, polymers impregnated with nanoparticles, carbon nanotubes, and compounds for mechanical, electrical, and biological applications were attempted using single double/triple jet electrospinning apparatus [11]. Table 3.1 provides a list of polymers (synthetic, natural, and biodegradable),

Table 3.1 A sample list of polymers (synthetic, natural, and biodegradable) and loading materials – ceramers, electro-ceramers, and bio-ceramers [8]

Polymers and Composites	Composition	Molecular weight	Loading	Applications/ comments
Synthetic	Polystyrene/Dichloroethane	299,000	CNT	Field Emission
	Nylon (6,12)/1,1,1,3,3,3 Hexafluoro-2-propanol	–	CNT	Flexible Substrates
	Poly(etheremide)/1,1,2-Tricholoroethane	30,000	CNT	Mechanical Strength – SoF
	Polyacrylonitrile (PAN)/ Dimethylformamide (DMF) (also dimethyl acetamide (DMAc))	114,000 M vol.	CNT ZnO	Strength Sensors (SoF)
	Polymethylmethacrylate (PMMA) in acetone, chloroform, Toluene	540,000	None	Porous fibers
	Polyvinylpyrrolidone/Ethanol/ DMF	58K–1.3M	Bi_2S_3	CT imaging
	Polyprrole in Acetonitrile (ACN), and dichloromethane (DCM)	–	TEA BF_4	Conducting Polymers – SoF
	Polyeurathane/DMF Polycarbonate/DMF	42,000	CNT	Protective Clothing
Natural	Lecithin/Dimethylformamide	–	None	Biomembranes
	Lecithin/Dimethylformamide	–	Bi_2S_3	CT imaging
Biodegradable	Poly (ε-caprolactone) (PCL)	–	PLA/PGA	Wound dressing
	Poly (L-lactide/co ε-caprolactone) (PLLA/CL)	150,000	PLA/PGA Collagen	Stents Tissue/Scaffold
	Poly (ethylene oxide)/water	400,000	PCL	Hollow Fibers
	Polyglcolide		PCL	
	Poly (ethylene oxide)/water	400,000	MoO_3, ZnO, TiO_2	Sensors
	Poly (L-lactide/co ε-caprolactone) (PLLA/CL)	150,000	Bi_2S_3 Collagen	CT imaging
	Poly (L-lactide/co ε-caprolactone) (PLLA/CL)	150,000	Metal nanoparticles	Targeted drug Delivery
Ceramers	Polyacrylonitrile (PAN)/ Dimethylformamide (DMF)	150,000 M vol.	TiO_2	Conductive & PV Polymer
	Polydiphenooxyphosphazene PDPP		SiO_2 (Y, Er)	Glass/ceramic

(continued)

Table 3.1 (continued)

Polymers and Composites	Composition	Molecular weight	Loading	Applications/ comments
Electro-optic polymers	Poly(acrylic acid) -polypyrene methanol (PAA-PM)/ Dimethylformamide	50,000		Optical sensors
	Poly (vinyl acetate)/Acetone		GeO_2	1D Opto-electronics
Bio-ceramers	Poly(l-lactic) acid, poly(glycolic) acid, Poly(l-lactide/glycolide)	70 K/>150 K /10–20 K	Collagen Mefoxin	Bone repair Antibiotics

Table 3.2 A list of polymers and loading materials under consideration

Polymer composition	Loading	Applications/comments
Nylon (6,12)/1,1,1,3,3,3 Hexafluoro-2-propanol	NRI NPs	Camouflage
Polyacrylonitrile (PAN)/ Dimethylformamide (DMF) (also dimethyl acetamide (DMAc))	NRI NPs/ZnO	Camouflage/Strength
Lecithin/Dimethylformamide	Ag	Biomembranes
Lecithin/Dimethylformamide	Bi_2S_3	CT imaging
Poly (ε-caprolactone) (PCL)	Turmeric	Wound dressing
Poly (L-lactide/co -caprolactone) (PLLA/CL)	Turmeric/Collagen	Tissue/Scaffold
Polymers including polyamide, polyviniyl alcohol, polyurethane, polypropylene, nylon-6, poly(viniyl chloride), polysulfone, poly(methyl methacrylate), cellulose acetate	Ag, TiO_2, Cu, ZnO, MgO, ZrO_2	Flame retardancy, UV protection, self cleaning, antistatic and antimicrobial activities
Polyacrylonitrile (PAN)/ Dimethylformamide (DMF)	MgO, TiO_2	Chem./Bio Decon

ceramers, electro-ceramers, and bio-ceramers that were studied and reported earlier [8]. The table lists their composition, molecular weight, approximate range of loading, and intended applications. The study was later extended for several new categories of materials, including negative refractive index (NIR) materials, bioactive compounds with powerful medicinal properties, and certain oxides to study phytotoxic and genotoxic effects. Table 3.2 provides a list of materials that are currently being tested with intended applications. Figure 3.2 shows a conceptual schematic of possible spectrum of applications that are potentially feasible [8].

We describe below applications of loaded nanofibers in context of chem.-bio sensors, filtration, and information and communication technologies in support of Command, Control, Communications, Computers and Intelligence (C4I). We also discuss features such as therapeutic capability that is either built-in or through

Fig. 3.2 Recommendation map listing technology readiness level vs. technological significance [13]

add-on layers. Filtration and decontamination capabilities include protection against ambient chemical/biological/radiological contaminants Additional capabilities, such as force protection (refers to mechanical properties that renders products mechanically strong, resilient, flexible), light weight, breathable, and with the ability to camouflage [12].

3.3 Data Mining and Predictive Analytics

During the last decade alone, new digital learning technologies have fundamentally modernised classical theory and practice of conducting research, articulating the path forward and future investments. Notwithstanding much progress, the synergy arising from the convergence of multiple disciplines offers boundless potential for transformational, revolutionary, and embryonic opportunities. Novel convergence methodologies will integrate and advance next generation solutions to current and future technological challenges. Convergence in research methodologies transforms the way research is conducted by overcoming specific barriers or filling existing knowledge gaps. Moore's law has made the storage and analysis of data scalable, creating opportunities for fine-grained data analytics. New data gathering tools have fostered precise and customized data management and analysis practices.

Multidisciplinary convergence and data analytics have exceptionally high potential for transforming the way state-of-the-art information is gathered, analysed, and leveraged to enable future advances and applications. The idea of "Converging Technologies" in conjunction with data analytics has considerably influenced scientific and social thinking about technological innovation space. To provide strategic assessment of revolutionary scientific breakthroughs in multidisciplinary environments based on the convergence, several data analytics platforms were developed [13]. Innovations lead to mapping technology roadmaps which are cautiously formulated based on extensive research, expert elicitation and networking approaches to project "future scenarios" realistically and epistemologically. Such roadmaps enable the development of transformative tools and methodologies that fill fundamental knowledge gaps and change culture in academia to foster collaboration, thus providing unique solutions. Synergy arising from converging technologies and research methodologies will leverage emerging and potentially transformative studies.

The objective of an independent investigation was to develop automated data mining tools – such as automated data analysis, mining and surveillance (ADAMS^{TM}), technology foresight and road-mapping (TechFARM^{TM}) and New and Emerging Science and Technology Trends and Surveillance (NESTTS^{TM}) to provide an independent confirmation to ensure that appropriate resources were allocated as this project progressed. Furthermore, such data mining tools, in addition to providing technological capabilities, also provide technological gaps, allowing us to focus on gaps to achieve desired capabilities. The scope and methodologies of these analytical tools is described elsewhere [13]. For the sake of continuity, scope of these tools is listed in Table 3.3.

Based on these tools, a list of recommendations was made, as shown in Fig. 3.2, indicating that wearable electronics and chem.-bio sensing platforms showed a significantly higher level of technology readiness level (TRL) and are of significant importance.

Within the framework of the recommendations of data analytics and anticipated characteristics of loading in nanofibers, a spectrum of potential applications was mapped and is represented in Fig. 3.3. The scope of applications considered here are textile for chem.-biological sensing/detection/protection and on-site and on-demand delivery of drugs from the uniform of soldiers in combat theatre.

Table 3.3 Data analytic tools developed and scope

TECH FARM Application No.: 85/287,943
Research and development of new products of others by providing technology foresight, assessment and forecasting services that apply a futures-oriented approach that exploits analytical methodologies, including heuristics, data-mining, scientometrics, modeling and simulation, and scenario development to provide systematic recognition and evaluation of new or existing science and technology solutions for the development of new products and written reports supplied in connection therewith in International Class 42.
NESTTS Application No. 85/287,940
Research services, namely, providing, tracking, analyzing and projecting new and emerging trends to provide awareness and means to mitigate threats and risks posed by chemical, biological, nuclear, radiological, high-yield explosive materials, and internet and intranet based communications for governments, non-governmental organizations, industries, and academic institutions for defense and security applications in International Class 42.
ADAMS: Application No. 85/287846
The mark ADAMS is an interactive tool that utilizes several data mining tools to analyze, evaluate, and determine emerging S&T trends based on quality of journals, recognition of individual authors and their international stature recognition, validation by heuristics, and other scientific databases to provide normalized and scientifically validated information on emerging trends in science and technologies.

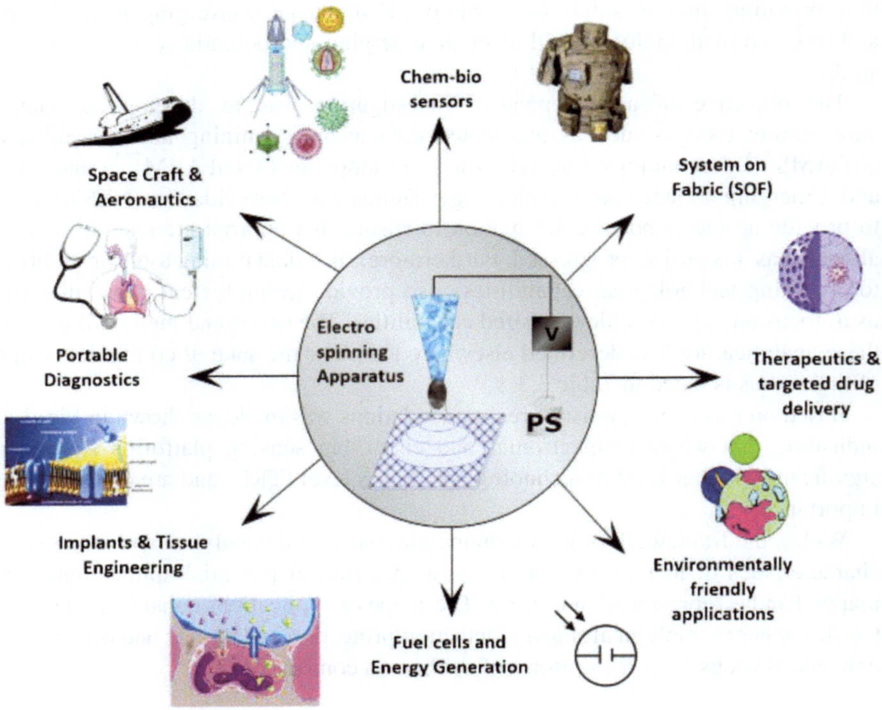

Fig. 3.3 Spectrum of potential applications

3.4 Characteristics and Bioactivities of Loaded Nanofibers in Support of Chem.-Bio Protection

Electrospun fibers are capable of forming a highly porous nanofibrous network. Due to their exceptionally large surface/volume ratio, the fibers have a very high surface energy, thus the resulting sensor platforms have very high sensitivity. Nanostructured morphologies further augment characteristics due to their reduced dimensionality [14]. Loading of various materials, such as carbon nanotubes, metal oxides, ceramers, etc. allow resulting fibers to be sensitive to chemical and biological environment. Electrospun nanofibrous materials, so produced are lightweight, breathable, provide protection against harsh environment and comfortable to wear. This is particularly critical to soldiers in combat theatre, where they have a great likelihood of experiencing environment contaminated with chem.-bio agents and residue arising from firearms. Although a discussion about chemical and biological warfare agents is the beyond the scope of this report, it is advisable to note some of the characteristics of chemical warfare agents (CWA) and biological agents (BA), in support of chemical-biological defence (CBD). Most CWAs are liquids that evaporate into vapours at varying rates and hence spread by evaporation, spray, explosion, or even small breezes to disperse. The agents are classified as vesicants and blood, nerve, or pulmonary agents. For biological agents, the Centre for Disease Control (CDC) classifies biological agents into categories according to their threat to national security. The top two categories are: Cat. A which result in high mortality and require enhanced preparedness, and Cat. B which results in moderate public health impact and requires surveillance to monitor spread [15].

3.4.1 Chem.-Biological Sensing Detection and Protection

Protective clothing that are currently in use may include activated charcoal impregnated with metal ions to physically adsorb chemical and biological agents. Metal oxide nanoparticles such as TiO_2 and MgO are added as potential catalysts for the decontamination of chem.-bio agents. Electrospun nanofibers may be fabricated from a wide range of polymers, either synthetic (poly-acrylonitrile (PAN), poly-styrene(PS), poly-vinylalcohol (PVA), poly-vinylpyrrolidone (PVP), poly-ethylenimine (PEI)) or natural (collagen, chitosan, gelatin) with loading of fillers such as metal salts, metal nanoparticles (NPs), carbon nanotubes (CNTs), graphene, ceramers, herbal bioactive compounds and fluorescent or photoluminescent markers to provide additional advanced functionalities.

Presently applied protective clothing against chemical and biological agents include the use of activated charcoal impregnated with metal ions, which physically adsorb agents thereby creating disposal hazards after its usage. The significant properties of these "adsorbents" are due to their high surface areas and high surface concentrations of reactive defect sites. However, there are disadvantages of modification of scaffold, mats, and e-spun textile with the nanoparticles, especially the ones that are not stabilized on the surface of the textiles. In addition to altering the physical and mechanical properties of the textile materials, there are linked toxic effects of nanoparticles on human health and is under independent investigation [16]. Furthermore, inhalation of those nanoparticles in high doses can result in pulmonary inflammation and allergic reactions, therefore the use of nanoparticles in powder form is difficult in protective clothing. To overcome this limitation, research effort was directed to incorporate nanoparticles with nanofibers by inclusion of nanoparticles with polymer solution before electrospinning.

Most chemical agents, as stated above, are released in the form of vapours, hence to detect such vapours/gases, several semiconducting metal oxides such as TiO_2, SnO_2, ZnO, WO_3, and other wide band gap metal oxides, well known for their ability to detect trace concentrations of various gases in air via charge transfer interactions between the sensor and chemisorbed species, were introduced in electrospun fibers. The nanowires act as sensitive, fast, stable and reproducible gas sensors that can be easily integrated into multi-component and integrated arrays. Investigation using TiO_2 nanofibers as a sensing membrane to detect traces of CO and NO_2 in air was conducted to fabricate ultrasensitive gas sensors. In a separate investigation, carbon filled polymeric matrices were used to sense various gases and detect presence of radiation by varying size and percolation in polymer matrix [17]. There are several investigations reporting acoustics wave [18], photoelectric [19], and optical sensors [20], however this investigation is focused primarily on resistive gas sensors. Some of the recent efforts are focused on the development of nanostructured sensors to achieve increased surface-to-volume ratios and reduced cross sections, offering more effective gas modulation of device resistance.

Electrospun nanofibers based biosensors reported in the literature mostly rely on electrochemical transduction. Sensing elements are largely enzymes, antibodies, and more scarcely using DNA strands or aptamers. The bio-receptors can be immobilized using various strategies, depending on both the recognition elements and their interfacial interactions. The most common methods used to generate bioreceptor-nanofibers hybrid assemblies consist of attaching of the biomolecules onto the fiber surface by physical or chemical sorption, covalent binding, cross-linking or entrapment in a membrane. This approach has been expansively utilized to immobilize enzymes, antibodies, DNA strands and aptamers. For enzyme biosensors, entrapping the bioactive molecules inside the nanofibers is accomplished by electrospinning a combination of enzymes and polymers. A selection of nano-taggants offer possibility to trace, track, and authenticate biological threats. For this investigation, we focused on nanofibers that are sensitive to chemical and biological environments and provide protection by way of filtration and decontamination. Nanofiber based materials suggest novel protecting clothing materials with comfortable, adsorptive and lightweight properties with active detoxyfying/decontaminating nanomaterials, primarily facilitated by their nature at nanoscale [14]. Several intended applications are listed in Table 3.4, also consistent with the priorities of defence agencies.

Table 3.4 Nanoscale filtration applications

Wearable Kidney
Water desalination
Water purification (collagen based ES)
Breathable NBS – protection clothing
Wearable anti-dehydration systems
Advanced rebreathers for indefinite-length underwater presence – artificial gills
Superior gas masks
MoS_2 nanoparticles – especially for VOCs

3.4.2 Bioactivities: Therapeutics and Drug Delivery

The high surface area combined with biocompatibility and biodegradable nature certain fibres offer tremendous promise for diverse biomedical applications from tissue engineering, targeted vaccine delivery, and non-thrombogenic materials for blood contacting applications. Biodegradable polymers are highly desirable for long-term drug delivery applications because they degrade in the body to biologically inert and compatible molecules. By incorporating a drug into biodegradable polymers, dosage forms of various shapes and sizes yield built-in timed-release or physiological condition based-release therapeutics. Secondary surgical procedures, common after the completion of dosing regimens are rendered moot post-delivery via such polymers.

Many biodegradable polymer chemistries were investigated, however, the most common and successful are polyesters biodegradable sutures. These polymers include poly(glycolide), poly(D,L-lactide) and their related copolymers poly(D,L-lactide-co-glycolide). Under suitable conditions, most synthetic and natural polymers demonstrate electrospinning; thus, allowing nanofiber preparation from nanoparticles or vaccine-impregnated polymers. We reported elsewhere that antibiotic loaded biodegradable and biocompatible polymeric electrospun nanofibrous membranes prevent post-surgery induced intra-abdominal adhesions by providing a physical barrier during wound healing process and at the same time serve a functional barrier against strains of microorganisms thereby reducing the local inflammatory response. Based on our previous investigations, we have prepared therapeutic patches using electrospun nanofibers using biocompatible and biodegradable polymers. The key to this methodology is, (a): loading the polymers with drugs and selection of appropriate polymers inducing release of the drugs from polymeric matrices; and (b): release of drugs either based on physiological condition or an external trigger mechanism, which would provide a linkage for the development of new functional barriers for protection applications. Related to this topic and reported elsewhere, we carried out several experiments with nanofibers which are sensitive to biomarkers for sensing, in particular, glucose, cholesterol, triglyceride, low density lipoprotein (LDL) and identifying symptoms of stress, influenza, and diabetes.

Use of biodegradable polymeric systems is an alternate and efficient route for vaccine administration, peripheral nerve regeneration, and tissue engineering, which finds interesting applications in the context of safety and security. Some relevant polymers include bio-degradable polymers which are specifically investigated for time-controlled release characteristics; copolymers with desirable hydrophilic/hydrophobic interactions; dendrimers for immobilization of enzymes, drugs, peptides, or other biological agents; and certain blends of hydrocolloids and carbohydrate-based polymers. These biomaterials with desirable functional groups are being created for innovative drug delivery systems, such as linings for artificial organs, as substrates for cell growth and as agents in drug targeting and immunology testing, as biomedical adhesives and bioseparation membranes, and as substances that are capable of mimicing biological systems. Some of the most promising medical applications include treatment of primary pulmonary hypertension and pulmonary arterial hypertension by time-controlled release templates impregnated with anticoagulants and calcium channel blockers; bioscaffold that mimic extracellular matrix topology [10], polyesters combined with phosphatidyl choline for biomimetic applications [21], intravascular stents from a blend of polyactide and trimethylene carbonate [22], dystrophin gene immobilized templates for curing aging skeletal muscles [23], cellulose based scaffold for cartilage tissue engineering [24], and esophageal tissue engineering [25].

3.4.3 *Information and Communications Technology*

Situation awareness is especially important in an environment where the information flow is relatively high and poor decisions may lead to serious consequences. Such is the case in combat theatre, where the actions are asymmetric, kinetic, and mostly unpredictable. SA is broken down into three segments: perception of the elements in the environment, comprehension of the situation, and projection of future status. Also, on an average, a soldier carries at least 60 pounds of gear. With an extended patrol, the weight is likely to double. Specialized warfighters, such as Automatic Riflemen, Combat Medics, and Special Operations can experience overall weight totals much higher.

Use of nanofibers with the embedded capability of transmitting signals and capability of storing power, offer options to reduce the weight and provide capability of projection of the elements of the environment – to the soldier and also to the unit commander. As shown in Tables 3.1 and 3.2, several loaded nanofibers were prepared which offer capability of receiving and transmitting of signal and storing power. The investigation is far from over however, initial results indicate different ICT options, as shown in Fig. 3.4a, b. Several other applications under consideration included outwear in support of force protection and optical cloaking – all of which offer promising and direct applications of loaded micro/nano fibres for safety and security.

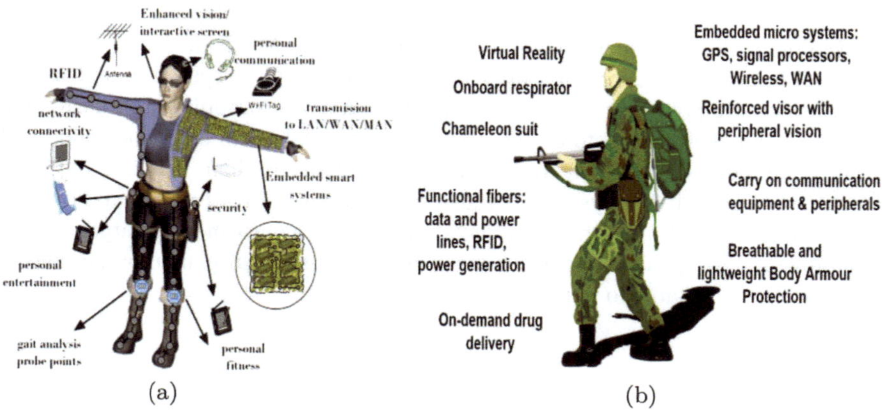

Fig. 3.4 (**a**): Use of nanofibers in information communication technologies, (**b**): Intended applications in soldier's uniform

3.5 Discussion and Future Prospects

Due to the growing interest in materials at nanoscale and associated novel and unique properties, research on nanofibers filled with nanomaterials fabricated using electrospinning process has increased significantly in the recent years. Using several innovative spinning techniques, inventive collector designs, new structures with innovative compositions, complex architectures, fibre morphologies, improved properties, and tuneable functional moieties have been produced. Electrospun nanostructured polymer structures of natural or synthetic origin have a multitude of possible applications in defence and security. For this investigation, the focus is on outerwear for soldiers which provides protection from chem.-bio environment, decontamination, vaccines and therapeutics and situational awareness using information and communication technologies. Development of new and smart protective textiles for future applications include, integration of nanoparticles with multiple threat sensing capability; and energy storing/harvesting devices built in the protective clothing. Smart protective textiles are still at the development phase and some of the materials are already quite close to commercial production. There are multitude of other viable applications, such as pharmaceuticals, optoelectronics, sensor technology, catalysis, and filtration. However, there are limitations, viz. nanofiber based protective textiles still require extensive research before large scale production. Life cycle analysis, as it relates to health and safety in all stages of the products, requires an extensive investigation and evaluation for its short and long-term health risks. Selecting a polymer for use as a degradable biomaterial must match the mechanical properties and the degradation rate to specific parameters. The greatest advantage of biodegradable polymers is that they are fragmented into biologically acceptable, metabolized and removable molecules. However, biodegradable materials do produce degradation by-products that must be endured with little to no adverse reactions within the biological environment. These degradation products must be tested thoroughly. Much of the development of novel materials in controlled drug delivery focuses on preparation and use of these responsive polymers with specifically designed macroscopic and microscopic structural and chemical features. These new biomaterials with desirable functional groups are being created by researchers who envision their use not only for innovative drug delivery systems but also as potential linings for artificial organs, as substrates for cell growth or chemical reactors, as agents in drug targeting and immunology testing, as biomedical adhesives and bio-separation membranes and as substances that are able to mimic biological systems.

References

1. Huang ZM, Zhang YZ, Kotaki M, Ramakrishna S (2003) A review on polymer nanofibers by electrospinning and their applications in nanocomposites. Compos Sci Technol 63:2223–2253
2. Chua KN, Lim WS, Zhang P, Lu H, Wen J, Ramakrishna S, Leong KW, Mao HQ (2005) Stable immobilization of rat hepatocyte spheroids on galactosylated nanofiber scaffold. Biomaterials 26:2537
3. Xang W, Drew C, Lee SH, Senecal KJ, Kumar J, Samuelson LA (2002) Electrospun nanofibrous membranes for highly sensitive optical sensors. Nano Lett 2(1):1273
4. Kim JS, Renekar DH (1999) Mechanical properties of composites using ultrafine electrospun fibers polymer composites. Polym Compos 20(1):124
5. Locher I, Kirstein T, Troster G (2004) Temperature profile estimation with smart textiles. In: Kiang K, Gorton J, Alert M (eds) Proceedings of 37th international symposium on microelectronics, IMAPS, p 204
6. E-Textile. Virginia Tech, Blacksburg. www.ccm.ece.vt.edu/etextiles
7. Rutledge S, Hohman MM, Shin MY, Brenner MP (2001) Electrospinning and electrically forced jets: I. Stability theory. Phys Fluids 13:2201; Electro-spinning and electrically forced jets: II. Apps. Phys Fluids 13:2221
8. Vaseashta A, Stamatin I (2007) Electrospun polymers for controlled release of drugs, vaccine delivery, and system-on-fibers. J Optoelectron Adv Mater 9(6):1606–1613
9. Lu Y, Chen S (2004) Micro and nanofab. of biodegradable polymers for drug delivery. Adv Drug Deliv Rev 56:1621
10. Kim TG, Park TG (2006) Biomimicking ECM: cell adhesive RGD peptide modified electrospun poly (D,L-lactic-co-glycolic acid) nanofiber mesh. Tissue Eng 12(2):221
11. Vaseashta A (2007) Controlled formation of multiple Taylor cones in electrospinning process. Appl Phys Lett 90:093115
12. Diao Z, Kraus M, Brunner R, Dirks J-H, Spatz JP (2016) Nanostructured stealth surfaces for visible and near-infrared light. Nano Lett 16(10):6610–6616
13. Vaseashta A (2014) Advanced sciences convergence based methods for surveillance of emerging trends in science, technology, and intelligence. Foresight – J Futur Stud Strateg Think Policy 16(1):17–36
14. Vaseashta A (2005) Nanostructured materials based next generation devices and sensors. In: Vaseashta A, Dimova-Malinovska D, Marshall JM (eds) Nanostructured and advanced materials. Springer, Dordrecht, pp 1–30
15. https://www.webmd.com/a-to-z-guides/secret-weapons-chemical-agents. Accessed 12 Jan 2017
16. Vaseashta A (ed) (2015) Life cycle analysis of nanoparticles – risk, assessment, and sustainability. Destech Publishers, New York, p 404. ISBN:978-1-60595-023-5

17. Arshak K, Cunniffe C, Moore E, Vaseashta A (2008) Anomalous behavior of carbon filled polymer composites based chemical and biological sensors. In: Vaseashta A (ed) Functionalized nanoscale materials, devices and systems. NATO security through science series B: physics and biophysics. Springer, Dordrecht, pp 321–324.
18. Ding B, Yamazaki M, Shiratori S (2005) Electrospun fibrous polyacrylic acid membrane-based gas sensors. Sens Actuat B-Chem 106:477–483
19. Yang M, Xie T, Peng L, Zhao Y, Wang D (2007) Fabrication and photoelectric oxygen sensing characteristics of electrospun Co doped ZnO nanofibers. Appl Phys A-Mat Sci Process 89:427–430
20. Luoh R, Hahn HT (2006) Electrospun nanocomposite fiber mats as gas sensors. Compos Sci Technol 66:2436–2441
21. Hillborn J (2005) Non-thrombogenic biodegrad-able polymers for the new generation of phospholipid analogues Priv. Com.
22. Medical device link. http://www.devicelink.com
23. Vaseashta A, Boskovic O, Webb A, Ozdemir N, Ozturk E (2006) Effect of dystrophin gene immobilized nano-structured therapeutic templates on aging skeletal muscles. In: Kassing R, Petkov P, Kulisch W, Popov C (eds) Functional properties of nanostructured materials. Springer, Dordrecht, p 511
24. Muller FA, Muller L, Hofman I, Greil P, Wenzel MM, Staudenmaier R (2006) Cellulose-based scaffold materials for cartilage tissue engineering. Biomaterials 27(21):3955
25. Zhu Y, Chian KS, Chan-Park MB, Mhaisalkar PS, Ratner BD (2006) Protein bonding on biodegradable poly (L-lactide-co-caprolactone) membrane for esophageal tissue engineering. Biomaterials 27:68

Chapter 4
Graphene Quantum Dots, Graphene Non-circular n–p–n-Junctions: Quasi-relativistic Pseudo Wave and Potentials

H. V. Grushevskaya, G. G. Krylov, S. P. Kruchinin, and B. Vlahovic

Abstract In our work, we build an atomic-like GQD-model and look for a GQD pseudopotential barrier, which is given by a set of well pseudopotentials for individual carbon atoms of the GQD. Numerical modelling of large-size GQDs has been performed in hydrodynamic approximation. It has been shown that pseudopotential removes degeneracy of energy levels for GQD-supercell and localizes valent electrons of the GQD-model on holes of n–p–n graphene-junction. Non-spherical symmetry of GQD wave functions leads to lifting of spin and valley degeneracy.

Keywords Graphene quantum dots · Graphene non-circular n–p–n-junctions · GQD-model · GQD pseudopotential barrier

4.1 Introduction

Nanoscale graphene quantum dots have the form of a quasicircle with a radius in the range from 3 [1] to 10 nm [2]. The parabolic confinement potential well has been used to calculate energy levels of a non-relativistic quantum dot [3–6]. But the results of such simulations describe only the low-lying part of the dot energy spectrum correctly.

H. V. Grushevskaya (✉) · G. G. Krylov
Physics Department, Belarusian State University, Minsk, Belarus
e-mail: grushevskaja@bsu.by

S. P. Kruchinin
Bogolyubov Institute for Theoretical Physics, Kiev, Ukraine

B. Vlahovic
North Carolina Central University, Durham, NC, USA

J. Bonča, S. Kruchinin (eds.), *Nanostructured Materials for the Detection of CBRN*, NATO Science for Peace and Security Series A: Chemistry and Biology,
https://doi.org/10.1007/978-94-024-1304-5_4

In the paper [8] using the scanning tunnelling microscope (STM) it has been managed to form large quantum dots of the type of circular n−p−n and p−n−p junctions of graphene with a radius of about $r_{\rm dot} \sim 100\,\rm nm$ and atomic-like distribution of the electron density has been established. In such an approach the equivalence has been assumed on the change of the problem of motion of holes and electrons in the potential well $(-\kappa r^2)$ and $(+\kappa r^2)$ respectively, to the problem on scattering by the barrier $(-\kappa r^2)$ and $(+\kappa r^2)$ for electrons and holes respectively with a sufficiently high repulsive potential at the distance L far from the boundary of the quantum point so that $L \gg r_{\rm dot}$. Naive considerations of the equivalence of such a change lead to the following unpleasant feature of the model as an appearance of false hole states (spurious states) at $r = 0,\ L$ (see section "Methods" in [8]) stipulated by the effective infiniteness of the well that is not a good choice for any Dirac problem. Usage of massless Dirac equation with a finite height step-like radial positive potential (barrier for electrons and well for holes) allows to perform correct relativistic simulations of electron density in the circular graphene n−p−n-junctions of the sizes 5.93 and 2.76 nm [1]. Despite the roughness of such calculations in comparison with the parabolic barrier, the theory and experiment should approximately coincide or, at least, qualitatively coincide if the assumption is valid on formation of a circular n−p−n-junction by mean addition of the step barrier to Dirac cone. But the relativistic simulation results satisfactory describe the only high energy level of the graphene quantum dots (GQDs), and the divergence between theoretical results and experimental data grows with the increase of GQD radius. Thus, relativistic barrier GQD models turn out to be inappropriate in the low-energy limit, and conversely, non-relativistic GQD models with a potential well do not work well in the high-energy limit.

Theory and experiment can be directly compared near graphene QD center and for following cases only: a scanning tunneling microscope (STM) tip radiuses $r_{\rm tip}$ are more than 70 nm or less than 20 nm. Electrically confining potentials have Gaussian shape for $r_{\rm tip} \leq 20\,\rm nm$ [7, 10]. The tight binding calculations (TB) in this case [8, 9] are applicable to the description of the behavior of wave functions in a QD-center only. Electrically confining potentials have cos-shape for $r_{\rm tip} \geq 70\,\rm nm$ [10]. Tight binding calculations (TB) in this case are applicable for the description of the wave functions in a QD center only. At present, there are no satisfactory methods for calculating GQDs and GQDs of large sizes.

In our work, we propose an atom-like GQD-model and look for a GQD pseudopotential barrier, which is given by a set well of potentials for distinct carbon atoms of the GQD.

4.2 An Atom-Like Model of Graphene Quantum Dot

A graphene quantum dot, consisting of $N_{\rm dot}$ carbon atoms is show schematically in Fig. 4.1a. Let a model GQD be considered as a "large atom" whose core i-th electrons are p_z-electrons of j-th C atoms, $j \neq i$. k-th p_z-electron of k-th C atom

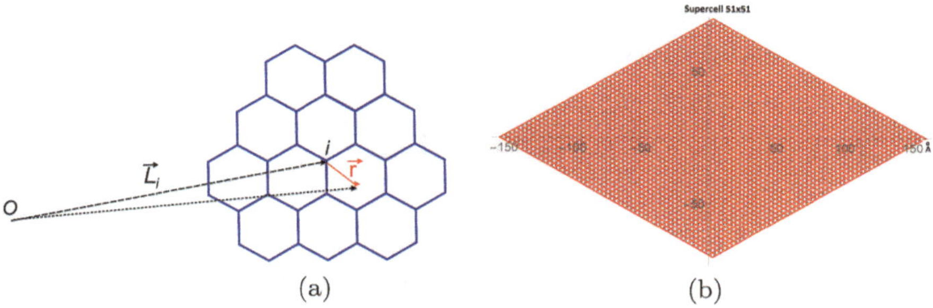

Fig. 4.1 (**a**) Scheme of graphene quantum dot: i-th atom C of GQD is located at i-th site with radius vector $\mathbf{L}_i$, and $\mathbf{r}$ is a radius vector of p_z-electron on respect to i-th lattice site. O is a reference point. (**b**) Graphene quantum dot super-cell from 51×51 primitive cells (100×150 Å)

plays a role of an external valent electron. Let k-th C atom is place in the lattice site with a radius-vector $\mathbf{L}_k$. The radius vector $\mathbf{r}$ will be calculated on respect to nearest lattice site and is a radius vector of the electron in an atom. Radius vector $\mathbf{X}_k$ of the valent electron of k-th atom reads $\mathbf{X}_k = \mathbf{L}_k + \mathbf{r}$.

Model graphene quantum dot has been constructed in the following way. Graphene primitive cell has basic vectors $\mathbf{b}_1 = a(3/2, \sqrt{3}/2), \mathbf{b}_2 = a(3/2, -\sqrt{3}/2)$ and two atoms (A and B) in the cell. Here a is the length of sp^2-hybridized C-C bond. We construct a supercell of the same symmetry type consisting of $(2n_1 + 1)(2n_2 + 1)$ primitive cells for $n_1 = 25, n_2 = 25$, that is shown in Fig. 4.1b.

Electronic band structure simulations in folding zone approximation gives the following set of eigenenergies $\pm\epsilon_{L_k}(q_i)$ and eigenstates $\psi_i^{(0)}(\mp q_i, \mathbf{r} + \mathbf{L}_k) = e^{\mp\imath\mathbf{q}_i\cdot(\mathbf{r}+\mathbf{L}_k)}u(\mathbf{r} + \mathbf{L}_k)$ for quasi-particle excitations of GQD:

$$\left\{\pm\epsilon_{L_k}(q_i),\ \psi_{L_k}^{(0)}\left(\mp\boldsymbol{q}_i, \mathbf{r} + \mathbf{L}_k\right)\right\}_{i,k=1,\ldots,N_{\text{dot}}}, \tag{4.1}$$

where $\mathbf{q}_i$ is the wavevector of i-th quasi-particle located in k-th lattice site, upper sign "+" is related to electrons and lower to "−" corresponds to holes.

Now, let us construct the pseudopotential for atoms C in GQD.

4.3 Pseudo-potentials for Atoms C in Quantum Dot

Let us consider n–p–n-junction. Fock operator describing non-paired valent electron in many-electron system can not be considered as a Hamiltonian one [11]. But the procedure of the secondary quantization restore the Hamiltonian property of the many-electron system by adding the operator of the "hole" $\hat{\epsilon}^{\dagger}$ to the Hartree-Fock Hamiltonian of m-th electron, scattering at a point $\mathbf{r}_i$ [12]:

$$\left[H(\mathbf{r}_i) + \hat{V}^{sc}(x_i) - \hat{\Sigma}^x(x_i)\right]\psi_m(x_i) = \left(\epsilon_m^{(0)} - \sum_{j=1}^{n} \hat{\epsilon}^{\dagger} P_j\right)\psi_m(x_i), \tag{4.2}$$

where $\hat{V}^{sc}(x_i)$ and $-\hat{\Sigma}^x(x_i)$ are operators of the Coulomb and exchange interactions respectively; $H(\mathbf{r}_i)$ is one-particle Hamiltonian without taking into account inter-electronic interactions, $\epsilon_m^{(0)}$ is an eigenvalue of $H(\mathbf{r}_i)$, $x_i = \{\mathbf{r}_i, \sigma_i\}$, σ_i is the spin of i-th electron, P_j is a projection operator, n is a number of electrons in a system.

Let us suppose that in the representation where the operator $\hat{\epsilon}^{\dagger}$ is a diagonal one, i-th electron with a radius-vector $\mathbf{L}_k + \mathbf{r}$ has been scattered on k-th atom with appearance of i-th hole with momentum $\mathbf{q}$, $i \neq k$ at Klein tunneling in the vicinity of the site with radius-vector $\mathbf{L}_k$. In this representation Eq. (4.2) is written as

$$\left[H(\mathbf{r}_i) + \hat{V}^{sc}(q\, x_i) - \hat{\Sigma}^x(q\, x_i)\right]\psi_m(q\, x_i) = \left(\epsilon_m^{(0)} - \sum_{j=1}^{n} \hat{\epsilon}^{\dagger} P_j\right)\psi_m(q\, x_i), \tag{4.3}$$

where $\hat{\epsilon}^{\dagger} = \tilde{\epsilon}_i(q)\hat{I}$, $\hat{I}$ is an operator unity in non-relativistic case or the identity matrix σ_0 for the Dirac equation. The inverse Fourier transformation

$$V_{i,L_k}(\mathbf{r})\psi_m(x_i) = \int e^{i\mathbf{q}\cdot(\mathbf{r}+\mathbf{L}_k)}\tilde{\epsilon}_i(q)\sigma_0\psi_m(q\, x_i)d\mathbf{q} \tag{4.4}$$

gives a nonrelativistic scattering potential. After calculating (4.4) we get

$$V_{i,L_k}(\mathbf{r}) = -2\pi\epsilon_{L_k}(q_i)\Theta\left(q_i - \frac{2\pi}{r}\right)\sigma_0 = -2\pi\epsilon_{L_k}(q_i)\Theta\left(\lambda_i - r\right)\sigma_0, \quad i \neq k; \tag{4.5}$$

where $q \to 0$, $r \leq a$, $\lambda_i = \frac{2\pi}{q_i}$. The potential (4.5), some well-known model potentials and a potential reconstructed based on experimental data are shown in Fig. 4.2. For a finite set of eigenenergies $\epsilon_{L_k}(q_i)$ the potential results to some staircase like potential which resemble the experimental one shown in Fig. 4.2b.

Further, a correct quasi-relativistic approximation of the pseudopotential for a graphene quantum dot will be constructed, and calculations of the valence orbitals of GQD will be performed based on the pseudopotential method.

Appearing after scattering of an electron on an atom hole with momentum $\mathbf{q}_i$ and radius-vector $\mathbf{L}_k + \mathbf{r}$ is always "frozen" in the vicinity of some lattice site with radius-vector $\mathbf{L}_k$. Due to the fact that such i-th hole together with electron belongs to

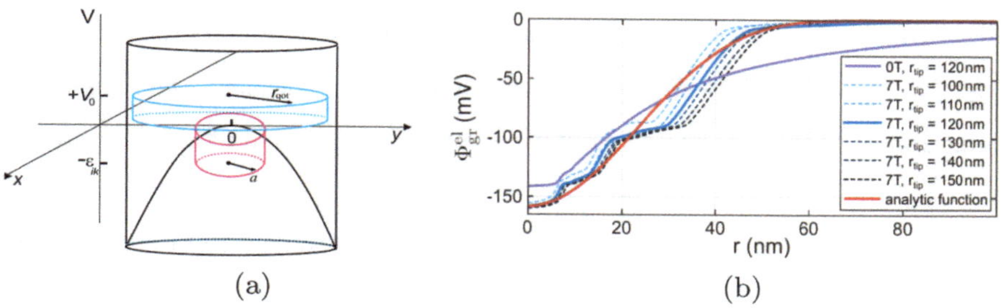

Fig. 4.2 (**a**) Model scattering potentials for the quantum dot (n–p–n junction): a circular well pseudo-potential $V_{i,L_k} = -\epsilon_{L_k}(q_i)\Theta(\lambda_i - r) \leq 0$ of the "frozen" atom with a size a (red curves); parabolic potential ($-\kappa r^2$) with a large repulsive potential outside the graphene quantum dot [8] (black curves); a circular barrier potential $V = +V_0\Theta(r_{\text{dot}} - r) \geq 0$ of radius r_{dot}, $V_0 > 0$ (blue curves) see e.g., [13] and references therein. (**b**) Reconstructed potential based on experimental data of [9]

"frozen" k-th atom, $i \neq k$, so in some sense internal "core" electron of the atom-like model GQD gains some additional binding energy of the electron-hole pair. Binding energy for every core electron is given by the set of quasi-particle energies (4.1). In the atom-like model GQD the valent electron is k-th electron which is scattered (binds) on its own k-th atom C. The valent electron is described by (4.3) in an approximation of the frozen core. In the approximation of the frozen core, matrix elements ϵ_{kc} of the hole energy operator $\hat{\epsilon}^{\dagger}$, $c \neq k$ is approximately defined by the difference between the energy level of k-th atom C with valent GQD-electron and energy levels of c-th atoms C with core GQD-electrons:

$$\epsilon_{kc} = \left[E_k^{(0)} - E_c^{(0)}\right], \quad c \neq k, \tag{4.6}$$

where $E_k^{(0)}$ is the energy of the valent electron, $E_c^{(0)}$ are eigenvalues of the equation

$$\left[H_{D,c} + V_{c,L_k}(\mathbf{r})\right]\psi_c(\mathbf{r} + \mathbf{L}_k) = E_c^{(0)}\psi_c(\mathbf{r} + \mathbf{L}_k), \quad c \neq k. \tag{4.7}$$

To get $E_k^{(0)}$, it is necessary to write down the equation for eigenvalues and eigenfunctions for valent electron

$$H_{D,k}\psi_k(\mathbf{r} + \mathbf{L}_k) + \sum_{c \neq k}\epsilon_{kc}P_c\psi_k(\mathbf{r} + \mathbf{L}_k) = \epsilon_m^{(0)}\psi_k(\mathbf{r} + \mathbf{L}_k). \tag{4.8}$$

Operator $\sum_{c \neq k}\epsilon_{kc}\psi_c(\mathbf{r} + \mathbf{L}_k)$ entering into (4.8) is a QDT pseudo-potential. The solution of the Eq. (4.8) can be obtained by the successive approximations/ In zero approximation we account of small differences between $\left[E_k^{(0)} - E_c^{(0)}\right]$, $k \neq c$: $\left[E_k^{(0)} - E_c^{(0)}\right] \approx \epsilon_m$ for the frozen core. Here ϵ_m is a constant. Then due to the arbitrariness of the zero energy point we assume $\epsilon_m = \epsilon_m^{(0)}$ and (4.8) is rewritten as

$$H_{D,k}\psi_k(\mathbf{r}+\mathbf{L}_k) = \epsilon_m^{(0)}\left[\psi_k(\mathbf{r}+\mathbf{L}_k) - \sum_{c\neq k}\epsilon_{kc}P_c\psi_k(\mathbf{r}+\mathbf{L}_k)\right]. \tag{4.9}$$

Let us designate the expression

$$\left[\psi_k(\mathbf{r}+\mathbf{L}_k) - \sum_{c\neq k}P_c\psi_k(\mathbf{r}+\mathbf{L}_k)\right] \tag{4.10}$$

through $\psi_{v,L_k}^{(0)}(\mathbf{r}+\mathbf{L}_k)$. It and (4.9) define a zero-order approximation for valent orbital

$$H_{D,k}\psi_{v,L_k}^{(0)}(\mathbf{r}+\mathbf{L}_k) = E_k^{(0)}\psi_{v,L_k}^{(0)}(\mathbf{r}+\mathbf{L}_k). \tag{4.11}$$

Substituting eigenvalues $E_k^{(0)}$ of (4.11) into (4.8), we find the equation for eigenfunctions and energy levels of k-th electron for

$$H_{D,k}\psi_k(\mathbf{r}+\mathbf{L}_k) + \sum_{c\neq k}\epsilon_{kc}\psi_c(\mathbf{r}+\mathbf{L}_k) = E_k^{(1)}\psi_k(\mathbf{r}+\mathbf{L}_k). \tag{4.12}$$

Taking into account the equality $i\hbar\frac{\partial}{\partial t}\psi_k(\mathbf{r}+\mathbf{L}_k) = E_k^{(1)}\psi_k(\mathbf{r}+\mathbf{L}_k)$. Then, the set

$$\left\{H_{D,k}\psi_k(\mathbf{r}+\mathbf{L}_k) + \sum_{c\neq k}\epsilon_{kc}\psi_c(\mathbf{r}+\mathbf{L}_k) = E_k^{(1)}\psi_k(\mathbf{r}+\mathbf{L}_k)\right\}_{k=1,r\to a}^{N_{qdots}} \tag{4.13}$$

can be considered in the continuous limit (hydrodynamic limit) $|\mathbf{L}_{k+1} - \mathbf{L}_k| \to 0$. Since the energy of electrons in the sites differs on the energy of quasi-particle excitation $\epsilon_{kc} = \epsilon_{c_i}$, Eq. (4.13) describes an electron moving in GQD:

$$\begin{aligned}&\left\langle \mathbf{R}_{\text{qdot}}\right| H_D^{\text{qdot}}\left|\Psi_v^{\text{qdot}}\right\rangle + \sum_{i=1,c_i\neq v}^{N_{\text{dot}}-1}\epsilon_{c_i}(\mathbf{R}_{\text{qdot}})\left\langle \mathbf{R}_{\text{qdot}}\right|\left.\Psi_{c_i}^{\text{qdot}}\right\rangle\left\langle\Psi_{c_i}^{\text{qdot}}\right|\left.\Psi_v^{\text{qdot}}\right\rangle \\ &= E_v^{\text{qdot}}\left\langle \mathbf{R}_{\text{qdot}}\right|\left.\Psi_v^{\text{qdot}}\right\rangle.\end{aligned} \tag{4.14}$$

Here

$$V_{\rm GQD} = \sum_{i=1, c_i \neq v}^{N_{\rm dot}-1} \left|\Psi_{c_i}^{\rm qdot}\right\rangle \epsilon_{c_i}(\mathbf{R}_{\rm qdot}) \left\langle\Psi_{c_i}^{\rm qdot}\right| \tag{4.15}$$

is a GQD pseudopotential,

$$\left\langle\Psi_{c_i}^{\rm qdot}\right| \left.\Psi_v^{\rm qdot}\right\rangle = \int \left\langle\Psi_{c_i}^{\rm qdot}\right| \left.\mathbf{R}'_{\rm qdot}\right\rangle \left\langle\mathbf{R}'_{\rm qdot}\right| \left.\Psi_v^{\rm qdot}\right\rangle d\mathbf{R}'_{\rm qdot}$$

is a scalar product, respectively, of these wave-functions $\left|\Psi_{c_i}^{\rm qdot}\right\rangle$ and $\left|\Psi_v^{\rm qdot}\right\rangle$ of core and valent electrons of GQD:

$$\left\langle\mathbf{R}_{\rm qdot}\right| \left.\Psi_j^{\rm qdot}\right\rangle \equiv \Psi_j^{\rm qdot}(\mathbf{R}_{\rm qdot}) \in \left\{\psi_{L_k}^{(0)}\left(\mp\mathbf{q}_j, \mathbf{r}+\mathbf{L}_k\right)\right\}_{k=1,\ r=a}^{N_{\rm qdot}}, \tag{4.16}$$

$\epsilon_{c_i}(\mathbf{R}_{\rm qdot}) \in \left\{\pm\epsilon_{L_k}(q_i)\right\}_{k=1,\dots,N_{\rm dot}}$; $\psi_{L_k}^{(0)}(\mp\mathbf{q}_i, \mathbf{r}+\mathbf{L}_k)$ and $\pm\epsilon_{L_k}(q_i)$ belong to the set (4.1), and for the valence electron the latter are an eigenvalue and a wave function for the same folding zone, and for the core electron they belong to different folding zones.

4.4 Simulation Methods

Quantization conditions can be obtained by "folding zones" applied in both directions allowing by quantum-dot symmetry. The quantization condition for the collective excitation with a wave-vector $\mathbf{k} = (k_x, k_y)$ for graphene nanotube with a chiral vector $\mathbf{C}$ reads

$$\mathbf{k} \cdot \mathbf{C} = 2\pi m, \quad m = 0, 1, \dots, N. \tag{4.17}$$

Analogous, for the quantum dot we have to choose the following vectors as the basis vectors $\mathbf{C}_1 = (2n_1 + 1)\mathbf{b}_1$, $\mathbf{C}_2 = (2n_2 + 1)\mathbf{b}_2$ of the supercell, this results to the following systems for quantized wavevectors

$$\begin{aligned} (2n_1 + 1)\mathbf{b}_1 \cdot \mathbf{k} = 2\pi m_1, \quad m_1 = 0, 1, \dots, N; \\ (2n_2 + 1)\mathbf{b}_2 \cdot \mathbf{k} = 2\pi m_2, \quad m_2 = 0, 1, \dots, N, \end{aligned} \tag{4.18}$$

where N^2 is a number of π(p$_z$)-electrons. Discrete set of energies is indexed by two integers m_1, m_2 and is based on graphene energy band spectrum $\epsilon(\mathbf{k})$. The last is used to find a distance from $\mathbf{k}_{mn} = (k_m, k_n)$ to a nearest Dirac point for a given

m, n. As a graphene model we use the earlier developed quasi-relativistic one in q^4 approximation for the exchange interactions (see detail in [14, 15]). The massless Dirac fermion model for graphene band structure is known to be good enough as a low energy excitations description up to 1 eV. For the quantum dot we have to work with the whole rather than the first Brillouin zone. This means accounting the **k** aside the Dirac cone apex in such a way to choose the nearest K-point used for energy evaluation. As an initial wave functions set for graphene quantum dot model we use the N^2 plane waves with appropriate $\mathbf{k}_{ij}$ and construct the lattice representation of this set in accord with the quantization condition (4.18). Spinor component of the waves are constructed as solutions of the appropriate Dirac-like equation within the framework of the graphene model [15].

Quantum Mechanics perturbation theory for non-orthogonal eigenfunctions sets should been used. Non-perturbed problem for the system (4.14) reads

$$H_0 \left|\psi_i^{(0)}\right\rangle = E_i^{(0)} \left|\psi_i^{(0)}\right\rangle \tag{4.19}$$

where the energy is determined as $E_i = E_i^{(0)} + \Delta E_i$. The solution of the perturbed problem

$$(H_0 + V) \left|\psi_i\right\rangle = E_i \left|\psi_i\right\rangle \tag{4.20}$$

for the state $\left|\psi_i\right\rangle$ is sought as a series on unperturbed eigenstates $\left|\psi_i^{(0)}\right\rangle$ in the form

$$\left|\psi_i\right\rangle = \sum_k c_{ik} \left|\psi_k^{(0)}\right\rangle. \tag{4.21}$$

After substitution (4.21) into (4.20) one gets

$$(H_0 + V) \sum_k c_{ik} \left|\psi_k^{(0)}\right\rangle = (E_i^{(0)} + \Delta E_i) \sum_k c_{ik} \left|\psi_k^{(0)}\right\rangle. \tag{4.22}$$

Scalar multiplying both sides of the last equation on $\left\langle\psi_m\right|$ and simplifying we get the generalized eigenvalue problem

$$\sum_k c_{ik} \left(V_{mk} - \Delta E_i \, S_{ij}\right) = 0 \tag{4.23}$$

where $V_{mk} = \left\langle\psi_m^{(0)}\right| V \left|\psi_k^{(0)}\right\rangle$, $S_{ij} = \left\langle\psi_i|\psi_j\right\rangle$ is the overlapping integral. The eigenenergies are given by

$$\mathrm{Det}\left(\hat{V} - \Delta E \, \hat{S}\right) = 0. \tag{4.24}$$

Simulations with pseudopotential (4.15) has been performed for lowest 40 levels. First order corrections for the energies and wave functions have been calculated.

4.5 Numerical Results and Comparison with Experiment for Electrostatically Confined QDs in Monolayer Graphene

Comparison of two spectra, initial one and the first perturbation theory results are shown in Fig. 4.3. As one can see, the perturbed spectrum is represented by five bands. Lowest one is a very narrow and consists of ten very near placed lines. Index of eigenstate, wave numbers and calculated energy levels in the folding zone approximation are given in the first three columns of the Table 4.1. As one can see from the table, some levels are degenerated. The resulting energy levels for first

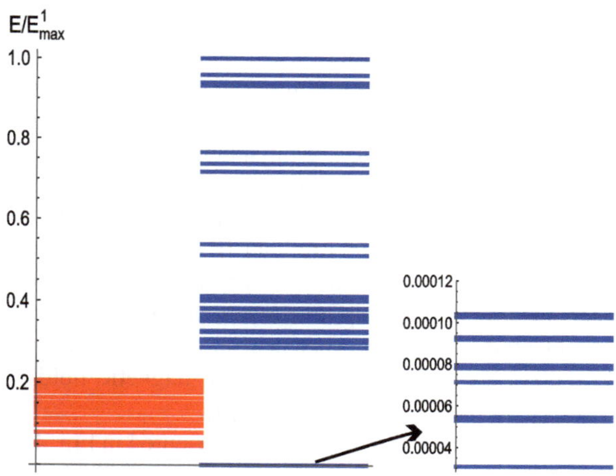

Fig. 4.3 Energy spectra for graphene quantum dot supercell from 51 × 51 primitive cells. Folding zone approximation (left); pseudo-potential approximation, first order perturbation theory result (right). Spectra are normalized on largest value of the right spectrum

Table 4.1 List of lowest eigenstates of the model graphene quantum dot

No.	$\mathbf{k}_i/\vert K_A\vert$	$E^{(0)}(\mathbf{k}_i)$, eV	$E_i^{(1)}$, eV
1	(0., –0.0522603)	0.682502	0.00044269
2	(0., 0.0522603)	0.787028	0.000762951
3	(–0.020115, –0.0871006)	**1.13688**	0.000785459
4	(0.020115, –0.0871006)	**1.13688**	0.00102502
5	(–0.020115, 0.0871006)	*1.37428*	0.00112078
6	(0.020115, 0.0871006)	*1.37428*	0.0011403
7	(0., –0.121941)	1.44642	0.00132088
8	(–0.04023, –0.121941)	**1.61086**	0.00133497
9	(0.04023, –0.121941)	**1.61086**	0.0014769
10	(–0.020115, –0.156781)	*1.80494*	0.00149691
11	(0.020115, –0.156781)	*1.80494*	4.10456

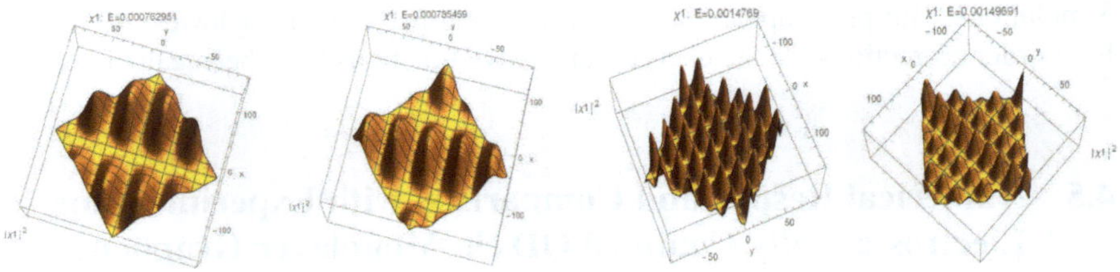

Fig. 4.4 Amplitude squares $|\chi|^2$ of strongly localized wave functions for different energy GQD-levels; energy values E are in insets to figures

order perturbation theory applied (again in sorted order) are shown in the fourth column of the Table 4.1. They already do not correspond to states with definite wave numbers $\mathbf{k}_i$ and are non-degenerated. Eigenfunctions for several lowest levels are shown in separate figures. As one can see electrons are localized on holes of n–p–n graphene junction. Figure 4.4 shows that every electron is confined by its hole but is not annihilated with it. Levels in Hartree–Fock approximations are ionization levels of the graphene QD.

The pseudopotential removes the degeneracy of the energy levels for the GQD-supercell and localizes the valent electrons of the GQD-model on the holes of the n–p–n graphene junction. Non-spherical symmetry of GQD wave functions in a magnetic field leads to lifting of spin and valley degeneration. Therefore, a tunneling current differential of the electrons pulled out from the holes by different electrical tip-induced fields yield more than 10 quadruplets of charging peaks for superlattice "graphene on boron nitride" in STM-experiments [9] Data in Table 4.1 and Fig. 4.3 with sign "–" are ten top hole levels of our n–p–n-junction model. These predicted by our model levels are very near in energy and are in the range from 0.00044269 to 0.00149691 eV. Their splitting in magnetic fields would explain the Coulomb staircase in the STM-experiments [9].

According to simulation results presented in Fig. 4.5a, b, there are weak localized energy levels, for example, with energies 0.00044269 and 0.00102502 eV, where electrons in GQD persist to be almost free. Thus, Klein paradox is absent in our model. Moreover, these weak localized states are wide ones that fit experimental data in Fig. 4.5c much better than theoretical curves in [8]. The weak localization explains the "tail" in distribution of hole density outside the quantum dot (see Fig. 4.5c). Besides, the presence of numerically revealed strongly localized GQD-states in Fig. 4.4 explains atomic-like distribution of electron density observed in STM-experiments [1, 8].

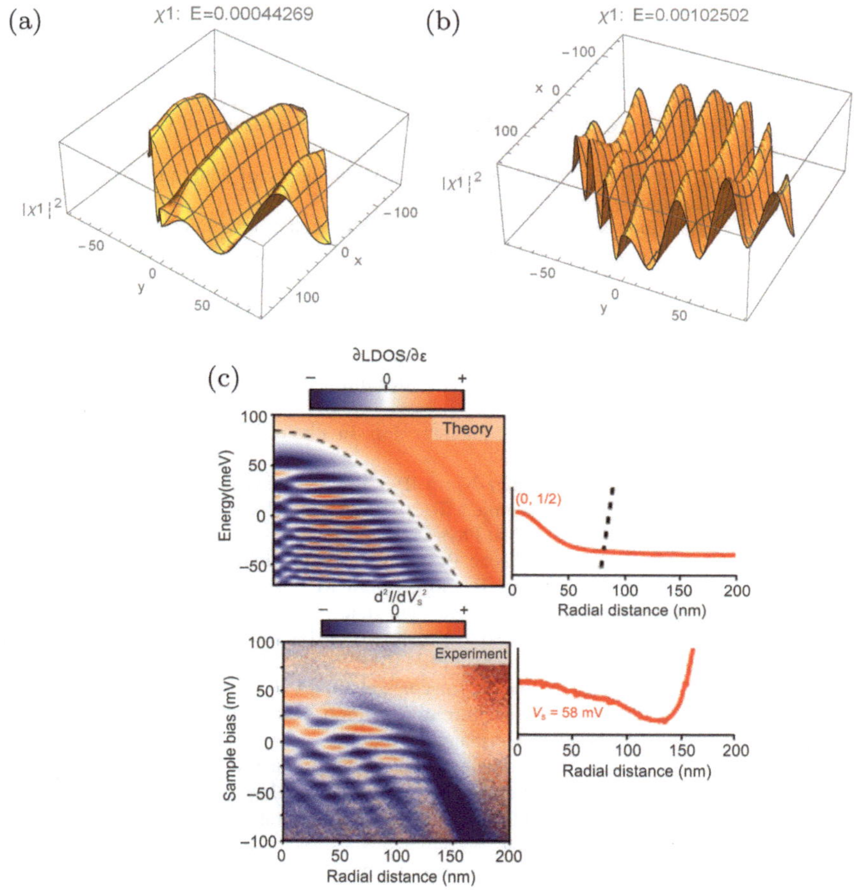

Fig. 4.5 Weak localized wave functions: amplitude square $|\Psi|^2$ for an energy GQD-level 0.00044269 (**a**) and 0.00102502 eV (**b**); (**c**) atom-like graphene QD with Gaussian potential: theory (up) and experiment (down) [8]

4.6 Conclusion

To summarize, the quasi-relativistic models of graphene n-p-n (p-n-p) junctions have been proposed. A supercell pseudopotential which electrically confines electrons (holes) in graphene quantum dot has been found. This potential has a finite depth.

References

1. Gutiérrez Ch, Brown L, Kim C-J, Park J, Pasupathy AN (2016) Klein tunnelling and electront rapping in nanometre-scale graphene quantum dots. Nat Phys 12:1069
2. Zhang J, Yu S-H (2016) Carbon dots: large-scale synthesis, sensing and bioimaging. Mater Today 19:382
3. Bogolubov NN, Soldatov NN, Kruchinin SP (2015) The theory of quantum dots in external magnetic field. Quantum Matter 4:352
4. Soldatov AV, Bogolyubov NN Jr, Kruchinin SP (2006) Method of intermediate problems in the theory of Gaussian quantum dots placed in a magnetic field. Condes Matter Phys 9(1):1
5. Repetsky SP, Vyshyvana IG, Molodkin VB, Kruchinin SP, Lizunov VV (2017) Influence of the adsorbed atoms of potassium on an energy spectrum of graphene. Metallofiz Noveishie Tekhnol 39:1017–1022
6. Kruchinin SP, Repetsky SP, Vyshyvana IG (2016) Spin-dependent transport of carbon nanotubes with chromium atoms. In: Bonca J, Kruchinin S (eds) Nanpmaterials for security. Springer, Dordrecht, pp 65–97
7. Giavaras G, Nori F (2012) Tunable quantum dots in monolayer graphene. Phys Rev B 85:165446
8. Lee J, Wong D, Velasco J Jr, Rodriguez-Nieva JF, Kahn S, Tsai H-Z, Taniguchi T, Watanabe K, Zettl A, Wang F, Levitov LS, Crommie MF (2016) Imaging electrostatically confined Dirac fermions in graphene quantum dots. Nat Phys 12:1032
9. Freitag NM, Chizhova LA, Nemes-Incze P, Woods CR, Gorbachev RV, Cao Y, Geim AK, Novoselov KS, Burgdörfer J, Libisch F, Morgenstern M (2016) Electrostatically confined monolayer graphene quantum dots with orbital and valley splittings. Nano Lett 16:5798
10. Maksym PA, Roy M, Craciun MF, Russo S, Yamamoto M, Tarucha S, Aoki H (2010) Proposal for a magnetic field induced graphene dot. J Phys Conf Ser 245:012030
11. Fock VA (1978) Fundamentals of quantum mechanics. MIR Publishing, Moskou
12. Krylova H, Hurski L (2013) Spin polarization in strongly correlated systems. LAP Lambert Academic Publishing. Saarbrücken, Germany
13. Kudryashov VV (2009) Exact wave functions of electron in a quantum dot with account of the Rashba spin-orbit interaction. Int J Nonlin Phen Compl Sys 12:199
14. Grushevskaya HV, Krylov G (2014) Quantum field theory of graphene with dynamical partial symmetry breaking. J Mod Phys 5:984
15. Grushevskaya HV, Krylov G (2015) Semimetals with Fermi velocity affected by exchange interactions: two dimensional Majorana charge carriers. Int J Nonlin Phen Compl Sys 18:266

Chapter 5
Magnetic Properties of *n*CoO/(1-*n*)ZnO Nanocomposites Obtained by Calcination

N. Guskos, G. Zolnierkiewicz, J. Typek, S. Glenis, D. Sibera, and U. Narkiewicz

Abstract Magnetic properties of the *n*CoO/(1-*n*)ZnO (n = 0.4, 0.50, 0.60 and 0.70) nanocomposites obtained by using traditional wet chemical synthesis method followed by calcination at 600 °C were investigated by dc magnetometry and magnetic resonance spectroscopy. XRD measurements revealed the presence of only two phases: ZnO (hexagonal nanocrystals with sizes in 64–300 nm range) and spinel phase Co_3O_4 (spheroidal nanocrystals with sizes in 14–21 nm range). Magnetic dc susceptibility measurements in 2–300 K range revealed dominating paramagnetic behavior in the whole temperature range and the presence of a strong temperature-independent component. With exception of $n = 0.70$ sample, no behavior connected with the expected phase transition to antiferromagnetic phase in Co_3O_4 and superparamagnetism was registered. Experimental results could be consistently explained by assuming that the most of high-spin Co^{2+} ions are involved in formation of antiferromagnetic pairs or clusters. Low intensity electron paramagnetic resonance spectra registered at RT were attributed to two different magnetic components – one involving paramagnetic Co^{2+} ions at Zn^{2+} sites in ZnO and the other due to Co^{2+} in Co_3O_4 phase or more probably the superparamagnetic resonance of Co_3O_4 nanoparticles. The former component dominates in nanocomposites with small concentration of cobalt, the latter in highly Co concentrated samples.

Keywords Magnetic properties · Zinc oxide nanocomposite · Cobalt oxides · Magnetic resonance

---

N. Guskos (✉) · G. Zolnierkiewicz · J. Typek
Institute of Physics, Faculty of Mechanical Engineering and Mechatronics, West Pomeranian University of Technology, Szczecin, Poland
e-mail: guskos@zut.edu.pl

S. Glenis
Department of Solid State Physics, Faculty of Physics, University of Athens, Athens, Greece

D. Sibera · U. Narkiewicz
Department of Chemical and Environment Engineering, West Pomeranian University of Technology, Szczecin, Poland

J. Bonča, S. Kruchinin (eds.), *Nanostructured Materials for the Detection of CBRN*, NATO Science for Peace and Security Series A: Chemistry and Biology,
https://doi.org/10.1007/978-94-024-1304-5_5

5.1 Introduction

Zinc oxide (ZnO) nanocomposites doped with different transition metal group ions have already demonstrated enormous potential in applications in microelectronic devices due to unique and specific physical and chemical characteristics [1–3]. Especially their various magnetic properties seems to be promising in novel and quickly developing area of spintronics. Particularly important is the room temperature ferromagnetism (RTFM), observed in in V-, Mn-, Fe-, Co-, Ni-, and Cu-doped ZnO, although reports of detected ferromagnetic (FM) properties are inconsistent [3–11]. This might be the result of the sensitivity of RTFM phenomenon to the preparation method of ZnO nanocomposites. It is not excluded that RTFM may arises from the precipitation of the secondary phases and not from the replacement of Zn ion by the transition metal ion in ZnO lattice. On the other hand, theoretical study suggested that the oxygen vacancies can make a significant contribution to RTFM by forming a bound magnetic polaron [16]. Consequently, the presence of intrinsic FM in transition metal doped ZnO is still an open topic in material science, mostly due to a low reproducibility of the results obtained from samples prepared by very different techniques.

Kuryliszyn-Kudelska et al. presented structural (XRD study) and magnetic properties (ac magnetic susceptibility and dc magnetometry) of Co doped ZnO nanocrystals prepared by two different methods: the microwave assisted hydrothermal synthesis and the traditional wet chemistry followed by calcination [12, 13]. Two series of samples in a wide range of CoO nominal concentrations were obtained: from 5 wt.% to 80 wt.% for the calcination method and from 5 wt.% to 60 wt.% for the hydrothermal method. The mean crystalline size of the magnetic phases (Co_3O_4 in calcined and $ZnCo_2O_4$ in hydrothermal samples) measured by X-ray diffraction (XRD) varied from 14 to 55 nm in samples obtained by the calcination method, while for samples obtained by the hydrothermal method it was in range from 33 to 77 nm. For ZnO phase the average size varied from 64 to 300 nm. The obtained results indicate that in both types of samples (hydrothermal or calcined) the crystalline size of the magnetic phases did not show any monotonous dependence. The crystalline size of ZnO decreased with the increase of nominal concentration of CoO for samples obtained by calcination, while for the second method of synthesis an opposite effect was observed. From scanning electron microscopy measurements two morphologies were distinguished: spherical and hexagonal nanograins. The hexagonal nanocrystals were assumed to correspond to the ZnO phase, while the spheroidal nanocrystals to the spinel phases (Co_3O_4 in samples produced by the calcination process and $ZnCo_2O_4$ by the hydrothermal method).

Magnetic study of these types of samples has shown two different types of magnetic behaviour: spin-glasslike and superparamagnetic behaviour, depending on the synthesis processes. In the high temperature range the Curie-Weiss behaviour was registered by ac magnetic susceptibility measurements. It was observed that the determined negative value of the Curie-Weiss temperature T depended strongly on the nominal content of cobalt oxide. It was also revealed that for samples synthesized by the calcination method the values of T increased with the increase of Co content, indicating enhancement of antiferromagnetic (AFM) interaction. On the other hand, for samples obtained by the hydrothermal method the opposite effect was detected indicating the breakdown of the predominance of AFM coupling with the increase of nominal Co ion content. Recently, the magnetic resonance investigation of a new series of $(CoO)_n/(ZnO)_{1-n}$ nanocomposites (where the Co composition index $n = 0.4, 0.5, 0.6$ and 0.7) prepared by the hydrothermal synthesis have shown in the spectra the presence of three main components: a broad line attributed to nanoparticles of $ZnCo_2O_4$ phase two narrow lines arising from Co^{2+} ions in ZnO nanoparticles, and a narrow component probably due to the Co_3O_4 phase [14].

Cobalt oxide (Co_3O_4) is a key phase that is formed during calcination synthesis in heavily Co doped ZnO samples. It crystallizes in the normal cubic spinal structure $Co^{2+}(Co^{3+})_2(O^{2-})_4$ (space group O_h^7) with Co^{2+} and Co^{3+} placed at tetrahedral (A-sites, magnetic moment 4.14 μ_B) and octahedral sites (B sites, diamagnetic), respectively [15, 16]. AFM coupling of A-sites ions brings about the AFM ordered phase below Neel temperature (observed in 30–40 K range). For bulk Co_3O_4 $T_N = 33$ K is reported and in the ordered magnetic state the magnetic moment on Co^{2+} sites is 3.26 μ_B at 4.2 K. It has been observed that in case of AFM nanoparticles the Neel temperature is reduced with respect to the bulk material and many new magnetic phenomena might appear (weak FM, spin canting, exchange bias effect, etc.) due to uncompensated surface or core spins [17–25]. An interesting study of magnetic properties of spinel Co_3O_4 (111) surface and its interface with ZnO (0001) has been reported by Kupchak et al. [26]. Spinel surface, containing Co^{2+} and Co^{3+} ions and terminated with either cobalt or oxygen ions were considered. Calculations have shown that Co^{3+} ions have non-zero magnetic moments at the surface and interface, leading to the FM ordering. Thus such magnetic ordering might be one of the possible origins of the magnetism in calcined samples It was also shown that the calcination temperature has a significant influence on magnetic resonance spectra of nanocrystalline cobalt oxide [27].

Recently, a new series of $(CoO)_n/(ZnO)_{1-n}$ nanocomposites synthesized by calcination method was created. It allowed to obtain more concentrated Co nanocomposites (up to $n = 0.7$). Magnetisation study of these samples will be described and discussed in this paper. Results of magnetic susceptibility measurements in zero field cooled (ZFC) and field cooled (FC) modes in n K temperature range as well as isothermal magnetisation in an external magnetic field up to 70 kOe will be analysed. Additionally, EPR/FMR (electron paramagnetic resonance/ferromagnetic resonance) method which is very useful in clarification of many unresolved problems in nano-magnetism of ZnO will be used. From the obtained experimental results the information about magnetic systems responsible for the observed characteristics and the involved magnetic interactions will be deduced.

5.2 Experimental

The nanocomposites of nCoO/(1 − n)ZnO ($n = 0.4$, 0.50, 0.60 and 0.70) were obtained by using traditional wet chemical synthesis method followed by calcination [15] A mixture of cobalt and zinc hydroxides from an aqueous solution of nitrites was obtained and the hydroxides were filtered, dried and calcined at 600 °C for 1 h. The detailed structural investigations by XRD revealed the presence of the crystalline phases of hexagonal ZnO and cubic Co_3O_4 The mean crystalline sizes of the magnetic phase (Co_3O_4) in $n = 0.4$, 0.50. 0.60, 0.70 samples were 17, 21, 14, and 15 nm, respectively The crystalline size of ZnO decreased with the increase of nominal concentration of CoO and for nanocomposites $n \geqslant 0.60$ the ZnO phase was not registered in XRD measurements.

Magnetization study was performed using a Quantum Design Magnetic Property Measurements System MPMS XL-7 with superconducting quantum interference device magnetometer in magnetic fields up to 70 kOe in the 2–300 K temperature range, in ZFC and FC modes The measurements of magnetic resonance spectra were performed on a conventional X-band ($\nu = 9.4$ GHz) Bruker E 500 spectrometer with 100 kHz magnetic field modulation. Samples were studied in 90–290 K temperature range.

5.3 Results and Discussion

Figure 5.1 shows the temperature dependence of dc magnetic susceptibility χ (defined as $\chi = M/H$) for three samples ($n = 0.40$, 0.50, 0.70) registered in an external magnetic field $H = 100$ Oe in ZFC mode. For sample $n = 0.60$ a very weak magnetic response was recorded with a strong diamagnetic component so this sample will not be discussed further due to very uncertain and not repeatable results. Temperature dependence of dc susceptibility of the three samples displayed in Fig. 5.1 shows a behavior typical for paramagnetic materials, with an intense increase of χ in the low temperature range. ZFC and FC modes produced similar $\chi(T)$ dependences for our samples. Absence of any indication of the paramagnetic-AFM transition that is usually observed for Co_3O_4 nanoparticles (in 30–40 K range) or a peak in ZFC mode related to the blocking temperature T_B is an unexpected phenomenon that needs explanation. Only for sample $n = 0.70$ there is an additional component visible in 20–30 K range superimposed on the dominating paramagnetic background. That additional component shows a local peak in $\chi(T)$ dependence at $T = 24$ K which may be interpreted either as the blocking temperature of Co_3O_4 nanoparticles or Neel temperature of the phase transition to AFM phase.

To verify if the Curie-Weiss law is fulfilled for our samples, Fig. 5.2 presents the temperature dependence of reciprocal susceptibility $\chi^{-1}(T)$. As there seems to be no relatively broad temperature range were the experimental points would aligned along the straight line, an additional term in $\chi^{-1}(T)$ fitting was added

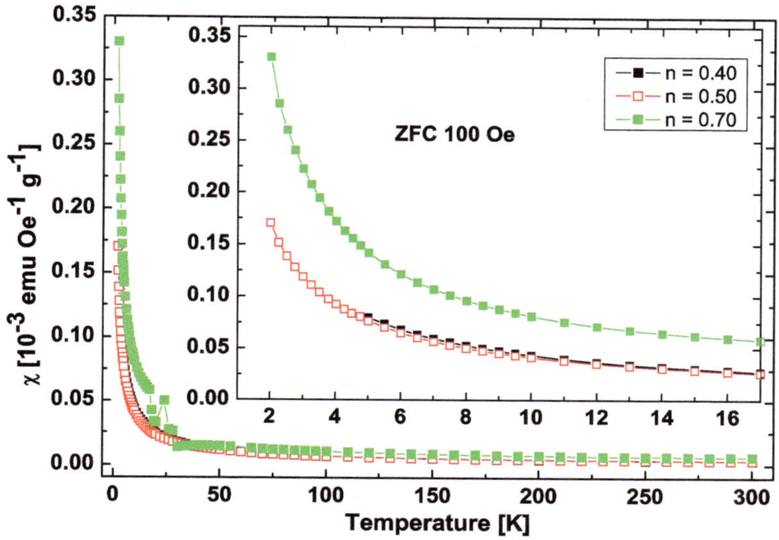

Fig. 5.1 Temperature dependence of the dc magnetic susceptibility $\chi(T)$ for three investigated samples in magnetic field $H = 100$ Oe measured in ZFC mode. The insert shows magnified view of $\chi(T)$ curves in the low temperature range

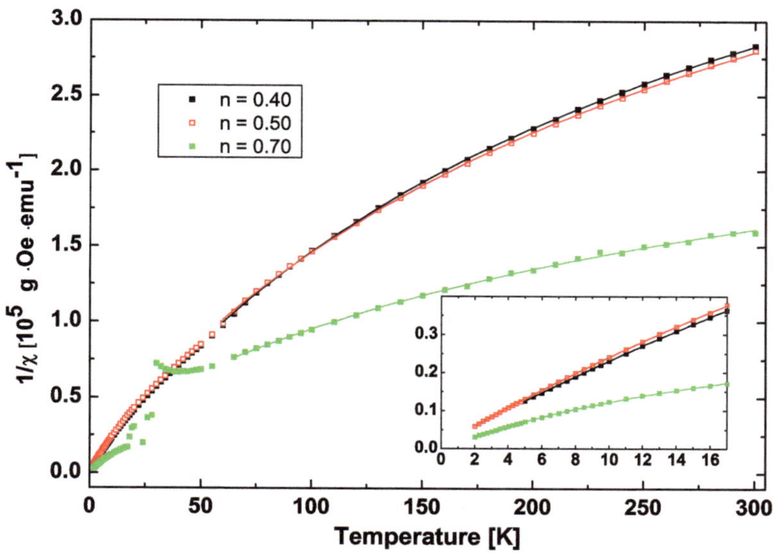

Fig. 5.2 Temperature dependence of the inverse magnetic susceptibility $\chi^{-1}(T)$ for three investigated samples in magnetic field $H = 100$ Oe measured in ZFC mode. The insert shows magnified view of $\chi^{-1}(T)$ curves in the low temperature range. Solid lines are the best least-squares fits to the Eq. (5.1)

Table 5.1 Values of the fit parameters in Eq. (5.1) in the high- and low-temperature ranges for the three investigated samples. High-temperature range: 50–290 K, low-temperature range: 2–16 K

Sample	Temperature range	$C\left[\frac{\text{emu}\cdot\text{K}}{\text{Oe}\cdot\text{g}}\right]$	$C\left[\frac{\mu_B}{\text{Co ion}}\right]$	T_0 [K]	$\chi_0\left[\frac{\text{emu}}{\text{Oe}\cdot\text{g}}\right]$
$n = 0.40$	High	$5.10(6) \times 10^{-4}$	0.87(1)	–2.2(0.8)	$1.85(2) \times 10^{-6}$
	Low	$4.10(3) \times 10^{-4}$	0.78(1)	–0.36(3)	$4.05(9) \times 10^{-6}$
$n = 0.50$	High	$5.30(9) \times 10^{-4}$	0.80(1)	–6.5(1.5)	$1.84(3) \times 10^{-6}$
	Low	$3.90(2) \times 10^{-4}$	0.68(1)	–0.36(2)	$4.32(9) \times 10^{-6}$
$n = 0.70$	High	$9.1(4) \times 10^{-4}$	0.88(1)	–28.9(3.6)	$3.4(1) \times 10^{-6}$
	Low	$5.60(6) \times 10^{-4}$	0.69(1)	+0.18(3)	$20.0(2) \times 10^{-6}$

that was independent on temperature. Thus the fitting function used for temperature dependence of dc susceptibility contained two terms

$$\chi(T) = \frac{C}{T - T_0} + \chi_0, \tag{5.1}$$

where C is the Curie constant, T_0 is the Curie–Weiss temperature, and χ_0 represents temperature independent susceptibility. Fitting experimental data with function (5.1) gave very satisfactory results only in case when it was done in two limited temperature ranges: high-temperature range (50–300 K) and low-temperature range (2–16 K). The solid lines in Fig. 5.2 are the best least-squares fits to the experimental data. The obtained values of the fit parameters (C, T_0, χ_0) are listed in Table 5.1. If the Curie constant is express in $\frac{\text{emu}\cdot\text{K}}{\text{mol}\cdot\text{Oe}}$ unit (molar Curie constant – C_{mol}) than the equation below enables calculation of the effective magnetic moment (in units of Bohr magnetons) for a single Co ion:

$$\mu_{\text{eff}} = \sqrt{\frac{3k_B C_{\text{mol}}}{N_A \mu_B^2}} \cong 2.824\sqrt{C_{\text{mol}}}, \tag{5.2}$$

where k_B is Boltzmann constant, N_A is Avogadro constant, and μ_B Bohr magneton. The obtained values of μ_{eff} for the three investigated samples are also listed in Table 5.1. In the high-temperature range μ_{eff} are slightly bigger (0.80–0.88 μ_B) than in the low-temperature range (0.68–0.78 μ_B) but still significantly smaller than expected for the high-spin Co^{2+} ion ($S = 3/2, g = 2.2, \mu \sim 4.2\mu_B$). This smallness of the average effective moment can be the result of AFM paring or clustering of majority of cobalt ions producing magnetic entities with a magnetic moment equal to zero. If this is the case, our experimental results (assuming $\mu_{\text{eff}} = 0.85\ \mu_B$) means that only 20% of all high-spin Co^{2+} ions is not involved in AFM pairs or clusters and forms paramagnetic centers.

The Curie–Weiss temperature T_0 conveys the information about the strength of the interaction between paramagnetic species and the type of that interaction (FM for positive, AFM for negative T_0 values). In the high-temperature range this parameter is negative what suggests AFM interaction between Co ions and its strength increase with Co content – the higher Co concentration, the stronger the AFM interaction. On the other hand, in the low-temperature range there seem to be only a very weak interaction that does not depend on Co concentration in our samples.

Temperature independent susceptibility χ_0 has much higher value in low-temperature range than in high-temperature range (Table 5.1). As the concentration dependence of χ_0 is concerned it seems to be constant for smaller Co concentrations and only displays substantial increase in $n = 0.70$ sample. Another factor that seems to play an important role in establishing the value of χ_0 is an average size of nanoparticles. Valeeva et al. have investigated the dependence of the Van Vleck paramagnetism (which is temperature independent) in nanocrystals of superstoichiometric titanium monoxide [28]. This investigation has established that this type of paramagnetism is inversely proportional to the average size of nanocrystals because of breakage of the symmetry of a local environment of the near-surface atoms of titanium and oxygen. Thus the Van-Vleck paramagnetism contribution due to atomic-vacancy disorder in superstoichiometric titanium monoxide nanocrystals, as well as in the stoichiometric composition, is proportional to a deviation of the degree of long-range order from its maximum value [28]. Temperature dependence of dc susceptibility measurements of our three samples seems to corroborate such conclusion (see Table 5.1). Sample $n = 0.70$ contains nanoparticles having the smallest average size (15 nm) and the biggest χ_0 term (3.4×10^{-6}), while sample $n = 0.5$ with the largest average size (21 nm) has the smallest χ_0 (1.84×10^{-6}). Thus our observations indicate that both the concentration of magnetic ion and the size of nanoparticles might influence the magnitude of the temperature independent susceptibility.

To examine if any FM component exists in our samples, the isothermal magnetization measurements at high (300 K) and low (2 and 5 K) temperatures in an external magnetic field up to 70 kOe were carried out. The results are presented in Fig. 5.3. As can be seen in the lower insert in this figure, a very insignificant hysteresis loop can be recognized at low temperature (2 and 5 K). The coercive field H_c is of the order of 20 Oe, and the remnant magnetization is not higher than 0.005 emu/g. These extremely small values indicate that only a very insignificant part of all magnetic ions (of the order of 10^{-5}) is involved in FM clusters. AFM nanoparticles can cause weak FM due to its uncompensated surface spins caused by the breaking of the symmetry and a reduction of the nearest neighbour coordination. At RT no evidence of FM component is to be seen.

Microstructure of Co doped ZnO nanocrystals seems to play an important role in establishing their magnetic properties [10]. Certain microstructural features can even cause the appearance of FM phase. It was noticed that samples with very homogenous microstructure can display FM, but they have a very small saturation moment, well below that expected for Co^{2+} ion. Conversely, samples showing a defective microstructure are paramagnetic and the increase of the Co content does not induce ferromagnetism [10]. Taking into account magnetisation results obtained for our samples it could be reasoned that their microstructure is very defective.

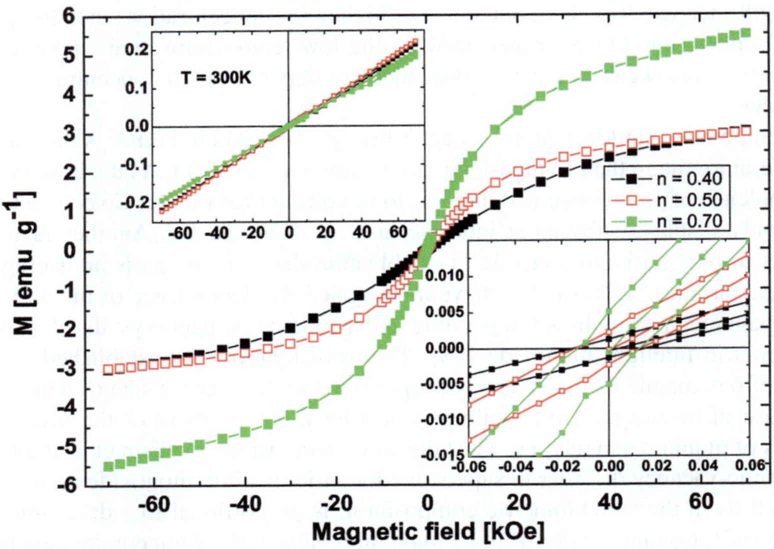

Fig. 5.3 Isothermal magnetisation $M(H)$ of three investigated samples at low temperature (5 K for $n = 0.40$, 2 K for $n = 0.50$ and 0.60 samples). The lower insert shows a magnified central part of the hysteresis loops, the upper insert shows $M(H)$ dependence registered at $T = 300$ K

Figure 5.4 presents magnetic resonance spectra of four studied samples at $T = 290$ K (top panel) and $T = 90$ K (bottom panel). RT spectra are not very intense (low signal/noise ratio), but they gain intensity as samples are cooled to liquid nitrogen temperature. The spectrum consist of two (or in fact three) lines that have been fitted by Lorentzian lineshape curves. Although this simple fitting method is very crude, it might provide required spectral parameters (A – amplitude, H_r – resonance field, and ΔH_{pp} – peak-to-peak linewidth) with accuracy that is satisfactory for introductory analysis of the registered spectra. The knowledge of these parameters allows to calculate also an effective g-factor (using equation $g = \frac{h\nu}{\mu_B H_r}$, where h is Planck constant and ν is microwave frequency) and another important quantity – the integrated intensity, $I = A(\Delta H_{pp})^2$, which is proportional to the magnetic susceptibility of the spin system on microwave frequency. Two lines are easily discerned in EPR spectra of all four samples taken at RT – one in lower magnetic field ~1500 Oe (it will be designated as line 1), the other in higher field ~3000 Oe (line 2). They were fitted by Lorentzian lines and two important spectral parameters: g-factor and the integrated intensity I have been calculated. The values of these parameters are listed in Table 5.2.

In Fig. 5.5 the total EPR integrated intensities of four samples at $T = 290$ and 90 K were displayed. Practically, the same behaviour of intensities as a function of the Co concentration index n is observed at these two different temperatures. An initial decrease of intensity with the increase of n and a final increase of I with further increase of Co content, suggests that the observed EPR spectra are due to

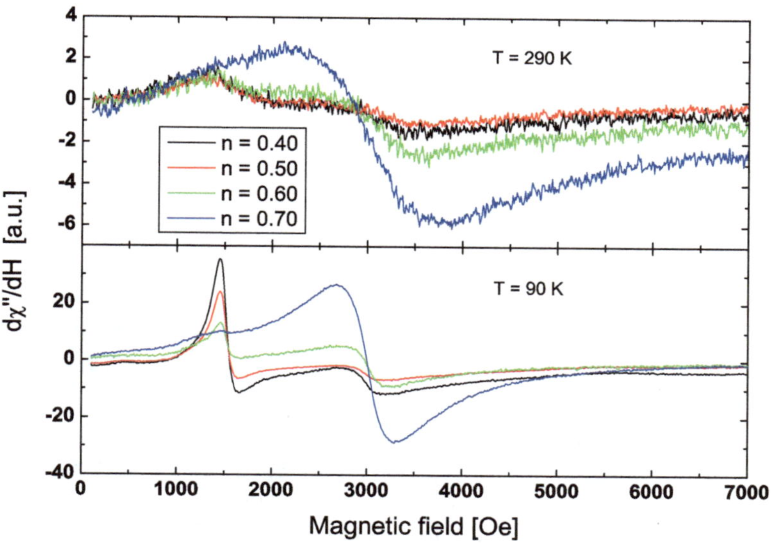

Fig. 5.4 Magnetic resonance spectra of four investigated samples at $T = 290$ K (top panel), $T = 90$ K (bottom panel). The spectra have been normalized to a unit mass of sample

Table 5.2 EPR parameters (g-factor and normalized integrated intensity) of the two lines (labelled as 1 and 2) spectrum registered at $T = 290$ K. All integrated intensities were normalized to the largest value in $n = 0.70$ sample

Sample	g-factor g_1	Integrated intensity I_1 [a.u.]	g-factor g_2	Integrated intensity I_2 [a.u.]
0.40	4.140	0.035	2.340	0.249
0.50	4.085	0.024	2.334	0.144
0.60	4.282	0.021	2.291	0.406
0.70	–	–	2.272	1.000

two types of paramagnetic centres. One centre (centre A) must be located in ZnO nanoparticles and its intensity will be proportional to the content of that phase in our samples, i.e. integrated intensity will decrease with an increase of Co concentration. The other centre (centre B) should show an opposite behaviour and thus must be associated with Co_3O_4 nanoparticles.

Two relatively narrow lines observed in $n = 0.40$, 0.50 and 0.60 samples, one in low magnetic field (line 1, $g \sim 4.2$) and the other near the resonance field 3 kOe (line 2, $g \sim 2.3$) show a very similar behaviour as a function of Co concentration – the higher the concentration of Co (bigger n index), the smaller the EPR amplitude of these two lines. It follows that they belong to the same magnetic component (centre A) and they are located in ZnO phase, which concentration diminishes with an increase of the composition index n. As a matter of fact they belong to the same powder-like EPR spectrum and the line 1 can be identified as

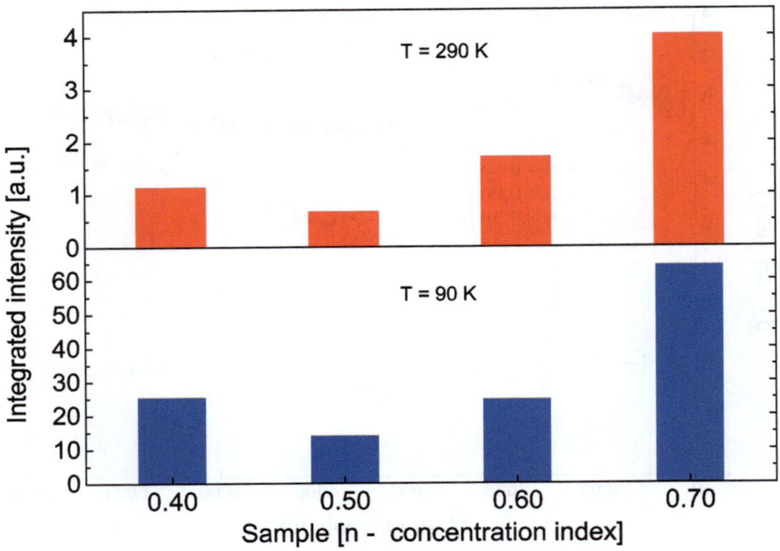

Fig. 5.5 Integrated intensities of four investigated samples at $T = 290\,\mathrm{K}$ (top panel), $T = 90\,\mathrm{K}$ (bottom panel)

the perpendicular component ($g_{\perp}$), while line 2 as the parallel component ($g_{||}$). Our measured g-factors are not very different from the values found for Co^{2+} in tetrahedral Zn^{2+}sites in bulk ZnO what further corroborates the proposed origin of the centre A in our samples. On the other hand the broader line 2 in $n = 0.70$ sample (and partially in sample $n = 0.60$) should be attributed to Co_3O_4 phase as its intensity increases strongly with the n index. Thus component B could be due to either Co paramagnetic defects located in Co_3O_4 or more probably to (mostly) agglomerated Co_3O_4nanoparticles and thus it might be identify with the superparamagnetic resonance line. In the latter case there should be a connection of this EPR line with a local maximum registered in the temperature dependence of ZFC susceptibility in 20–30 K range.

5.4 Conclusions

Magnetization and magnetic resonance studies of a series of nCoO/(1-n)ZnO nanocomposites with composition index $n = 0.4$, 0.5, 0.6, and 0.7 synthesized by calcination at 600 °C have provided interesting magnetic characteristics of these materials. In all studied samples the dominating magnetic response comes from the paramagnetic system consisting of a small part of all Co^{2+} ions (about 20%), the rest of cobalt ions form AFM pairs or clusters that are magnetically silent. This might be the result of highly defective microstructure of our samples. EPR study has shown that the Co^{2+} ions could also be found at Zn^{2+} sites in ZnO phase. Only for sample $n = 0.70$, with a high concentration of cobalt oxide, an additional magnetic component appears in magnetometric and magnetic resonance studies. It might be related to the superparamagnetic Co_3O_4 nanoparticles which produce an intense line in EPR spectrum at RT and a local maximum in ZFC $\chi(T)$ dependence at the blocking temperature $T_B = 24$ K. The blocked Co_3O_4 nanoparticles might be responsible for a very weak FM component detected at low temperatures in form of a hysteresis loop in $M(H)$ dependence.

References

1. Wang ZL (2004) Zinc oxide nanostructures: growth, properties and applications. J Phys Condens Matter 16:R829
2. Chien CH, Chiou SH, Guo GY, Yao YD (2004) Electronic structure and magnetic moments of 3D transition metal-doped ZnO. J Magn Magn Mater 282:275–278
3. Yates K (2011) Diluted magnetic oxides: current status and prospects. In: Nasirpouri F, Nogaret A (eds) Nanomagnetism and spintronics. Fabrication, materials, characterization and applications. Word Scientific Publishing Ltd., Singapore
4. Lawes G, Risbud AS, Ramirez AP, Seshadri R (2005) Absence of ferromagnetism in Co and Mn substituted polycrystalline ZnO. Phys Rev B 71:045201
5. Ueda K, Tabota H, Kamai T (2001) Magnetic and electric properties of transition-metal-doped ZnO films. Appl Phys Lett 79:988
6. Guskos N, Typek J, Zolnierkiewicz G, Wardal K, Sibera D, Narkiewicz U (2011) Magnetic resonance study of nanocrystalline ZnO nanopowders doped with Fe_2O_3 obtained by hydrothermal synthesis. Rev Adv Mat Sci 29:142–149
7. Guskos N, Glenis S, Zolnierkiewicz G, Typek J, Berczynski P, Guskos A, Sibera D, Narkiewicz U (2012) Magnetic properties of $ZnFe_2O_4$ ferrite nanoparticles embedded in ZnO matrix. Appl Phys Lett 100:122403
8. Typek J, Wardal K, Zolnierkiewicz G, Guskos N, Sibera D, Narkiewicz U (2014) Magnetic study of 0.20(Fe_2O_3)/0.80(ZnO) nanocomposite. J Magn Magn Mater 361:12–18
9. Typek J, Wardal K, Zolnierkiewicz G, Szymczyk A, Guskos N, Narkiewicz U, Piesowicz E (2016) Magnetic studies of 0.7(Fe_2O_3)/0.3(ZnO) nanocomposites in nanopowder form and dispersed in polymer matrix. Mater Sci Poland 34:286–296
10. Martínez B, Sandiumenge F, Balcells L, Arbiol J, Sibieude F, Monty C (2005) Structure and magnetic properties of Co-doped ZnO nanoparticles. Phys Rev B 72:165202
11. Rao CNR, Deepak FL (2005) Absence of ferromagnetism in Mn- and Co-doped ZnO. J Mater Chem 15:573-9578
12. Kuryliszyn-Kudelska I, Hadzic B, Sibera D, Romcevic M, Romcevic N, Narkiewicz U, Łojkowski W, Arciszewska M, Dobrowolski W (2013) Magnetic properties of ZnO(Co) nanocrystals. J Alloys Compd 561:247251
13. Kuryliszyn-Kudelska I, Dobrowolski W, Arciszewska M, Romčević N, Romčević M, Hadžić B, Sibera D, Narkiewicz U, Łojkowski W (2013) Transition metals in ZnO nanocrystals – magnetic and structural properties. Sci Sintering 45:31–48
14. Typek J, Guskos N, Zolnierkiewicz G, Sibera D, Narkiewicz U (2017) Magnetic resonance study of Co-doped ZnO nanomaterials: a case of high doping. Rev Adv Mat Sci 50:7687
15. Roth WL (1964) The magnetic structure of Co_3O_4. J Phys Chem Solids 25:1–10
16. Hill AH, Harrison A, Ritter C, Yue W, Zhou W (2011) Neutron powder diffraction and magnetic studies of mesoporous Co_3O_4. J Magn Magn Mater 323:226–231

17. Takada S, Fujii M, Kohiki S, Babasaki T, Deguchi H, Mitome M, Oku M (2001) Intraparticle magnetic properties of Co_3O_4 nanocrystals. Nano Lett 1:379–382
18. Li S, Bi H, Zhang F, Du Y, Jiang X, Yang C, Yu Q, Zhu Y (2004) Anomalous magnetic properties in Co_3O_4 nanoparticles covered with polymer decomposition residues. J Appl Phys 95:7420–7422
19. Dutta P, Seehra MS, Thota S, Kumar J (2008) A comparative study of the magnetic properties of bulk and nanocrystalline Co_3O_4. J Phys Condens Matter 20:015218
20. Farhadi S, Safabakhsh J, Zaringhadam P, Farhadi S (2013) Synthesis, characterization, and investigation of optical and magnetic properties of cobalt oxide (Co_3O_4) nanoparticles. J Nanostruct Chem 3:69
21. Wang Y, Yank CM, Schmidt W, Spliethoff B, Bill E, Schuth F (2005) Weakly ferromagnetic ordered mesoporous Co_3O_4 synthesized by nanocasting from vinyl-functionalized cubic $Ia3d$ mesoporous silica. Adv Mater 17:54–56
22. Kruchinin S, Dzezherya Yu, Annett J (2006) Imteractions of nanoscale ferromagnetic granules in a London superconductors. Supercond Sci Technol 19:381–384
23. Dzezherya Yu, Novak IYu, Kruchinin S (2010) Orientational phase transitions of lattice of magnetic dots embedded in a London type superconductors. Supercond Sci Technol 23:1–5
24. Kruchinin S, Nagao H (2012) Nanoscale superconductivity. Inter J Mod Phys B 26:1230013
25. Kruchinin S, Nagao H, Aono S (2010) Modern aspect of superconductivity: theory of superconductivity. World Scientific, Singapore, p 232
26. Kupchak I, Serpak N, Chkrebtii A, Hayn R (2017) Surface magnetism and electronic structure: ZnO(0001)/Co_3O_4(111). arXiv:1704.07148 [cond-mat.mtrl-sci]
27. Guskos N, Typek J, Maryniak M, Zolnierkiewicz G, Podsiadly M, Arabczyk W, Lendzion-Bielun Z, Narkiewicz U (2006) Effect of calcination and structural additives on the EPR spectra of nanocrystalline cobalt oxides. Mater Sci Poland 24:1095–1102
28. Valeeva AA, Nazarova SZ, Rempel AA (2016) Dependence of Van Vleck paramagnetism on the size of nanocrystals in superstoichiometric TiO_y. J Exp Theor Phys 122:722–726

Chapter 6
Photocatalytic Properties of TiO_2 Thin Films Doped With Noble Metals (Ag, Au, Pd and Pt) for Water Decontamination

C. Moslah, G. A. Mousdis, M. Kandyla, G. Petropoulou, and M. Ksibi

Abstract Access to clean water is a major problem for many people. The use of contaminated drinking-water causes hundreds of thousands deaths every year. Even in the developed countries chemical residuals from the industry or human activities or even from the disinfection process can cause several health problems. There are many methods to purify contaminated water. Recently, semiconductor photocatalytic process has shown a great potential as a sustainable treatment technology. It is low-cost, environmentally friendly and in accordance with the "zero" waste scheme. The ability of the photocatalytic process to oxidize and remove persistent organic compounds and microorganisms in water has been widely demonstrated. Although a lot of work has been done and some commercial devices have been prepared, there are still many scientific challenges.

Keywords Photocatalysis · Water purification · TiO_2 · Photocatalytic reactors · Mineralization

---

C. Moslah
NHRF–National Hellenic Research Foundation, Theoretical and Physical Chemistry Institute-TPCI, Athens, Greece

Laboratory of Environment Engineering and Ecotechnology, ENIS, University of Sfax, Sfax, Tunisia

G. A. Mousdis (✉) · M. Kandyla · G. Petropoulou
NHRF–National Hellenic Research Foundation, Theoretical and Physical Chemistry Institute-TPCI, Athens, Greece
e-mail: gmousdis@eie.gr

M. Ksibi
Laboratory of Environment Engineering and Ecotechnology, ENIS, University of Sfax, Sfax, Tunisia

J. Bonča, S. Kruchinin (eds.), *Nanostructured Materials for the Detection of CBRN*, NATO Science for Peace and Security Series A: Chemistry and Biology,
https://doi.org/10.1007/978-94-024-1304-5_6

6.1 Introduction

6.1.1 *Water Purification*

Safe and readily available water is important for public health, whether it is used for drinking, domestic use, food production or recreational purposes. Due to the rapid development of industrialization, population growth and increasing demand, the access to clean water has become an issue worldwide. Today 1.1 billion people lack access to safe drinking water supply; as a consequence 1.8 million people die from diarrheal disease each year with the vast majority children under 5 years old, caused mainly by polluted water. The problem is expected to worsen in the short future, due to the increasing water contamination from the overwhelming discharge of micropollutants and contaminants into the natural water cycle [1]. Simple techniques for treating water could save a huge number of lives each year [2]. Furthermore improved management of water resources and better quality of water can boost economic growth and contribute to poverty reduction.

Even at the developed countries with modern water purification systems the drinking water is sometimes polluted by chemical substances or even byproducts of the disinfection process (such as trihalomethanes and haloacetic acids) that can cause serious health problems. Also, during the last decades, an emerging issue is the pollution of water from bioactive trace pollutants such as, pharmaceuticals and ingredients of personal care products for which there is no information about their impact to the humans and to the environment. Furthermore because of their low concentrations in water (ppb or ppt) and their complex chemical structure, common technologies used for cleaning of drinking water are not efficient enough to complete removal [3].

Many cases of organic biorefractory compounds pollute the water supplies and remain for many years (e.g. the Methyl Tertiary Butyl Ether a carcinogenic compound [4]). The removal of these soluble refractory organic compounds is both tedious and expensive [5].

A major cause of water pollution is the industrial, agriculture and human wastes that are dumped untreated into bodies of clean water especially at the developing countries. For that reason methods to neutralize or remove the pollutants from the wastes are needed. To solve this problem, various low-cost and high efficiency methods have been adopted to purify the water.

The water pollutants can be separated into 3 main categories:

- Macroscopic (e.g. plastics, papers etc.,)
- Biological (Microorganism, Bacteria, toxins etc.)
- Chemicals (petroleum, pharmaceuticals, insecticides, herbicides, paints, fibers, plastics, solvents and volatile organic compounds). Most of these compounds are toxic and biorefractory, resisting to microbial degradation and remaining in the water for many years. There are also some inorganic pollutants (e.g. heavy metals).

There are mainly methods for purification of the water:

- Physical (filtration, sedimentation, adsorption distillation, etc.,)
- Chemical (chlorination, coagulation, flocculation, etc.,)
- Biological (aerobic and anaerobic biological processes)
- Physicochemical (UV-radiation, Photocatalysis)

The best method depends on several factors like the type and quantity of the pollutants, the use and the quantity of water, the power consumption, the byproducts etc. Usually they use a combination of methods. In most of the cases (e.g. physical methods), they transferring the pollutants to other phases, but still remain concentrated and not being completely "destroyed", or generate toxic secondary pollutants (e.g. chlorination) that are mutagenic and carcinogenic to human health [6, 7].

6.1.2 Photocatalytic Method for Water Decontamination

Heterogeneous photocatalysis [8] is an innovative method, involving utilization of light along with a catalytic solid to oxidize the water chemical pollutants and disinfect from the microorganisms. In recent years this method has shown a great potential as a low-cost, environmentally friendly, sustainable and low-cost, water treatment technology that is in accordance with the "zero" waste scheme [9]. Another important advantage of photocatalysis is that oxidation works even at very low concentrations (even at ppt) because the contaminant is attracted strongly to the surface of the catalyst.

Photocatalysis involves the acceleration of photoreaction in the presence of a semiconductor catalyst. Several semiconductors (TiO_2, ZnO, Fe_2O_3, CdS, ZnS) can act as photocatalysts. It is widely used to describe the process in which the acceleration of a reaction occurs when a semiconductor, interacts with light of sufficient energy (or of a certain wavelength) to produce reactive transitory oxidizing species (i.e. H_2O_2, $OH^{\cdot}$, $O_2^{\cdot -}$, e^-, O_3). These species can oxidize, up to mineralization, the organic pollutants including refractory compounds, as well as pathogens and water disinfection by-products.

Among the semiconductors, titanium dioxide (TiO_2) has received the greatest interest for photocatalytic applications [10] due to its hydrophilicity, high reactivity, reduced toxicity, chemical stability and low cost [11].

TiO_2 exists as three different polymorphs; anatase, rutile and brookite [12]. Titanium dioxide is typically an n-type semiconductor due to oxygen deficiency [13]. The band gap is 3.2 eV for anatase, 3.0 eV for rutile, and ~3.2 eV for brookite. According to the experimental results, anatase has better photocatalytic action than rutile [14].

Fig. 6.1 Photo-induced formation mechanism of electron-hole pair in a TiO_2 semiconductor

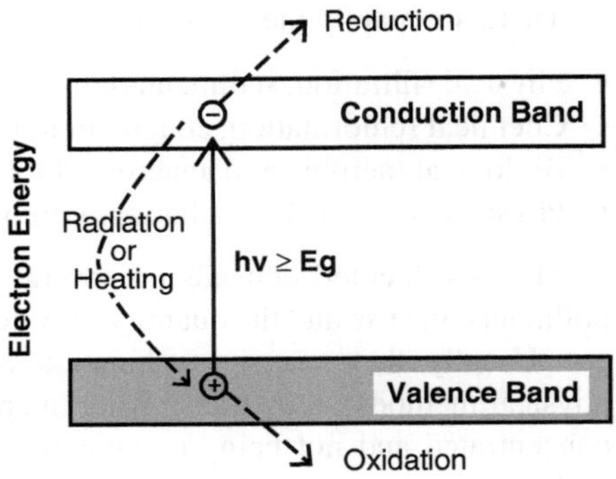

6.1.3 Electronic Processes in TiO_2 Photocatalysis

Photocatalysis does not imply the action of light as a catalyst in a reaction but photon-assisted generation of catalytically active species [15]. The 1st step of photocatalysis is the absorption of a photon with energy greater than the band gap of TiO_2 (in the case of anatase TiO_2 with a band gap 3.2 eV, light with wavelengths $\leqslant$387 nm is required). The absorbed energy excites an electron from the valence band to the empty conduction band in femtoseconds, leaving behind a positive hole in the valence band (Fig. 6.1, Eq. (6.1)).

$$TiO_2 + hv \rightarrow h^+ + e^-. \tag{6.1}$$

These charge carriers can be [16]:

1. recombined radiatively or non-radiatively, dissipating the energy as light or heat,
2. trapped as O^- and Ti^{3+} at defect sites in the TiO_2 lattice,
3. migrated to the catalyst surface and initiate redox reactions with adsorbates.

The photogenerated trapped holes (h^+) at the semiconductor surface can oxidize the adsorbed H_2O, OH^- or surface titanol groups and produce free hydroxyl radicals $OH^{\cdot}$, (Eqs. (6.2), (6.3)) but also direct oxidation of the organics by trapped holes has also been reported.

$$H_2O_{(ads)} + h^+ \rightarrow OH^{\cdot} + H^+, \tag{6.2}$$

$$OH^-_{(ads)} + h^+ \rightarrow OH^{\cdot}. \tag{6.3}$$

The photogenerated trapped electrons e^- react with the adsorbed O_2 which is reduced to form superoxide radical anion $O_2^{\cdot -}$, (Eq. (6.4)) that may further react with H^+ to generate hydroperoxyl radical $HO_2^{\cdot}$ species, (Eq. (6.5)). These reactive oxygen species may also contribute to the oxidative pathways such as the degradation of a pollutant (Eqs. (6.6) and (6.7)) or they can give H_2O_2 (Eq. (6.8)) that can produce more hydroxyl radicals $OH^{\cdot}$ (Eqs. (6.9), (6.10) and (6.11))

$$e^{-} + O_{2(ads)} \rightarrow O_2^{\cdot -}, \tag{6.4}$$

$$O_2^{\cdot -} + H^{+} \rightarrow HO_2^{\cdot}, \tag{6.5}$$

$$O_2^{\cdot -} + C_x|H_y \rightarrow \ldots \rightarrow CO_2 + H_2O, \tag{6.6}$$

$$HO_2^{\cdot} + C_x|H_y \rightarrow \ldots \rightarrow CO_2 + H_2O, \tag{6.7}$$

$$2HO_2^{\cdot} \rightarrow H_2O_2, \tag{6.8}$$

$$H_2O_2 + h\nu \rightarrow 2OH^{\cdot}, \tag{6.9}$$

$$H_2O_2 + O_2^{\cdot} \rightarrow OH^{\cdot} + O_2 + HO^{-}, \tag{6.10}$$

$$H_2O_2 + e^{-} \rightarrow OH^{\cdot} + HO^{-}. \tag{6.11}$$

The hydroxyl radical $OH^{\cdot}$ is a highly reactive specie with stronger oxidation power than ordinary oxidants normally used in the oxidation process, in water (oxidation potential: 2.80 eV) [17] (for comparison the O_3 that is the strongest oxidizing compound used has 2.07 eV). It can oxidize organics RH by abstraction of protons producing organic radicals $R^{\cdot}$ (Eq. (6.12)), which are highly reactive and can be further oxidized to aldehydes, ketones or acids by ring opening reactions, the organics can be completely detoxified by full conversion to CO_2, water and in the case of substituted organics, inorganic salts if the treatment continue [18]. It can oxidize even non-biodegradable organic pollutants such as $CHCl_3$.

$$OH^{\cdot} + RH \rightarrow H_2O + R^{\cdot}. \tag{6.12}$$

The overall process consists from the following steps:

1. Transfer of the organic pollutant(s) from the liquid phase to the TiO_2 surface.
2. Adsorption of the organic pollutant(s) onto the TiO_2 surface.
3. Surface activation of TiO_2 by photon.
4. Photocatalysis reaction for the adsorbed substance on the TiO_2 surface.
5. Desorption of the intermediate(s) (e.g. B) from the TiO_2 surface.
6. Transfer of the reaction products from the TiO_2 surface to liquid.

The total velocity of the reactions depends on the slowest step. Depending on the case, the slowest step can be different (e.g. in the case of very low concentrations of pollutants the slowest step is the 1st. In the case of higher concentrations the slowest are the reaction steps 2, 3, 4, 5. For these reasons the ideal system depends on the sample).

In general there are 2 methods to increase the efficiency of the photocatalyst:

1. to increase the surface, using nanoparticles of the catalyst or mesoporous phases,
2. to increase the yield of the photocatalysis by doping the catalyst with sensitizers,

6.1.4 *Forms of the Catalyst*

Today the best results were obtained with TiO_2 nanoparticles due to the large surface area-to-volume ratio which promotes the efficient charge separation and trapping at the physical surface [19].

The most common photocatalyst is the Degussa P-25 TiO_2 catalyst that is used as a slurry, but then an additional process step is added; the post-separation of the catalysts. The removing of the catalyst nanoparticles has to be done to avoid the loss of catalyst and not pollute the water with the catalyst. The separation of catalyst nanoparticles is difficult and expensive with many technical problems [20].

A solution of this problem could be the immobilization of the nano catalysts on to different substrates (e.g. activated carbon [21, 22], onto a membrane [23], mesoporous clays [24], fibers [25], or glass).

6.1.5 *Doping Catalysts*

The TiO_2 needs radiation with wavelengths lower than 400 nm and in most cases we use lamps to activate it. A solution will be to broaden the photoresponse of the catalyst and use the abundance of outdoor solar irradiation.

That can be done by the introduction of a sensitizer (Metals, metal ions dyes [26] etc.).

There are the following mechanisms (Fig. 6.2)

1. Dye sensitization: The photon is absorbed by the dye molecule and the excited electron is transferred to the conduction band of TiO_2. But dye sensitization is usually accompanied by desorption and recombination of the electron hole during the photoreaction [27].

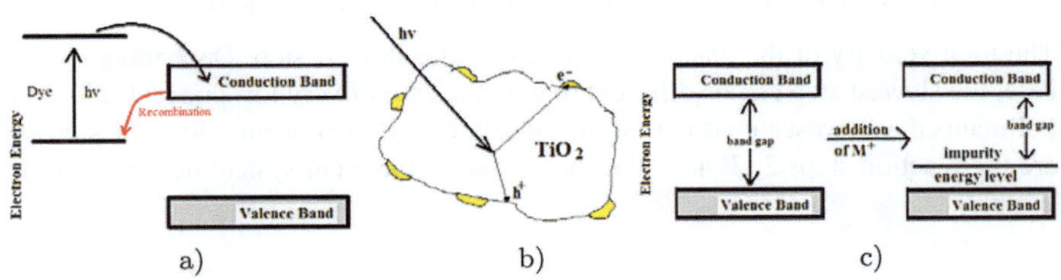

Fig. 6.2 Mechanisms of enhanced photocatalysis when doped (**a**) dye, (**b**) metals, (**c**) crystal doping

2. Noble metal sensitization: Addition of noble metals with Fermi level lower than TiO_2 catalyst (e.g. Au, Pt, Pd etc) photo-excited electrons can be transferred from conduction band to the metal particles, while the holes remain on the TiO_2, resulting in the efficient separation and consequently to stronger photocatalytic reactions [28].
3. Crystal doping: By addition of a metal ion in a lattice site of the titania, we introduce defect sites and incorporate energy levels inside the band gap.

6.1.6 Synthetic Methods for TiO_2 Films

Various techniques have been used for the deposition of TiO_2 films, such as doctor-blade technique [29], chemical vapour deposition [30], electrophoretic deposition [31], flame-aerosol [32], plasma enhanced chemical vapor deposition (PECVD) [33], sputtering deposition [34] etc., but the most common technique is the sol–gel method combined with the spin or dip coating methods.

The sol–gel technique is a widely used method for the deposition of thin films because it presents many advantages such as the utilization of very simple equipment, the possibility of work in normal atmospheric conditions, the high homogeneity of the final films. The sol–gel solutions are composed by a metal alkoxide, an alcohol and water. The metal alkoxide undergoes a partial hydrolysis (6.13) and polymerization (6.14) process. The reactions that take place are:

$$Ti(OR)_4 + 4H_2O \rightarrow Ti(OH)_4 + 4ROH, \tag{6.13}$$

$$2Ti(OH)_4 \rightarrow (OH)_3Ti\text{-}O\text{-}Ti(OH)_3 + H_2O. \tag{6.14}$$

The polymerization process continuous further, until the formation of an oxide network. The solutions can be deposited on the substrate by spinning, or dipping. After this, the sintering of the sample is required for the oxide conversion (removal of solvent and residual organics) and the film densification.

In our work we prepared thin films of TiO_2 on glass, undoped and doped with metals and we studied their photocatalytic properties for purification of water polluted by organic substances.

6.2 Experimental

In our approach we studied the effect of metal doping on the photocatalytic purification properties of TiO_2 film on glass.

The substrates, microscope glass slides ($2.5 \times 3\,cm^2$), were cleaned by soaking for 24 h in a sulforochromic bath, then washed copiously with distilled water and kept in isopropanol until used.

Thin films were prepared using a Laurell WS-400BX-6NPP spin-coater.

6.2.1 Au Precursor

$HAuCl_4$ solution in ethanol was prepared by dissolution metal Au in $HNO_3/3HCl$. After the complete dissolution of the metal the solution was dried under vacuum and the remaining powder was dissolved in a specific volume of ethanol to obtain a 0.3 M solution.

6.2.2 Pt Precursor

H_2PtCl_6 solution in ethanol was prepared by dissolution metal Pt in $HNO_3/3HCl$. After the complete dissolution of the metal the solution was dried under vacuum and the remaining powder was dissolved in a specific volume of ethanol to obtain a 0.3 M solution.

6.2.3 Pd Precursor

A fresh solution was prepared by dissolving $PdCl_2$ in ethanol to prepare a 0.3 M solution. The solution was used immediately.

6.2.4 Ag Precursor

The $AgNO_3$ solution was prepared by dissolving $AgNO_3$(0.25 M) in distilled water

6.2.5 Ti Precursor

The Ti-precursor solution was prepared by mixing 2 different solutions. The “A” solution was prepared by dissolving 1 mL of titanium (IV) isopropoxide in 5 mL of 2-methoxyethanol. The “B” solution was made by mixing 0.15 mL of HNO_3 and 0.25 mL of H_2O.

6.2.6 Preparation of the Films (Fig. 6.3)

The working solution was prepared by mixing the solutions A with solution B the mixture was stirred at room temperature for 10 min and in the case of doped samples

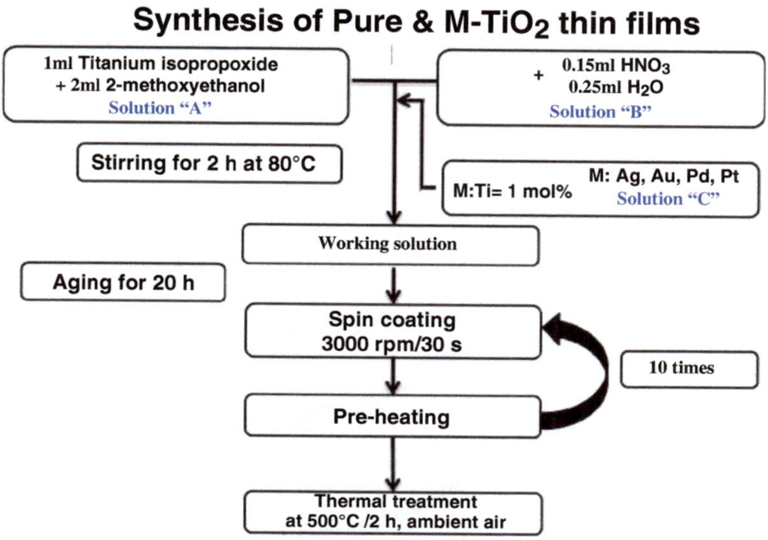

Fig. 6.3 Flow chart of sol-gel spin coating deposition process for TiO_2 and TiO_2-1% M films (M = Au, Ag, Pd, Pt)

the appropriate quantity of the metal precursor (solution C) was added to obtain 1% metal to titanium molar ratio. The mixture was stirred at 80 °C for 2 h and aged for 20 more hours at room temperature.

Doped and undoped thin films of TiO_2 were deposited on the glass substrates by spin coating at room temperature, with a rate of 3000 rpm for 30 s. After each spin-coating step, the films were heated on a hot plate at 250 °C in air for 5 min to remove the solvents this procedure was repeated 10 times for each film. Subsequently, the films were annealed in a furnace at 500 °C for 2 h to obtain TiO_2 crystalline thin films.

6.3 Results and Discussion

6.3.1 Films Characterization

The films were characterized by Structural analysis [Philips X'Pert Pro MPD X-Ray Diffraction (XRD) system, Cu-Ka radiation, $k = 1.541837$ [Å], Raman spectra were recorded by the Renishaw inVia Raman Microscope equipped with a HeNe laser at 633 nm, and chemical composition analysis of the films was done with a FEI scanning electron microscope, equipped with an energy dispersive X-ray spectrometer (EDX). Optical Transmittance spectra for the UV-Vis spectral

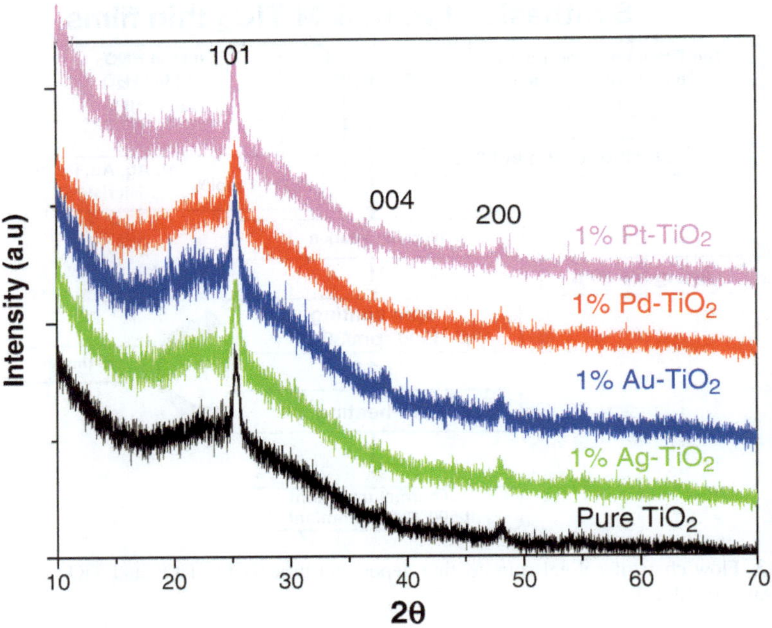

Fig. 6.4 XRD patterns of pure TiO_2 and 1% M-TiO_2 films (M = Ag, Au, Pt, Pd)

region (300–800 nm) were recorded by a Perkin Elmer UV/VIS/NIR Lambda 19 spectrophotometer. The film thickness was measured by an Alpha-Step IQ Surface Profiler.

6.3.1.1 Structural Properties

Figure 6.4 shows the XRD patterns of pure and 1% M-TiO_2 films. As we can see all the samples have characteristic peaks at 25.3°, 37.8° and 48.1° (2θ) degrees that corresponds to the (hkl) values of TiO_2 anatase phase (101), (004) and (200) crystal planes (JCPDS Card no. 21-1272), respectively, concluding that all the samples crystallize to anatase phase with a preferred orientation along the (101) direction. Moreover, no additional diffraction peaks belonging to M species are observed, probably due to low concentration [35].

Rietveld refinements were carried out using a tetragonal structure with the space group $I4_1$/amd to derive the lattice parameters of the films (Table 6.1). The lattice parameters were calculated using the unit cell refinement method in the PANalytical X'Pert HighScore Plus software and the average crystallite size, ranging between 9 and 13 nm (Table 6.1), was evaluated by applying the Scherrer formula [36]. As we can see there is no significant change of lattice parameters when doping with Ag, Au, Pd, Pt, indicating that there is no possible substitution of Ti^{4+} by metallic ions.

Table 6.1 Rietveld refinements results of pure TiO_2 and M-TiO_2 thin films

TiO_2 films	Crystallite size (nm)	Lattice parameters		Lattice volume
		a (Å)	c (Å)	V (Å^3) (b)
Pure	11	3.785	9.536	136.600
1% Ag-TiO_2	13	3.789	9.431	135.397
1% Au-TiO_2	12	3.778	9.445	134.844
1% Pd-TiO_2	9	3.780	9.606	137.265
1% Pt-TiO_2	11	3.790	9.469	136.032

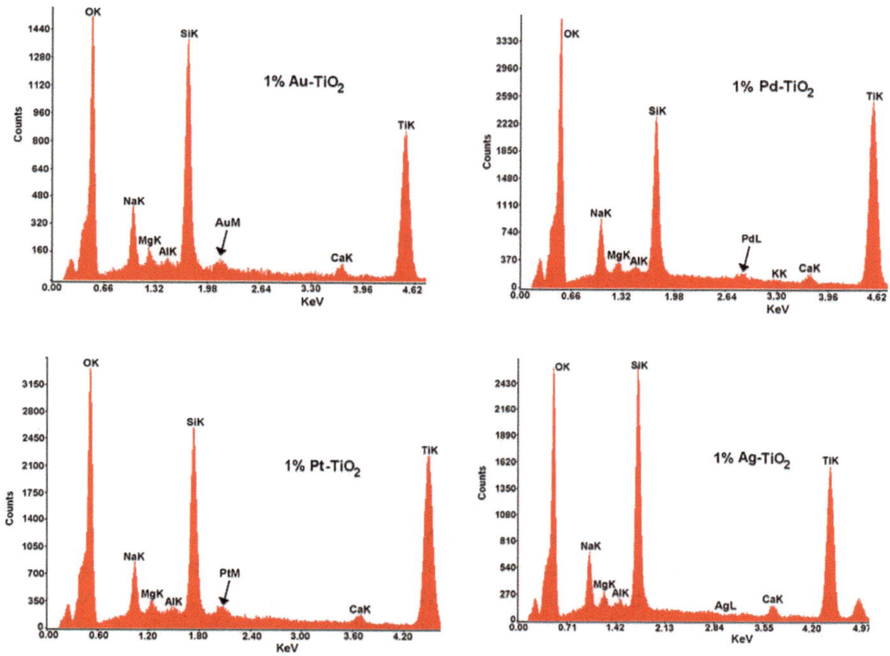

Fig. 6.5 EDX spectra of 1% M-TiO_2 films (M = Ag, Au, Pt, Pd)

6.3.1.2 EDX Analysis

The EDX spectra in Fig. 6.5 show the presence of high Ti and O content in the films. They also confirm the presence of small amount of Au, Pt, and Pd to the doped films.

In the case of Ag-TiO_2 film EDX spectrum does not show any Ag traces. This is because during the sol-gel synthesis the Ag nanoparticles are preferentially segregated on the surface of the film, thus limiting the detection with EDX [37].

The Si, Na, Al, Ca, and Mg peaks observed on the EDX spectra are due to the glass substrates on which the M-TiO_2 films are grown.

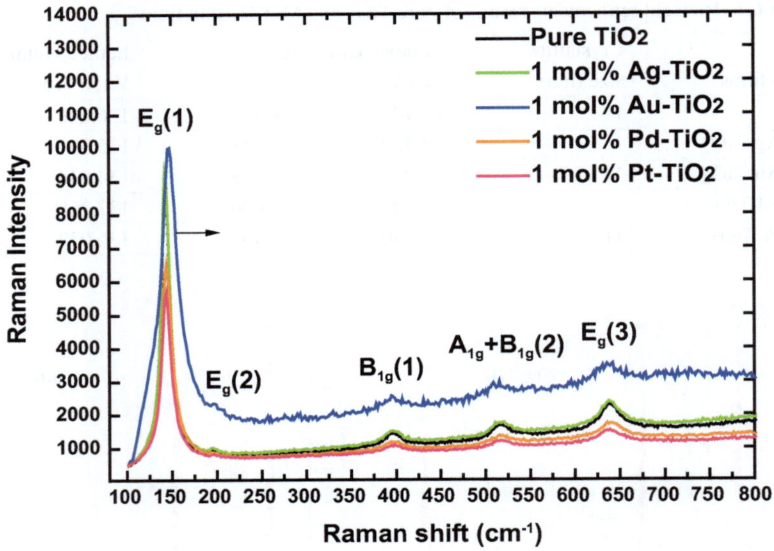

Fig. 6.6 Raman spectra of pure and 1% M-TiO_2 films (M = Ag, Au, Pt, Pd)

6.3.1.3 Raman Spectra

The crystallographic unit cell of anatase phase of TiO_2 is body centered (space group $I4_1/amd$) and contains two primitive unit cells, each of which contains two formula units of TiO_2. According to the factor group analysis, six modes, $A_{1g} + 2B_{1g} + 3E_g$, are Raman active and three modes, $A_{2u} + 2E_u$ are infrared active. One vibration, B_{2u}, will be inactive in both infrared and Raman [38]. Figure 6.6 shows the Raman spectra of pure and 1% M-TiO_2 films (M = Ag, Au, Pt, Pd) films.

All the spectra show a peak at 143 cm^{-1} ($E_g(1)$), a weak peak at 197 cm^{-1} ($E_g(2)$), and three medium intensity peaks at 397 cm^{-1} ($B_{1g}(1)$), 516 cm^{-1} (that is a doublet of A_{1g} + $B_{1g}(2)$ modes), and 638 cm^{-1} ($E_g(3)$). These values are in agreement with the literature [38].

6.3.1.4 Transmittance Spectra

As we can see at Fig. 6.7 all the films (doped and undoped) are highly transparent in the visible region with a sharp absorption edge at ~350 nm shows. All the spectra show interference fringes, due to reflections from the front and back side of the films indicating very smooth surface of the film [39].

The frequencies of these oscillations in these spectra depend on the film thickness, which is shown in Table 6.2.

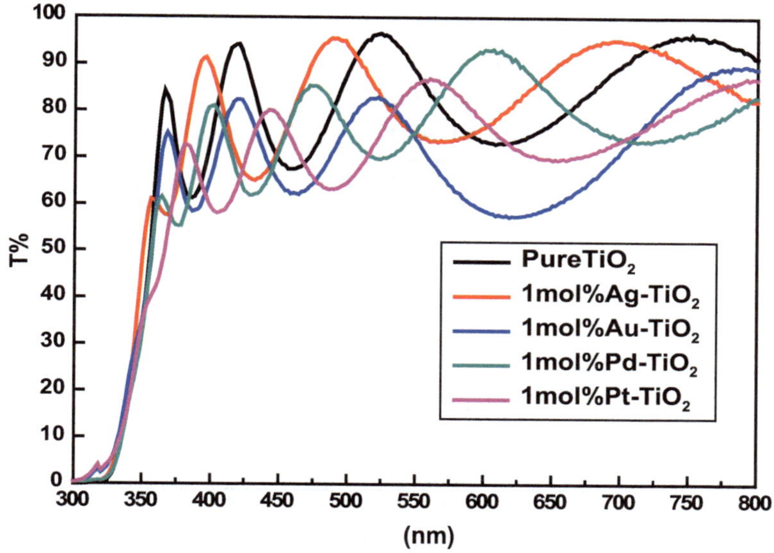

Fig. 6.7 Transmission spectra of pure and 1% M-TiO_2 films (M = Ag, Au, Pt, Pd)

Table 6.2 Transmittance, E_g and thickness of pure TiO_2 and 1 mol% M-TiO_2 films

TiO_2 films	Transmittance (%) (475–575 nm)	Eg(eV)	Thickness (nm)
Pure	96	3.46	242
1% Au-TiO_2	83	3.33	347
1% Pd-TiO_2	85	3.40	363
1% Pt-TiO_2	87	3.23	352

The optical band gap was estimated by employing the Tauc's model [40] in the high absorption region: $(\alpha h\nu)^{1/2} = A(h\nu - E_g)$, where α is the absorption coefficient, $h\nu$ is the photon energy, E_g is the band gap energy and A is an optical constant.

Plotting $(\alpha h\nu)^{1/2}$ versus $(h\nu)$, a straight line is obtained for a certain photon energy range, the extrapolation of this straight line intercepts the x-axis at the value of the indirect optical E_g (Fig. 6.8).

Table 6.2 contains the estimated values of the optical E_g. We observe that, when Au and Pt dopant ions are added into the TiO_2 thin films, E_g red-shifts from 3.46 eV to lower E_g values

We observe that the band gap of doped films is decreasing which is in accordance with the literature [37, 41–43]. The reduction of the band gap of TiO_2 in the presence of a noble metal can be attributed to the Strong metal support Interaction (SMSI) [44] because TiO_2 is a well-known material exhibiting such phenomenon [45].

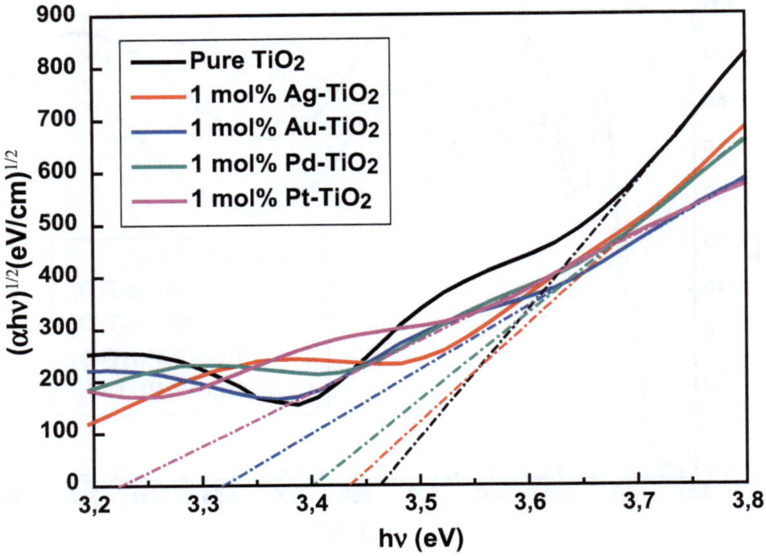

Fig. 6.8 Tauc's plots of pure and 1% M-TiO_2 films (M = Ag, Au, Pt, Pd)

6.3.2 Photocatalytic Properties

The photocatalytic activity of pure and doped TiO_2 films was evaluated by studying the decolorization of Methylene Blue under UV irradiation. A 4 W low-pressure Hg lamp (Hanovia Ltd.) was used as a light source, which mainly emits light centered at 254 nm. Two identical thin films of pure TiO_2 or M-TiO_2 on glass substrates ($2.5 \times 3\,cm^2$), were placed in a container filled with 10 mL of 2 ppm MB aqueous solution (Fig. 6.9). Before the UV lamp was turned on, the solution was left in the dark for 30 min in order to reach adsorption-desorption equilibrium with the photocatalyst. The photocatalytic activity of the films was evaluated by measuring the decolorization of MB after 5 h of UV irradiation using a Perkin Elmer UV/VIS/NIR Lambda 19 spectrophotometer. The change in the concentration of MB was estimated by the maximum absorbance peak of MB at $\lambda_{max} = 664$ nm. The degradation efficiency ($D\%$) of MB was calculated after 5 h of irradiation of the samples, by the following expression:

$D\% = [(C_0 - C_{5h})/C_0]^*100$ and based on the Beer–Lambert law (A = eCL);

$D\% = [(A_0 - A_{5h})/A_0]^*100$ where C_0, A_0, C_{5h} and A_{5h} is the initial concentration of MB and its absorbance initial and after 5 h respectively.

Figure 6.10a shows the spectra of MB solution every 30 min during the experiment with the pure TiO_2 sample and the Fig. 6.10b the change of MB concentration with regard to the time of photocatalysis. As we can see in the beginning (from

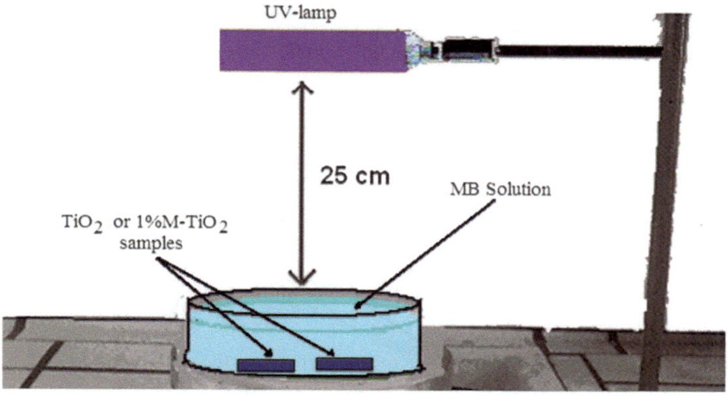

Fig. 6.9 Scheme of the set up used for the photocatalysis measurements

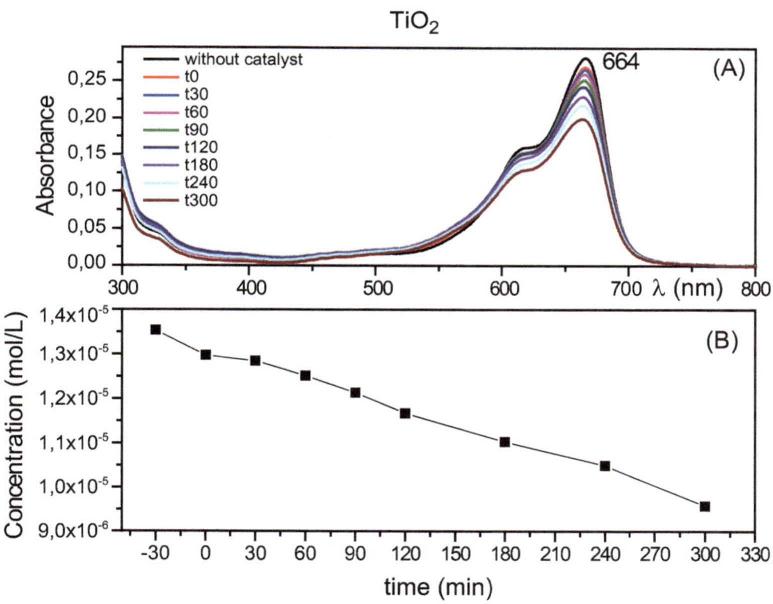

Fig. 6.10 (**a**) Absorption spectra of MB during photocatalysis for different time periods (**b**) Plot of MB concentration to time of photocatalysis

−30 to 0 min) there is a phase when the MB is adsorbed on the surface of the sample, subsequently the photocatalytic reaction starts with a linear diminution of the concentration vs the time.

Figure 6.11 shows the absorbance spectra of MB after 5 h of UV irradiation in the presence of TiO_2 or M-TiO_2 films, as well as the MB spectrum before the irradiation. As we can see around 15% of the MB dye was decolorized by direct

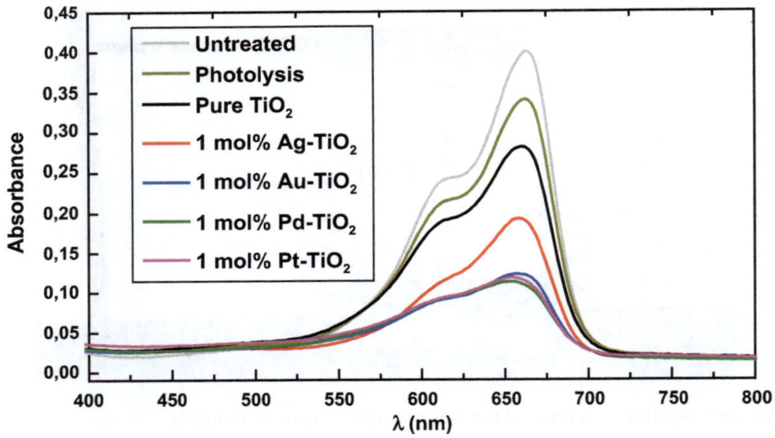

Fig. 6.11 Absorbance spectra of 2 ppm MB solutions, before and after 5 h irradiation, in the absence of catalysts (photolysis) and in the presence of pure TiO_2 and 1% M-TiO_2 thin-film catalysts

UV photolysis, without the presence of any photocatalyst. In the presence of pure TiO_2 the degradation efficiency is $D\% = 30\%$. In the case of M-TiO_2 films, the decolorization of MB increases significantly with the Pd-TiO_2 shows the highest photocatalytic activity, with $D\% = 72\%$, Au-TiO_2 ($D\% = 67\%$) and Pt-TiO_2 ($D\% = 70\%$) show similar photocatalytic activity to Pd-TiO_2 while Ag-TiO_2 performs less effectively ($D\% = 52\%$).

In order to explain the photocatalytic behavior shown in this work, we suggest that two aspects should be considered: first, the narrowing of the band gap due to the SMSI phenomenon and second due to the caption of photo-excited electrons to the metal particles reducing the recombination of e^-/h^+ pairs contributing to the stronger photocatalytic reactivity.

The lower photocatalytic activity of Ag-TiO_2 can be explained by the lower work functions of Ag (4.6 eV) in comparison to other metals (Pt (5.64 eV), Pd (5.12 eV), and Au (5.1 eV)). When two materials of different work functions are in contact, an energy barrier for electron transfer is created at the junction. This is sometimes known as the Schottky barrier. The transfer of electrons is facilitated from a material of lower work function [in our case the TiO_2 (4.2 eV)] to one that has a higher work function [46].

6.4 Conclusions

Pure and M-doped TiO_2 thin films were prepared by the sol-gel/spin coating method. XRD data and Raman spectra reveal the presence of the anatase only crystalline phase for all TiO_2 thin films. Energy-dispersive spectroscopy (EDX)

confirmed the presence of elemental Au, Pt, and Pd signals in the corresponding M-TiO_2 films, while no Ag peak was detected, due to Ag migration near the surface of TiO_2. Tauc's plots from transmittance data shows a narrowing of the band gap due to SMSI phenomenon. The Pd-doped TiO_2 film showed the highest photocatalytic activity in MB decolorization. Au-TiO_2 and Pt-TiO_2 films performed similar to Pd-TiO_2 as photocatalysts, while Ag-TiO_2 performed less effectively probably due to the lower work function of Ag.

Acknowledgements The research was financially supported by the Tunisian Ministry of Higher Education and Scientific Research and the National Hellenic Research Foundation (NHRF), through a Bilateral Scientific Cooperation. The authors would like to thank N.S. Tagiara for Raman analysis and Dr. K. Giannakopoulos for EDX measurements.

References

1. Wintgens T, Salehi F, Hochstrat R, Melin T (2008) Emerging contaminants and treatment options in water recycling for indirect potable use. Water Sci Technol 57:99
2. World Health Organization (2007) Combating waterborne disease at the household level. WHO Document Production Services, Geneva. ISBN:978-92-4-159522-3
3. Suárez S, Carballa M, Omil F, Lema JM (2008) How are pharmaceutical and personal care products (PPCPs) removed from urban wastewaters? Rev Environ Sci Biotechnol 7:125
4. World health organization (2005) Methyl tertiary-butyl ether (MTBE) in drinking-water. Background document for development of WHO Guidelines for drinking-water quality. World Health Organization, Geneva. (WHO/SDE/WSH/05.08/122)
5. Viessman W Jr, Hammer MJ, Perez ME, Chadik PA (2009) Water supply and pollution control, 8th edn. Prentice Hall, Upper Saddle River. ISBN:10:0132337177
6. Becher G (1999) Drinking water chlorination and health. Acta Hydrochem Hydrobiol 27:100
7. Lu J, Zhang T, Ma J, Chen Z (2009) Evaluation of disinfection by-products formation during chlorination and chloramination of dissolved natural organic matter fractions isolated from a filtered river water. J Hazard Mater 162:140
8. Fox M (1988) Photocatalytic oxidation of organic substances. In: Schiavello M (eds) Photocatalysis and environment: trends and applications. Kluwer Academic, Boston
9. Chong M-N, Jin B, Chow Ch, Saint Ch (2010) Recent developments in photocatalytic water treatment technology: a review. Water Res 44:2997
10. Pelaez M, Nolan NT, Pillai SC, Seery MK, Falaras P, Kontos AG, Dunlop PSM, Hamilton JWJ, Byrne JA, O'Shea K, Entezari MH, Dionysiou DD (2012) A review on the visible light active titanium dioxide photocatalysts for environmental applications. Appl Catal B Environ 125:331
11. Hoffmann MR, Martin ST, Choi W, Bahnemann DW (1995) Environmental applications of semiconductor photocatalysis. Chem Rev 95:69
12. Nolan NT, Seery MK, Pillai SC (2009) Spectroscopic investigation of the anatase-to-rutile transformation of sol-gel-synthesized TiO_2 photocatalysts. J Phys Chem C 113:16151
13. Wisitsoraat A, Tuantranont A, Comini E, Sberveglieri G, Wlodarski W (2009) Characterization of n-type and p-type semiconductor gas sensors based on NiO_x doped TiO_2 thin films. Thin Solid Films 517:2775
14. Luttrell T, Halpegamage S, Tao J, Kramer A, Sutter E, Batzill M (2014) Why is anatase a better photocatalyst than rutile? – Model studies on epitaxial TiO_2 films. Sci Rep 4:4043
15. Testino A, Bellobono IR, Buscaglia V, Canevali C, D'Arienzo M, Polizzi S, Scotti R, Morazzoni F (2007) Optimizing the photocatalytic properties of hydrothermal TiO_2 by the control of phase composition and particle morphology. A systematic approach. J Am Chem Soc 129:3564

16. Cozzoli PD, Comparelli R, Fanizza E, Curri ML, Agostiano A (2003) Photocatalytic activity of organic-capped anatase TiO_2 nanocrystals in homogeneous organic solutions. Mater Sci Eng C 23:707
17. Lee S-Y, Park S-J (2013) TiO_2 photocatalyst for water treatment applications. J Ind Eng Chem 19:1761
18. Augugliaro V, Bellardita M, Loddo V, Palmisano G, Palmisano L, Yurdakal S (2012) Overview on oxidation mechanisms of organic compounds by TiO_2 in heterogeneous photocatalysis. J Photochem Photobiol C 13:224
19. Nagaveni K, Sivalingam G, Hegde MS, Madras G (2004) Solar photocatalytic degradation of dyes. High activity of combustion synthesized nano TiO_2. Appl Catal B Environ 48:83
20. Molinari R, Caruso A, Argurio P, Poerio T (2008) Degradation of the drugs Gemfibrozil and Tamoxifen in pressurized and de-pressurized membrane photoreactors using suspended polycrystalline TiO_2 as catalyst. J Memb Sci 319:54
21. Lee DK, Kim SC, Cho IC, Kim SJ, Kim SW (2004) Photocatalytic oxidation of microcystin-LR in a fluidized bed reactor having TiO2-coated activated carbon. Sep Purif Technol 34:59
22. Kruchinin SP, Repetsky SP, Vyshyvana IG (2016) Spin-depent transport of carbon nanotubes with chromium atoms. In: Bonca J, Kruchinin S (eds) Nanomaterials for security. Springer, Dordrecht, pp 65–97
23. Lee HS, Im SJ, Kim JH, Kim HJ, Kim JP, Min BR (2008) Polyamide thin-film nanofiltration membranes containing TiO_2 nanoparticles. Desalination 219:48
24. Mogyorús K, Dékány I, Fendler JH (2003) Preparation and characterization of clay mineral intercalated titanium dioxide nanoparticles. Langmuir 19:2938
25. Tetsuya T, Ko T, Yasuyuki M, Masayuki S (2013) Improving the photocatalytic properties of rayon fibers containing a titanium dioxide photocatalyst through enzymatic treatment. Text Res J 83:1615
26. Hussein FH, Alkhateeb AN (2007) Photo-oxidation of benzyl alcohol under natural weathering conditions. Desalination 209:350
27. Zhang X, Veikko U, Mao J, Cai P, Peng T (2012) Visible-light-induced photocatalytic hydrogen production over binuclear Ru^{II}-bipyridyl dye-sensitized TiO_2 without noble metal loading. Chem Eur J 18:12103
28. Ni M, Leung MKH, Leung DYC, Sumathy K (2007) A review and recent developments in photocatalytic water-splitting using TiO_2 for hydrogen production. Renew Sust Energ Rev 11:401
29. Matsui H, Okad K, Kawashim T, Tetsuya E, Tanabe N, Kawano R, Watanabe M (2004) Application of an ionic liquid-based electrolyte to a 100×100 mm sized dye-sensitized solar cell. J Photochem Photobiol A 164:129
30. Siefering KL, Griffin GL (1990) Growth kinetics of CVD TiO_2: influence of carrier gas. J Electrochem Soc 137:1206
31. Zhang Y, Wan J, Ke Y (2010) A novel approach of preparing TiO_2 films at low temperature and its application in photocatalytic degradation of methyl orange. J Hazard Mater 177:750
32. Wang ZM, Sahle-Demessie E, Aly Hassan A, Perrett C (2012) Surface structure and photocatalytic activity of nano-TiO_2, TiO_2 thin film for selective oxidation. J Environ Eng 138:923
33. Sobczyk-Guzenda A, Pietrzyk B, Szymanowski H, Gazicki-Lipman M, Jakubowski W (2013) Photocatalytic activity of thin TiO_2 films deposited using sol–gel and plasma enhanced chemical vapor deposition methods. Ceram Int 39:2787
34. Liu B, Zhao X, Zhao Q, Li C, He X (2005) The effect of O_2 partial pressure on the structure and photocatalytic property of TiO_2 films prepared by sputtering. Mater Chem Phys 90:207
35. Zhong JB, Lu Y, Jiang WD, Meng QM, He XY, Li JZ, Chen YQ (2009) Characterization and photocatalytic property of Pd/TiO_2 with the oxidation of gaseous benzene. J Hazard Mater 168:1632
36. Cullity BD, Stock SR (2014) Elements of X-ray diffraction. Pearson, Harlow. ISBN:1292040548

37. Bensouici F, Souier T, Dakhel AA, Iratni A, Tala-Ighil R, Bououdina M (2015) Synthesis, characterization and photocatalytic behavior of Ag doped TiO_2 thin film. Superlattices Microst 85:255
38. Balachandran U, Eror NG (1982) Raman spectra of titanium dioxide. J Solid State Chem 42:276
39. Nakaruk A, Ragazzon D, Sorrell CC (2010) Anatase–rutile transformation through high-temperature annealing of titania films produced by ultrasonic spray pyrolysis. Thin Solid Films 518:3735
40. Tauc J, Grigorovici R, Vancu A (1966) Optical properties and electronic structure of amorphous germanium. Phys Status Solid (B) 15:627
41. Mani Rahulan K, Ganesan S, Aruna P (2011) Synthesis and optical limiting studies of Au-doped TiO_2 nanoparticles. Adv Nat Sci Nanosci Nanotechnol 2:025012
42. Rodionov VE, Shnidko IN, Zolotovsky A, Kruchinin SP (2013) Electroluminescence of Y_2O_3:Eu and Y_2O_3:Sm films. Mater Sci 31:232
43. Vlaskina S, Kruchinin S, Rodionov V et al (2016) Nanostructures in silicon carbide crystals and films. Inter J Mod Phys B 30(13):1042015
44. Tauster SJ (1987) Strong metal-support interactions. Accounts Chem Res 20:389
45. Alshammari A, Bagabas A (2015) Semiconductor supported gold nanoparticles for photodegradation of Rhodamine B. Int J Chem Mol Eng 9:47
46. Tan TTY, Yip CK, Beydoun D, Amal R (2003) Effects of nano-Ag particles loading on TiO_2 photocatalytic reduction of selenate ions. Chem Eng J 95:179

Chapter 7
Electrical Measurements of the Dimensions of Nanostructures

W. Nawrocki and Yu. M. Shukrinov

Abstract We propose to use measurements of electrical resistance for estimation of the geometric dimensions of nanostructures made of conductive materials. Transistors, resistors and conductive paths inside integrated circuits are fabricated in a form of thin films of progressively smaller size. At present (2017) microprocessors of fabrication technology of 10 nm have been manufactured. The technology of 7 nm will be implemented next year and the technology of 5 nm is expected in 10 years. The dimension of 10 or 5 nm refers to the length of a gate of a MOSFET transistor inside the chip. For the next generation of nanostructures, the 3D nanostructures, their scaling will be continued and measurements of them will be also necessary.

Keywords Measurements on nanostructures · Conductance quantization

7.1 Introduction

Studies of nanostructures and the measurements of them in the fabrication process can be helpful in development of the semiconductor industry. Miniaturization of transistors in integrated circuits is limited by numerous physical effects and manufacturing technology. For 60 years silicon of monocrystalline structure (referred as mono c-Si) has been a base material for the manufacturing of transistors and digital integrated circuits (IC) done on industrial scale. Digital integrated circuits are produced either in TTL or CMOS technology. The measure of miniaturization transistors inside of integrated circuits is described by a physical gate length of MOSFET transistors. For instance, the 14-nm technology (firstly applied in 2014

W. Nawrocki (✉)
Faculty of Electronics and Telecommunications, Poznan University of Technology, Poznan, Poland
e-mail: waldemar.nawrocki@put.poznan.pl

Yu. M. Shukrinov
Join Institute for Nuclear Research, BLTP, Dubna, Russia

J. Bonča, S. Kruchinin (eds.), *Nanostructured Materials for the Detection of CBRN*, NATO Science for Peace and Security Series A: Chemistry and Biology,
https://doi.org/10.1007/978-94-024-1304-5_7

Table 7.1 Data of integrated circuits according to the report of the international technology roadmap for semiconductors, NAND Flash (Edition 2015 V2 [3])

Year		2015	2016	2020	2024	2030
2D NAND Flash uncontacted poly 1/2 pitch	nm	15	14	12	12	12
3D NAND minimum array 1/2 pitch	nm	80	80	80	80	80
Number of word lines in one NAND string	–	32	32–48	64–96	128–192	384–512
Product highest density (2D or 3D)	–	256 G	384 G	768 G	1.5 T	4 T
3D NAND number of memory layers	–	32	32–48	64–96	128–192	384–512

by Intel, the Intel 14 nm Broadwell microprocessor) refers to the 14 nm length of a gate in CMOS transistors inside of IC microprocessor. The latest forecast of the development of the semiconductor industry (ITRS 2015) predicts that sizes of electronic devices in ICs circuits will be smaller than 5 nm in the next 10 years – Table 7.1 [1]. It is worth to mention that the new generation of integrated circuits are 3D circuits [1]. Also for transistors and the 3D integrated circuits the scaling process will continue. Then, measurements of structures of nanometer size dimensions will be is necessary in manufacturing as well in development phase.

7.2 Quantization of Electrical Conductance

7.2.1 Theory

Electric proprieties of electronic devices or paths with sizes in nanometers are described by quantum theories of conductance. Figure 7.1 presents a picture of a nanowire – the constriction in an electrical conductor with dimensions W(width), H (thickness) and L (length). The quantum unit of electrical conductance $G_0 = 2e^2/h$ was predicted by R. Landauer in his theory of electrical conductance [1, 2]. Parameters characterizing a nanowire (or other nanostructure) are a Fermi wavelength λ_F and an electron mean free path l_e. For metals like gold the Fermi wavelength ($\lambda_F \approx 0.52$ nm) is much shorter than the mean free path l_e ($l_{e\mathrm{Au}} = 38$ nm) (Table 7.2).

For semiconductors the Fermi wavelength depends on the intensity of doping. If the length of the system is shorter than the free electron path, the impurity scattering is negligible, so the electron transport is ballistic. If a nanostructure has the cross-section dimensions comparable with the Fermi wavelength λ_F, and its length L is less than l_e, the system can be regarded as one-dimensional (1-D), the electron regarded as a wave, and one can expect quantum effects – see Fig. 7.1. A long Fermi wavelength in Si determined the limit for the 1st step of conductance quantization in conducting paths inside of ICs.

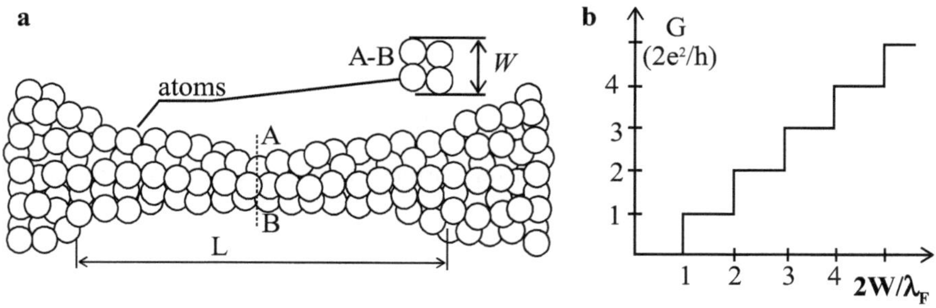

Fig. 7.1 Conductance quantization in a nanowire (conductor with length $L < l_e$ and width W comparable with the length of Fermi wave λ_F): (**a**) a nanowire formed by atoms; (**b**) conductance quantization G versus width W

Table 7.2 Fermi wavelength, λ_F, and electron mean free path, l_e, for metals (Au, Cu, Al) and doped silicon (doping from 10^{16} to 10^{19} cm^{-3}) at room temperature

Material		Si	Si	Si	Si	Au	Cu	Al
Doping, n_o	cm^{-3}	10^{16}	10^{17}	10^{18}	10^{19}	no	no	no
Fermi wave-length, λ_F	nm	93	43	20	9.3	0.52	0.46	0.36
Electron mean free path, l_e	nm	from 37 to 120 [5]				38	40	19

The total electrical conductance of a nanowire is given by formula (7.1)

$$G = \frac{2e^2}{h} N, \tag{7.1}$$

where N is the number of transmission channels.

For 2-D system, with thickness $H \leqslant \lambda_F$, N depends on the width of the wire,

$$N = \text{Int}(2W/\lambda_F), \tag{7.2}$$

where Int(D) means the integer of the real part of D number.

The same formula for G and N in 2-D nanostructures was published by Sharvin in 1965 [4] and quoted by Beenakker [5]. Sharvin analyzed the conductance in 3-D nanosystems as well. According to the Sharvin theory for 3-D nanosystem with a cross section A the conductance G is described by Eq. (7.3), e.g. thickness H and thickness W, $A = H \times W$, both H and $W \geqslant \lambda_F$

$$G = \frac{2e^2}{h} \frac{k_F^2 A}{4\pi}, \tag{7.3}$$

where k_F is the Fermi wavevector, $k_F = 2\pi\lambda_F$, A – area of the cross section of nanostructure at its narrowest point.

Thus, the number of transmission channels in the 3-D nanosystem is

$$N = \mathrm{Int}\frac{\pi\, A}{\lambda_F^2}. \tag{7.4}$$

However, defects, impurities and irregularities of the shape of the conductor can induce scattering, then conductivity is given by the Landauer equation:

$$G = \frac{2e^2}{h}\sum_{i,j=1}^{N} t_{ij}, \tag{7.5}$$

where t_{ij} denotes probability of the transition from jth to ith state. In the absence of scattering Eq. (7.4) is reduced to Eq. (7.1).

7.2.2 Measurements of Conductance Quantization

The experimental setup consists of a pair of metallic wires (they form a nanowire), a digital oscilloscope, a motion control system (not shown in the figure) and a PC, see Fig. 7.2a. Instruments are connected in one system using the IEEE-488 interface. The resistor $R_p = 1\,\mathrm{k}\Omega$, connected in series to the wires, is used.The circuit is fed by the constant voltage V_s. Measurements of current $I(t)$ are performed. Conductance is determined by current I accordingly to:

$$G = I\frac{1}{V_s - IR_p}. \tag{7.6}$$

Transient effects of making contact or breaking contact give time dependent current. The voltage V_p on the resistor R_p was measured with a computer controlled oscilloscope. The piezoelectric device is used to control the backward and forward

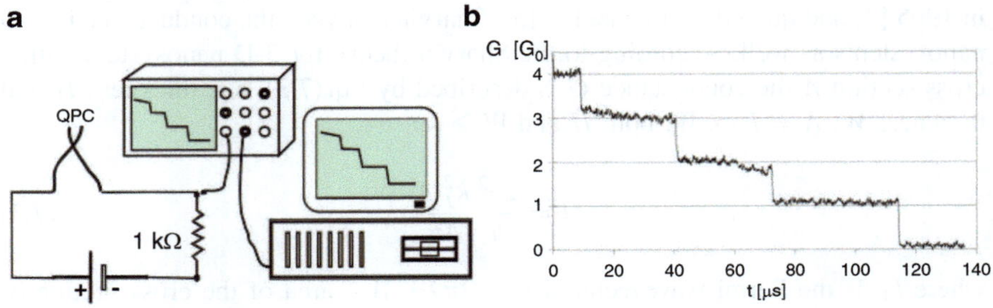

Fig. 7.2 Measurements of electrical conductance in nanowires (QPC): **(a)** set-up; **(b)** recorded process of conductance quantization

movement of the macroscopic wires between which nanowires are created (QPC – quantum point contact). A high voltage amplifier controlled by a digital function generator supplies the piezoelectric device. Both electrodes of the QPC (see Fig. 7.2) are made of wire 0.5 mm in diameter or of small pieces of metal. The conductance was measured between two metallic electrodes, moved to contact by the piezoelectric tube actuator. The oscilloscope was triggered by a single pulse. All experiments were performed at room temperature and at ambient pressure [6, 7].

In order to compare our results with those published before by other groups, the first experiment was performed with gold wires. The quantization of electric conductance (with G_0 units) does not depend either on the kind of metal nor on temperature. Despite this the purpose of studying quantization for different metals was to observe how properties of the metal affect the contacts between wires. Therefore, we have investigated the conductance quantization in nanowires for three nonmagnetic metals (gold, copper and tungsten) and for magnetic metals (cobalt and nickel).

For nonmagnetic metals, the conductance quantization in units of $G_0 = 2e^2/h = 7.75 \times 10^{-5}$ [A/V] $= (12.9\,\text{k}\Omega)^{-1}$ was previously observed for the following nanowires: Au–Au, Cu–Cu, Au–Cu, W–W, W–Au, W–Cu. The quantization of conductance in our experiment was evident. All characteristics showed the same steps equal to $2e^2/h$. We observed two phenomena: quantization occurred when breaking the contact between two wires, and quantization occurred when establishing contact between the wires. The characteristics were only partially reproducible; they differ in number and height of steps, and in the time length. The steps corresponded to 1, 2, 3 or 4 quanta. It should be stated that quantum effects were observed only for some of the characteristics recorded. The conductance quantization has been so far more pronouncedly observable in gold contacts.

7.3 Estimation of Geometric Dimensions of a Nanostructure by Electrical Measurements

7.3.1 The Method

The electrical conductance of nanostructures within the smallest atomic size is changed stepwise as a function of the cross section area of a nanostructure at its narrowest point. The effect of quantization of electrical conductance (resistance) with the quantum $G_0 = 2e^2/h$ occurs for nanostructures under the conditions for nanostructures given by Landauer in his new approach to electrical of conductance.

This effect can be used to measure the width of nanostructures (samples), or rather to estimate it. The method proposed here is based on electrical measurements – Fig. 7.3. Estimation of the width of the 2-D nanostructure consists in measuring its electrical resistance R (conductance $G = 1/R$) and calculating the number of transmission channels N. The width W of the 2-D nanostructures is calculated using the Eq. (7.7) obtained from the transformation of Eq. (7.2)

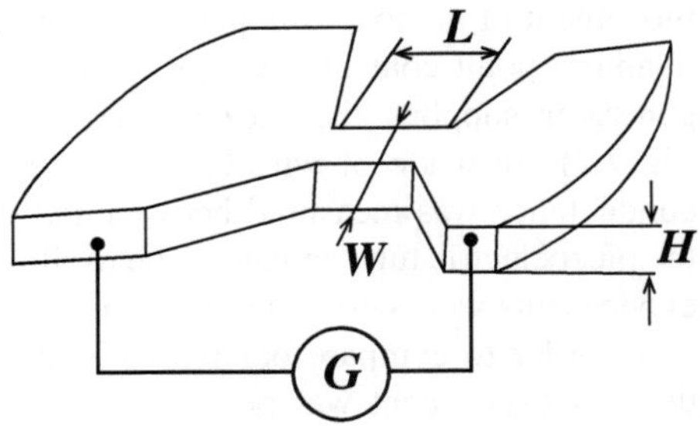

Fig. 7.3 Measurement of a width of a nanostructure by an electrical method

$$W = (N \times \lambda_F)/2. \tag{7.7}$$

Estimation of the area of the cross section A of the 3-D nanostructure at its narrowest point consists again in measuring its conductance G and calculating the number of transmission channels N. The area A of the 3-D nanostructures is calculated using the Eq. (7.8) obtained from the transformation of Eq. (7.4)

$$A = \frac{N\lambda_F^2}{\pi}. \tag{7.8}$$

7.3.2 *Technical Conditions*

The technical conditions necessary to carry out the measurements of the geometric dimensions of nanostructures using electrical conductance measurements are: that the nanostructure is made of electrical conductor; that the nanostructure (sample) has a constriction (narrow throat) between two wide terminals; that the length of the constriction is shorter than the mean free path l_e in sample material. The thickness of the nanostructure is comparable with the Fermi wavelength λ_F. This means the thickness of one or numerous atoms (as it is in graphene).

In order to measure dimensions of a nanostructure one should know three important parameters of the materials to be measured: the Fermi wavelength λ_F, the electron mean free path l_e and the lattice space a_n. Values of λ_F and l_e depend on carriers concentration, n_o. Equations (7.9), (7.10) and (7.11) describe λ_F and l_e [8–10].

$$\lambda_F = \frac{h}{\sqrt{2m\,E_F}} = \frac{2.03}{\sqrt[3]{n_o}}, \tag{7.9}$$

$$E_F = \frac{h^2}{8m}\left(\frac{3n_o}{\pi}\right)^{2/3}, \tag{7.10}$$

$$l_e = \sqrt[3]{\frac{3}{\pi}} \frac{\sigma}{G_0 n_o^{2/3}} = \sqrt[3]{\frac{3}{\pi} \frac{\sigma}{n_o^2 G_0}}, \qquad (7.11)$$

m – mass of electron, n_o – carriers concentration, E_F – Fermi energy, σ – conductivity of material of a sample.

Values of the Fermi wavelength λ_F and the electron mean free path l_e for three metals (Au, Cu, Al) and doped silicon (doping from $n_o = 10^{15}$ 1/cm^3 to 10^{20} 1/cm^3) are presented in Table 7.2 [11].

The nanowire of the smallest size forms a chain of atoms connected one to another by path, $L < l_e$. The chain creates one transmission channel for free electrons, accordingly to the Landauer theory. Thus, the conductance $G_0 = 7.75 \times 10^{-5}$ [A/V] of the nanowire indicates that the nanowire contains only a single atom at its narrowest point. The cross-sectional area of such a nanowire is the area of a circle defined by $2R_a$, where R_a is the radius of the atom. For example, for gold: the mean free path $l_{e\mathrm{Au}} = 38$ nm (at 293 K), the atomic radius $R_{a\mathrm{Au}} = 0.174$ nm.

7.3.3 Discussion

The chain of atoms forming a gold nanowire with conductance of G_0 contains from one to about 90 atoms (38/0.408 nm $\approx$ 90 atoms in the chain). The lattice space in Au, $a_n = 0.408$ nm. The upper limit of the number of atoms in the nanowire chain can‘t be given exactly. The reason of this is that only the *mean* free path of electrons in the material. Thus, a real free path length in gold may be e.g. 20 or 50 nm instead of 38 nm. One can note that a chain of atoms containing about 30 atoms or more is a nanostructure which is mechanically unstable.

The conductance of $2G_0$ or NG_0 of a nanowire means that in the nanowire there exist two or N transmission channels, respectively. The width of the nanowire at its narrowest point is W.

One can note that the same conductance of a two samples of nanostructure may show different values of width W (for 2-D) or area of cross section A (for 3-D). For example, the resistance of Au sample and Al sample is the same, $R = 6.2\ \mathrm{k}\Omega$, thus $G = 1/(6.2\ \mathrm{k}\Omega)$. Using the (7.1), the number of transmission channels in both samples, $N = G/G_0 = 6$. According to the Eq. (7.7) and taken $\lambda_{F-\mathrm{Au}} = 0.52$ nm and $\lambda_{F-\mathrm{Al}} = 0.36$ nm we obtain the values of the width: $W_{\mathrm{Au}} = 1.56$ nm and $W_{\mathrm{Al}} = 1.08$ nm.

The mean free path depends on the temperature T because of the conductivity dependence on T. For metals the mean free path at low temperatures is longer than at room temperature. In semiconductor material, like Si, Ge or GaAs, the carriers concentration n_o, presented in formulas (7.9), (7.10) and (7.11), depends on the temperature as well. Thus, for semiconductors both physical parameters (λ_F and l_e) depend on the temperature. As a result of these temperature dependencies, the nanostructure (sample) that meets Landauer's conditions for ballistic electron transport at the T_1 temperature does not meet them after heating up to T_2 (metals) or after cooling down to T_3 (semiconductors). The conductance of the sample, in which electron transport is no longer ballistic, is rapidly decreasing

7.4 Conclusions

The method of measurements of geometrical dimensions, presented above, is relatively simple and the results of measurement are more accurate when nanostructures is smaller. Electrical measurements of geometric sizes of nanostructures are potentially very important for nanotechnology, where other methods of measurement (e.g. STM) are much more complex and expensive.

Acknowledgements The authors are grateful for funding their scientific cooperation with the Bogoliubov – Infeld Programme fund. The paper was prepared with the financial support of the project No. 08/83/DSPB/4724 at Poznan University of Technology, Poland.

References

1. Landauer R (1957) Spatial variation of currents and fields due to localized scatteres in metallic conduction. IBM J Res Dev 1:223
2. Landauer R (1989) Conductance determined by transmission: probes and quantised constriction resistance. J Phys Condens Matter 1:8099
3. The International Technology Roadmap for Semiconductors (2016) Edition 2015. www.itrs.net/reports
4. Sharvin YV (1965) A possible method for studying Fermi surfaces. Zh Eksp Teor Fiz 48:984
5. Beenakker C, van Houten H (1991) Quantum transport in semiconductor nanostructures. Solid State Phys 44:1
6. Nawrocki W, Wawrzyniak M, Pajakowski J (2009) Transient states in electrical circuits with a nanowire. J Nanosci Nanotechnol 9:1350
7. Nawrocki W (2015) Introduction to quantum metrology. Quantum standards and instrumentation. Springer, Heidelberg
8. Ashcroft NW, Mermin ND (1976) Solid state physics. Hartcourt College Publisher, Orlanado
9. Kruchinin S (2016) Energy spectrum and wave function of electron in hybrid superconducting nanowires. Int J Mod Phys B 30(13):1042008
10. Kruchinin S, Pruschke T (2014) Thermopower for a molecule with vibrational degrees of freedom. Phys Lett A 378:157–161
11. Gall D (2016) Electron mean free path in elemental metals. J Appl Phys 119:085101

Chapter 8
Thermoelastic Phase Transformations and Microstructural Characterization of Shape Memory Alloys

O. Adiguzel

Abstract Shape memory effect is a peculiar property exhibited a certain alloy systems with special chemical compositions in the β-phase fields of alloys and other materials like polymers. Successive martensitic transformations, thermal induced and stress induced martensitic transformations, govern shape memory effect in shape memory alloys. Martensitic transformations are structural phase transformations, and thermal induced martensitic transformation occurs as martensite variants with lattice twinning in crystallographic or atomic scale in materials on cooling below martensite finish temperature. Twinned martensite structures turn into detwinned martensite structure by means of stress induced transformation by deforming plastically in a strain limit in martensitic condition. Shape memory alloys are in the fully martensitic state below martensite finish temperature with fully twinned structure and can be easily deformed through variant reorientation/detwinning process. Thermal induced martensitic transformation is lattice-distorting phase transformation and occurs as martensite variants with the cooperative movement of atoms by means of shear-like mechanism. Martensitic transformations occur by two or more lattice invariant shears on a {110}-type plane of austenite matrix, as a first step, and the transformed region consists of parallel bands containing alternately two different variants. Copper based alloys exhibit this property in metastable β-phase region, which has bcc-based structures at high temperature parent phase field. Lattice invariant shears are not uniform in these alloys, and the ordered parent phase structures martensitically undergo the complex layered structures.

Keywords Thermoelastic phase transformations · Shape memory effect · Twinned and detwinned martensites

---

O. Adiguzel (✉)
Department of Physics, Firat University, Elazig, Turkey
e-mail: oadiguzel@firat.edu.tr

J. Bonča, S. Kruchinin (eds.), *Nanostructured Materials for the Detection of CBRN*, NATO Science for Peace and Security Series A: Chemistry and Biology,
https://doi.org/10.1007/978-94-024-1304-5_8

8.1 Introduction

Shape memory alloys take place in a class of functional materials by exhibiting a peculiar property called shape memory effect. This property is characterized by the recoverability of desired shape on the material at different conditions. Shape memory effect is linked with martensitic transformation, and comprises a reversible transition from product martensite to parent austenitic phase.

Martensitic transformations are first order lattice-distorting phase transformations and occur with the cooperative movement of atoms by means of lattice invariant shears in the materials on cooling from high temperature parent phase region. When the copper based beta-phase alloys are cooled below a critical temperature called martensite start temperature, M_s, the martensitic transformation occurs and martensite forms as plates in groups of variants. It enables the shape memory alloys to deform under low stresses by variant coalescence because the total shape change on transformation becomes nearly zero for the group [1]. Martensitic transformation is evaluated by the structural changes in microscopic scale. Shape memory effect is based on martensitic transformation, and microstructural mechanisms are responsible for shape memory behaviour. In particular, the twinning and detwinning processes are essential as well as martensitic transformation in reversible shape memory effect [1, 2]. Thermal induced martensite occurs by means of a shear-like mechanism as multivariant martensite in self-accommodating manner and consists of lattice twins. Also, this martensite is called twinned martensite or multivariant martensite.

Shape memory alloys can be deformed plastically in low temperature martensitic condition, and recover the original shape on heating over the austenite finish temperature. The material cycles between the deformed and original shapes on cooling and heating in reversible shape memory case. By applying external stress, the martensitic variants are forced to reorient into a single variant leading inelastic strains, and deformation of shape memory alloys in martensitic state proceeds through a martensite variant reorientation or detwinning of twins [1, 2]. The twinning occurs with internal stresses, while detwinning occurs with the external stresses. The basic mechanism of phase transformation and shape memory effect in crystallographic level is schematically illustrated in Fig. 8.1 [2]. The deformed material recovers the original shape in bulk level, and crystal structure turns into the parent phase structure on first heating over the austenite finish temperature in both reversible and irreversible memory cases. The parent phase structure returns

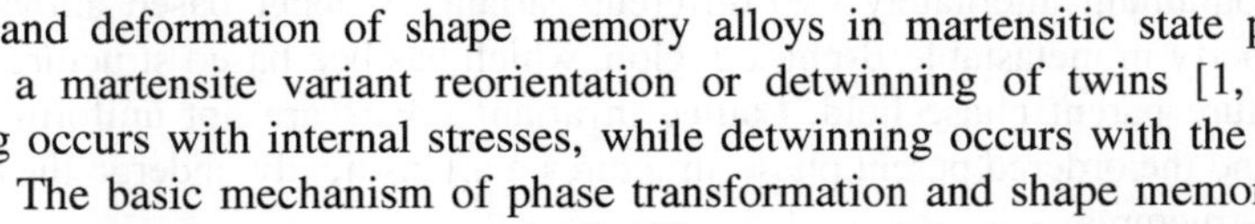

Fig. 8.1 Schematic illustration of the mechanism of the shape-memory effect: (**a**) atomic configuration on {110} – type planes of parent austenite phase, (**b**) twinned martensite phase occurring thermally on cooling, (**c**) detwinned martensite occurring with deformation [2]

to the multivariant martensite structures in irreversible shape memory effect on cooling below the martensite finish temperature; in contrast, the material returns to the detwinned martensite structure in reversible shape memory case.

Shape memory alloys exhibit another property called superelasticity (SE), which is performed by mechanical stress. Shape memory alloys can be deformed just over austenite finish temperature, and recover the original shape on releasing the stress in superelastic manner.

Shape memory effect is performed in a temperature interval depending on the forward (austenite → martensite) and reverse (martensite → austenite) transformation, on cooling and heating, respectively. Super-elasticity is performed in the parent austenite phase region, just over Austenite finish temperature, A_f. Superelastic materials are deformed in the parent phase region and, shape recovery is carried out instantly and simultaneously upon releasing the applied stress. This property exhibits rubber like behaviour or classical elastic material behaviour. Stress-strain behaviour is different in two cases, shape memory effect and super-elasticity. The maximum strain rate is reached to 8% in superelasticity [2]. Deformation is performed plastically in product martensitic condition in shape memory case, and the material keeps the deformed shape releasing the mechanical stress. The shape memory processes are performed thermally after this deformation. Meanwhile, the material is deformed in parent in parent phase region, and the material recover the original shaper after releasing the external stress. Superelasticity proceeds with only mechanical treatment. This behaviour is also result of martensitic transformation which is induced by applying external stress only in mechanical manner. With this stress, parent austenite phase structures turn into the fully detwinned martensite, and very large strain is observed; and a complete shape recovery is observed upon unloading, therefore, the material behavior resembles elasticity. This strain is maximum 8% in NiTi alloys [2].

The loading and releasing paths demonstrating the shape memory and superelastic effects in stress-strain diagram are shown in Fig. 8.2.

Copper based alloys exhibit this property in metastable β-phase region, which has B2 or DO3 – type ordered lattice at high temperature, and these structures martensiticaly turn into layered complex structures with lattice twinning process, on cooling from high temperature austenitic phase region.

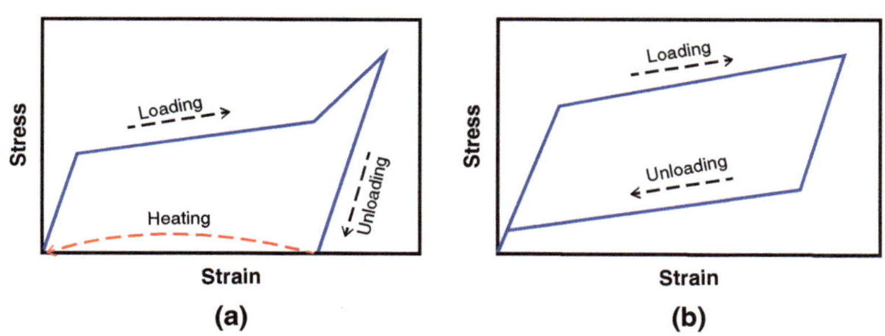

Fig. 8.2 Stress-strain diagram of shape memory effect and superelasticity [2]. (**a**) Shape memory effect. (**b**) Superelasticity

Martensitic transformations occur in a few steps with the cooperative movement of atoms less than interatomic distances by means of lattice invariant shears on {110}-type planes of austenite matrix which is basal plane of martensite. First one is Bain distortion, and second one is lattice invariant shear. Bain distortion consists of an expansion of 26% parallel to the ⟨001⟩-type axes, called Bain axes, and compression of 11% parallel to the ⟨110⟩ and ⟨1$\bar{1}$0⟩-type directions, perpendicular to Bain axes. The compressed directions are [110] and [1$\bar{1}$0] directions for [001] – Bain axes. For the illustration of Bain distortion, an fct unit cell is delineated in matrix, and this cell undergoes to the corresponding fcc unit cell with Bain distortion [3]. Lattice invariant shears occur with cooperative movement of atoms less than interatomic distances on {110}-type close packet plans of austenite matrix. The lattice invariant shears occurs, in two opposite directions, ⟨110⟩-type directions on the {110}-type basal planes and this type of shear can be called as {110} ⟨110⟩-type mode and has 24 variants in self-accommodating manner [4–9]. Lattice invariant shear and twinning is not uniform in copper based ternary alloys and gives rise to the formation of unusual complex layered structures called long period layered structures such as 3R, 9R or 18R depending on the stacking sequences on the close-packed planes of the ordered lattice. The periodicity and unit cell is completed through 18 layers in case of 18R martensite.

8.2 Experimental Details

In the present contribution, two copper based ternary shape memory alloys were selected for investigation; Cu-26.1%Zn 4%Al and Cu-11%Al-6%Mn (in weight). The martensitic transformation temperatures of these alloys are over the room temperature and both alloys are entirely martensitic at room temperature. Specimens obtained from these alloys were solution treated for homogenization in the β-phase field (15 min at 830 °C for CuZnAl alloy and 20 min at 700 °C for CuAlMn alloy), then quenched in iced-brine to retain the β-phase and aged at room temperature after quenching (both alloys).

Powder specimens for X-ray examination were prepared by filling the alloys. These specimens were then heated in evacuated quartz tubes at 830 °C for 15 min and immediately quenched into iced-brine for homogenization. X-ray diffraction profiles were taken from the quenched specimens using Cu-K_α radiation with wavelength 1.5418 Å. The scanning speed of the Geiger counter was chosen as 2°, 2θ/min for the diffractograms.

Specimens for TEM examination were prepared from 3 mm diameter discs and thinned down mechanically to 0.3 mm thickness. These specimens were heat-treated for homogenization at 830 °C for 15 min and quenched into iced-brine to obtain β-type martensite. The quenched disc-shaped specimens were electropolished in a Struers Tenupol-2 instrument at −20 °C in a solution of 20% nitric acid in methanol, and examined in a JEOL 200CX electron microscope operated at 160 kV.

8.3 Results and Discussion

X-ray powder diffractograms and electron diffraction patters were taken from CuZnAl and CuAlMn samples. An x-ray powder diffractogram taken from the long term aged CuAlMn alloy sample is shown in Fig. 8.3. This diffractogram has been indexed on the monoclinic M18R basis. On the other hand, electron diffraction pattern were also taken from the samples these alloys, the details were given elsewhere [3, 6]. Two electron diffraction patterns taken from the quenched samples of Alloy1 and 2 are also shown in Fig. 8.4a, b.

X-ray diffractograms and electron diffraction patterns reveal that both alloy have the ordered structure in martensitic condition, and exhibit superlattice reflections. A series of X-ray powder diffractograms and electron diffraction patterns were taken from both CuZnAl and CuAlMn alloy samples in a large time interval and compared with each other. It has been observed that some changes occur at the diffraction angles and peak intensities on the x-ray diffractograms with aging duration. These

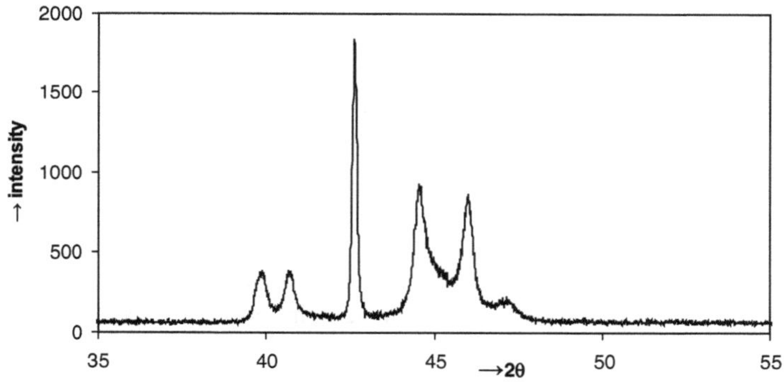

Fig. 8.3 An x-ray diffractogram taken from the long term aged CuAlMn alloy sample

Fig. 8.4 Electron diffraction patterns taken from CuZnAl and CuAlMn alloy samples

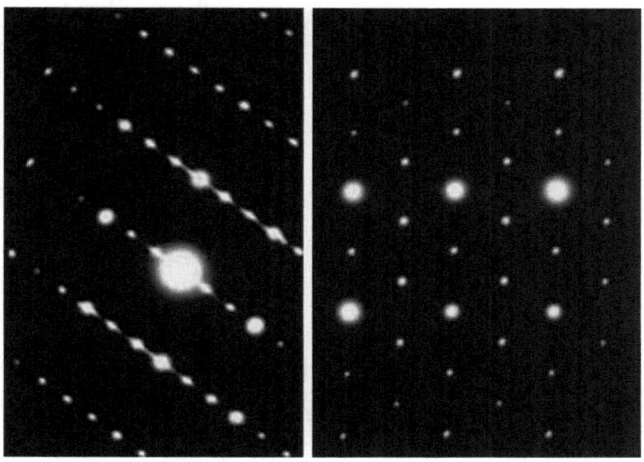

changes occur as rearrangement or redistribution of atoms in the material, and attribute to new transitions in diffusive manner [3, 6, 12]. The ordered structure or super lattice structure is essential for the shape memory quality of the material. In the shape memory alloys, homogenization and releasing the external effect is obtained by ageing at β-phase field for adequate duration.

On the other hand, post-quench ageing and service processes in devices affect the shape memory quality, and give rise shape memory losses. These kinds of results lead to the martensite stabilization in the reordering or disordering manner.

Although martensitic transformation has displacive character, martensite stabilization is a diffusion controlled phenomena, and this result leads to redistribution of atoms on the lattices sites. Stabilization is important factor and causes to memory losses, and changes in main characteristics of the material; such as, transformation temperatures, and x-ray diffraction peak location and intensities.

The martensitic phase in copper-based β-phase alloys is based on one of the $\{110\}\beta$ planes of parent phase called basal plane for martensite. The (110) basal plane which has a rectangular shape in parent phase is subjected to hexagonal distortion and undergoes a hexagon. This process is schematically illustrated in Fig. 8.5.

The powder specimens were aged at room temperature after quenching process, and a series of x-ray diffractograms have been taken from both of the alloy samples in a large time interval. Although all of the diffractograms exhibit similar characteristics, some changes have been observed in peak intensities and diffracted angles on the diffractograms with aging duration. These changes are attributed to new transitions which have diffusive character. It means that some neighbour atoms change locations.

In particular, some of the neighbour peak pairs have moved toward each other. It is interesting that miller indices of these plane pairs provide a special relation:

$$\left(h_1^2 - h_2^2\right)/3 = \left(k_2^2 - k_1^2\right)/n,$$

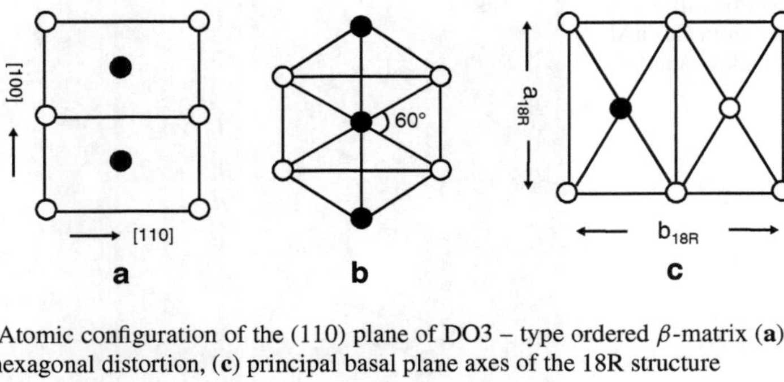

Fig. 8.5 Atomic configuration of the (110) plane of DO3 – type ordered β-matrix (**a**) before and (**b**) after hexagonal distortion, (**c**) principal basal plane axes of the 18R structure

where $n = 4$ for 18R martensite [3]. These plane pairs can be listed as follow; (1 2 2)-(2 0 2), (1 2 8)-(2 0 8), (1 2 10)-(2 0 10), (0 4 0)-(3 2 0). This result can be attributed to a relation between interplane distances of these plane pairs and rearrangement of atoms on the basal plane. In these changes, atom sizes play important role.

The different sizes of atomic sites lead to a distortion of the close-packed plane from an exact hexagon and thus a more close-packed layered structure may be expected. In the disordered case, atom sizes can be taken nearly equal, and martensite basal plane becomes an ideal hexagon. Although martensitic transformations are diffusionless, the post-martensitic transitions have the diffusive character because this transition requires a structural change and this also gives rise to a change in the configurationally order.

Metastable phases of copper-based shape memory alloys are very sensitive to the ageing effects, and any heat treatment can change the relative stability of both martensite and parent phases [12–14]. Martensite stabilization is closely related to the disordering in martensitic state. Structural ordering is one of the important factors for the formation of martensite, while atom sizes have important effect on the formation of ordered structures [10, 11, 13].

8.4 Conclusion

It can be concluded from the above results that the copper-based shape memory alloys are very sensitive to the ageing treatments. X-ray diffraction profiles and electron diffraction patterns reveal that both alloys exhibit super lattice reflections inherited from parent phase due to the displacive character of martensitic transformation. X-ray diffractograms taken in a long time interval show that diffraction angles and intensities of diffraction peaks change with the aging time in martensitic condition. In particular, some successive peak pairs come close each other. These changes lead to the martensite stabilization in the redistribution or disordering manner, and stabilization proceeds by a diffusion-controlled process. The martensite stabilization is a diffusion controlled phenomena and leads to redistribution of atoms on the lattices sites, although martensitic transformation has displacive character.

The basal plane of martensite turns into a hexagon by means of Bain distortion with martensite formation on which atom sizes have important effect. In case the atoms occupying the lattice sites have the same size, the basal plane of martensite becomes regular hexagon; otherwise the deviations occur from the hexagon arrangement of the atoms. The above mentioned peaks come close each other in the disordered case, and occur separately in the ordered case. The changes in the diffraction angles of the selected plane pairs can be a measure of the ordering degree in martensite.

References

1. Ma J, Karaman I, Noebe RD (2010) High temperature shape memory alloys. Inter Mater Rev 55:257–315
2. Sun L et al (2012) Stimulus-responsive shape memory materials: a review. Mater Des 33: 577–640
3. Adiguzel O (2013) Phase transitions and microstructural processes in shape memory alloys. Mater Sci Forum 762:483–486
4. Zhu JJ, Liew KM (2003) Description of deformation in shape memory alloys from DO3 austenite to 18R martensite by group theory. Acta Mater 51:2443–2456
5. Sutou Y et al (2005) Effect of grain size and texture on pseudoelasticity in Cu–Al–Mn-based shape memory wire. Acta Mater 53:4121–4133
6. Adiguzel O (2012) Martensitic transformation and microstructural characteristics in copper based shape memory alloys. Key Eng Mater 510–511:105–110
7. Uhera T, Tamai T (2006) An atomistic study on shape memory effect by shear deformation and phase transformation. Mech Adv Mater Struct 13:197–204
8. Ermakov V, Kruchinin S, Fujiwara A (2008) Electronic nanosensors based on nanotransistor with bistability behaviour. In: Bonca J, Kruchinin S (eds) Proceedings of the NATO ARW "Electron transport in nanosystems". Springer, pp 341–349
9. Dzezherya Y, Novak IY, Kruchinin S (2010) Orientational phase transitions of lattice of magnetic dots embedded in a London type superconductors. Supercond Sci Technol 23:105011–105015
10. Guo YF et al (2007) Mechanisms of martensitic phase transformations in body-centered cubic structural metals and alloys: molecular dynamics simulations. Acta Mater 55:6634–6641
11. Aydogdu A, Aydogdu Y, Adiguzel O (2004) Long-term ageing behaviour of martensite in shape memory Cu–Al–Ni alloys. J Mater Proc Technol 153–154:164–169
12. Li Z, Gong S, Wang MP (2008) Macroscopic shape change of Cu13Zn15Al shape memory alloy on successive heating. J Alloys Compd 452:307–311
13. De Castro F, Sade M, Lovey F (2012) Improvements in the mechanical properties of the 18R ↔ 6R high-hysteresis martensitic transformation by nanoprecipitates in CuZnAl alloys. Mater Sci Eng A 543:88–95
14. De Castro Bubani F et al (2016) Numerical simulations of the pseudoelastic effect in CuZnAl shape-memory single crystals considering two successive martensitic transitions. Smart Mater Struct 25:1–11

Chapter 9
Nanostructured SnO_2 as CBRN Safety Material

V. Grinevych, V. Smyntyna, and L. Filevska

Abstract The present review briefly reflects tin dioxide applications for safety devices for the last 2 years as convenient, cheap, widespread material with suitable physical and chemical properties. The usage of nanoscale SnO_2 forms are considered for several types of devices such as: gas sensors of conductometric type, electrochemical sensors, sensors on the SPR effect, material for electrodes of lithium-ion batteries and solar cells, together with catalytic applications for decomposition of pollutants.

Keywords Tin dioxide · Gas sensing · Electrode material · Catalysts

9.1 Introduction

Tin dioxide, used for many years for various applications, is a well known material for researchers. It may seem that its properties have long been studied [1], the possibilities are exhausted and this material has any prospects. However, mastering the nanotechnologies has expanded the possible applications for many long-known materials, including tin dioxide. This position is supported due to tin dioxide active mentioning in the latest 10–12 years in a number of scientific reviews on gas sensors' [2–5], catalytic, and electrode materials [6, 7], as well as reviews on the manufacture, properties and application of various forms of this material [8–13]. There is also a work where nano tin dioxide is applied as sorbent in medical purposes [14].

This review of works (authors' inclusive) over the past 2 years supports the idea of nano tin dioxide as a cheap widely available material with convenient physicochemical properties for the devices of safety production.

V. Grinevych (✉) · V. Smyntyna · L. Filevska
Odessa I.I. Mechnikov National University, Odessa, Ukraine
e-mail: grinevich@onu.edu.ua

J. Bonča, S. Kruchinin (eds.), *Nanostructured Materials for the Detection of CBRN*,
NATO Science for Peace and Security Series A: Chemistry and Biology,
https://doi.org/10.1007/978-94-024-1304-5_9

Some tin dioxide applications (not complete) as a CBRN material are given below:

Sensors material: flammable and toxic gas sensors, heavy metal ion sensors, sensors of biologically dangerous objects.

Electrode end Catalytic applications: Li-ion batteries, solar cells, hydrogen and ozone production, the decomposition of pollutants.

Tin dioxide mentioning data in literature over the past year is given below (NCBI databases PubMed Central® (PMC) – free full-text archive of biomedical and life sciences journal literature in the U.S.; National Institute of Health, National Library of Medicine (NIH/NLM)): Total number of articles about tin dioxide is ~448, among them: nano SnO_2 – 231, sensors on SnO_2 – 186, SnO_2 electrode – 308, from them for solar cells – 118, for Li-Ion Battery – 63.

Over the past 10 years, more than a dozen scientific reviews on its properties, applications, research and developments with its use have been published.

9.2 Tin Dioxide in Sensor Applications

The spectrum of sensory materials for ensuring the safety of industrial, domestic, natural environments and the detection of toxic and dangerous components is currently estimated at tens. Among them, a whole class of metal oxides, which have been used for many years. One of the first places in this class belongs to tin dioxide. Its sensory properties, reduced to nanoscale, are much more effective.

The number of ways to register the environment compositions deviations from the standard has also increased. The conductometric methods used for more than 40 years were complemented by optical ones (control of direct absorption/reflection of light, a luminescence, use of methods of automatic calculation and comparison of optical characteristics of materials in the process of measurement), plasmon resonance and other methods. It was found that in the most of them the nano tin dioxide was very productive as a sensitive or matrix element.

9.2.1 Tin Dioxide as the Conductometric Type Adsorptive and Sensitive Element for Gas Sensors

An active study of tin dioxide gas sensitivity mechanisms to various polluting gases, as H_2, Cl_2, SO_2, and NH_3, NO, NO_2, CO, CO_2, CH_4, and C_3H_8, ethanol, acetone, etc., was applied for gas sensors production. The main role in sensitivity belongs to the active forms of oxygen (O_2^-, O^-, O^{2-}) adsorbed on the surface of the sensor material. The SnO_2, due to its chemical characteristics, adsorbs oxygen precisely in such active forms.

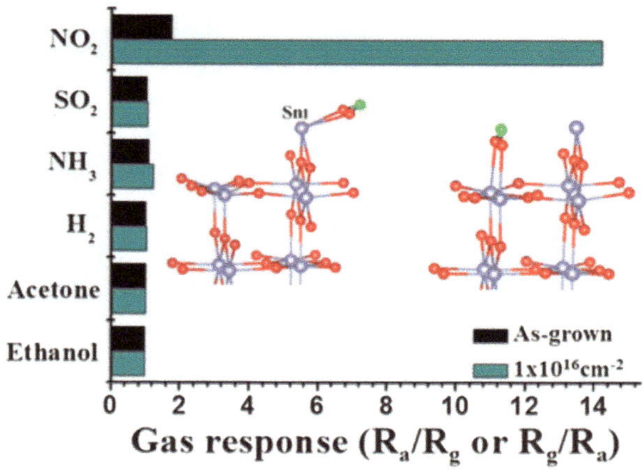

Fig. 9.1 Comparison of the sensitivity of nanowires of tin dioxide to various gases [15]. The inset shows the crystalline structure of SnO_2 nanowires

The work [15] is devoted to improvement of selectivity to NO_2 in SnO_2 nanowires by means of He ions radiation with different ion flux density. The maximum sensitivity to NO_2 shows the nanowire under ion flux density of 1×10^{16} ions/cm^2. The greatest number of surface defects in the form of interstitial tin was detected by photoluminescence analysis and X-ray photoelectron spectroscopy in the wires. Just these defects created by ionic radiation provide the adsorption of NO_2. The molecular modeling for the irradiated surface of SnO_2 (110) was carried out in the work. The Fig. 9.1 shows the comparison of sensitivity of tin dioxide nanowires to various gases and the crystalline structure of SnO_2 nanowires.

Theoretical studies of gas sensitivity mechanisms of SnO_2 (110) surface to mono and dioxide of nitrogen are conducted in work [16]. The phase diagram of the SnO_2 (110) surface in contact with O_2 and NO gas environment was determined by means of ab initio thermodynamic method. It was found that the fully reduced surface containing the bridging and in-plane oxygen vacancies was under oxygen-poor conditions, while the fully oxidized surface containing the bridging oxygen atoms and the oxygen dimer is under oxygen-rich conditions. The stoichiometric surface was proved to be the most stable. NO-rich conditions were formed by NO adsorption on the most stable surfaces only in the presence of oxygen.

Researchers from the Hanoi University of Science and technologies [17] made metal-oxide nanowires (NW) sensors for chlorine monitoring of ZnO, WO_3, and SnO_2 by on-chip growth technique with chemical vapor deposition method. The adsorptive response (R_{Cl2}/R_{air}) of SnO_2 nanowires sensor detecting 50 ppb Cl_2 at 50 °C was the largest and amounted about 57 units. Figure 9.2 shows general picture of nanowires sensor and sensitivity to chlorine comparison for metal-oxide nanowires of ZnO, WO_3, and SnO_2.

Tin dioxide was chosen by China researchers as a material for creation various three-dimensional nanostructures for sounding of gases [18]. By means of single-stage hydrothermal approach and the subsequent thermal annealing 3D nanostructures with various hierarchical morphology and, respectively, various

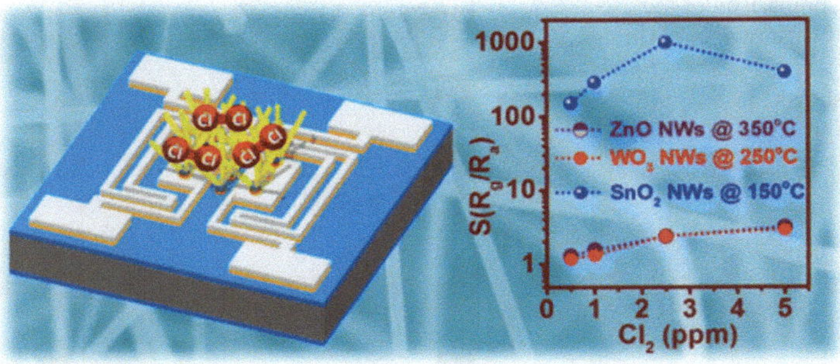

Fig. 9.2 General picture of nanowires sensor together with sensitivity comparison to chlorine of metal-oxide ZnO, WO_3, and SnO_2 nanowires [17]

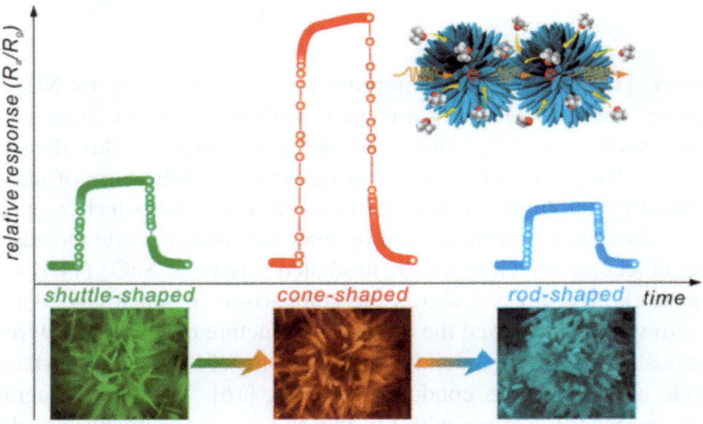

Fig. 9.3 A comparison of sensitivity to acetone of 3D nanostructures with various hierarchical morphology [18]

gas-sensitive properties were created. As it appeared the three-dimensional SnO_2 of cone-shaped hierarchy nanostructure shows the maximum sensitivity about 175 in relation to 100 ppm of acetone. A good sensitivity was shown by these structures also to some other organic compounds. As the microstructural analysis showed, the best sensitivity of cone-shaped hierarchical nanostructures is provided by more active superficial defects and discrepancies formed in the course of thermal recrystallization. A comparison of sensitivity to acetone of 3D nanostructures with various hierarchical morphology is shown at Fig. 9.3.

For improvement of sensor properties, increase in selectivity, the tin dioxide was doped by Cu [19, 20], Zn, Mn [21], Au [22], Sb and other additives. The study [19] of pure and copper-doped nanowires of tin dioxide, synthesized by thermal evaporation showed a good incorporation of copper atoms into the tetragonal rutile

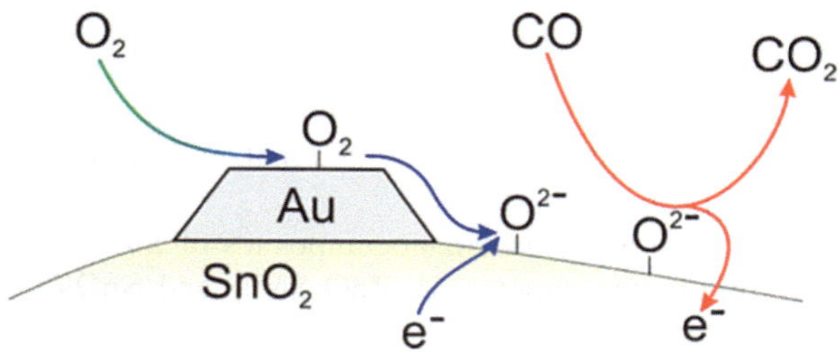

Fig. 9.4 Scheme of additional adsorption of oxygen on gold with its transfer to the surface of tin dioxide [22]

lattice of SnO_2. The undoped sensors based on SnO_2 nanowires showed a high sensitivity to gaseous ethanol, which was improved by the addition of Cu. In the other research [20], nano SnO_2in the form of nanosheets and nanodisc-like structures doped by 1% of Cu showed highly improved properties of sounding for CO in comparison with pure SnO_2 nanostructures and excellent selectivity to CO with insignificant hindrances by CH_4, CO_2 and NO_2. In the work [21] the SnO_2 doped by Cu (II) ions showed the fastest response time to 100 ppm of ethanol (9.7 sec. in comparison with 12.4 sec. for not doped SnO_2). At the same time, SnO_2 containing 2.91% of Mn showed 2.5 times higher response at detection of 100 ppm of ethanol in comparison with not doped material.

In [22] the sensitivity of tin dioxide covered with gold particles to gaseous CO was studied. The mechanism of sensitivity increasing was associated with additional adsorption of oxygen on gold with its transfer to the tin dioxide surface. Negatively charged oxygen ions on tin dioxide surface give additional states for interaction with the detected gas, thus increasing productivity of a sensitive layer. Figure 9.4 shows scheme of additional adsorption of oxygen on gold with its transfer to the surface of tin dioxide.

Tin dioxide is applied in the gas analysis not only as independent gas-sensitive material, but also as a constituent in composites, heterostructures and complex nanostructures for gas sounding. At the same time, its main role in a complex material is a gas-sensitive role, since it due to its structure changes its electrical parameters when interacting with the external environment [23]. Other materials provide selectivity by means of their catalytic features.

The work [24] is devoted to the research of such gas-sensitive nanoheterostructures as TiO_2/SnO_2 for H_2 control. The structures themselves were nanocrystalline TiO_2/SnO_2 n–n heterojunctions and have been obtained using flame spray synthesis from nanopowders of pure SnO_2, 90 mol % SnO_2/10 mol % TiO_2, 10 mol % SnO_2/90 mol % TiO_2 and pure TiO_2. It is interesting that with a long recovery time of SnO_2-rich samples their H_2 detection threshold is lower than 1 ppm and their large responses is over the whole measuring range. Researchers systematized the data on sensor response value, working temperature of gas-sensitive nanoheterostructures of TiO_2/SnO_2 and correlating with these data on structure and methods of material production. Such systemizing gave information on heterostructures in the form of nanopowders, nanofibres, nanotapes, nanowires, thin and thick films, ceramics, and systems: core-shell, the coral-like structures obtained by very different methods. Among the production methods there are a co-deposition and mechanical mixing, sol-gel and hydrothermal methods, atomic layer-by-layer deposition and thermal evaporation, together with metal-organic chemical vapor deposition.

The heterostructured nanocomposites of SnO_2-ZnO were studied in the works [25, 26]. In the paper [25], ZnO-SnO_2 and SnO_2 nanoparticles were fabricated by hydrothermal synthesis for gas sensor to H_2 and CO. Heterostructures showed hypersensibility to the detected gases in comparison with undecorated SnO_2 nanoparticles. This result, together with the nano grain size, is explained by additional number of reaction states at the heterojunctions ZnO-SnO_2 nanoparticles where there can be noticable electron transfer between the compound nanostructures and the absorbed oxygen species.

Group of researchers [26] studied sensitive properties of SnO_2-ZnO composite nanostructured thin films series with different amounts of SnO_2 (from 0 to 50 wt %) for highly toxic and flammable gases (CO, CO_2, CH_4, and C_3H_8). Samples were deposited on a miniaturized porous alumina transducer using the sol-gel and dip coating method. Improved sensing was achieved for the ZnO (98 wt %) – SnO_2 (2 wt %) composite as compared to the sensors containing only the pristine oxides. Actually, only 2 wt % of tin dioxide was the suitable quantity for ensuring the best sensitivity of heterostructures. The sensor of the mentioned composition showed the highest sensitivity to carbon monoxide (min. 5 ppm).

The successful combination of dimensional effect of nanostructure and transit properties of n–n heterojunctions SnO_2-ZnO is used by authors [27] for creation gas-sensitive one-dimensional (1D) n–n SnO_2-ZnO heterostructures perspective for highly effective sensors of organic amine compounds. The heterostructures were ZnO nanowires (80–100 nm in diameter, 12–16 μm long) coated with layers of SnO_2 nanoparticles (about 4 nm) by effective solvothermal treatment followed by calcination at 400 °C. The size of tin dioxide nanoparticle, comparable with the Debye length in this material (~3 nm), provides almost complete particle depletion by electrons due to various surface adsorbed oxygen species $O^{\delta-}(O_2^-, O^-$ and $O^{2-})$. Additional depletion is created at the SnO_2/ZnO heterointerface due to different work functions of ZnO (5.2 eV) and SnO_2 (4.9 eV). The energy band structures and gas sensing mechanisms of the SnO_2-ZnO NW heterostructure sensor is shown at Fig. 9.5.

A greater amount of oxygen can be adsorbed by the wire 1D structures both due to a lower agglomeration tendency and to a larger surface to volume ratio. All this facilitates the gas diffusion through a sensor and promotes active superficial reactions. The above mentioned features of one-dimensional (1D) n–n nanowire heterostructures of SnO_2-ZnO allowed to create on their basis a sensor of organic amine compounds with the improved characteristics: high speed of reaction and recovery, good selectivity and excellent reproducibility for n-butylamine vapors.

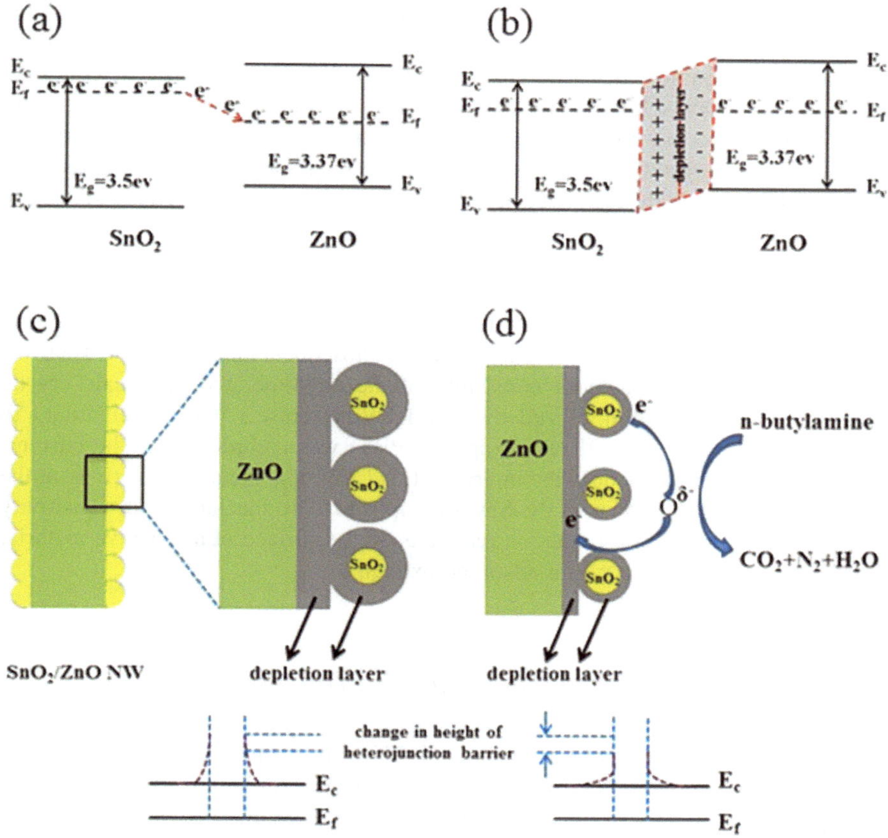

Fig. 9.5 (**a**, **b**) The energy band structures and (**c**, **d**) gas sensing mechanisms of the SnO_2-ZnO NW heterostructure sensor [27]

9.2.2 Tin Dioxide as Electrochemical Sensors of Organic Compounds and Heavy Metals

Enough often tin dioxides are used in electrochemical type sensors for heavy metals and dangerous organic compounds detection in various environments. Researchers from China [28] has conducted a complex of experimental and theoretical studies (adsorption/desorption tests, density-functional theory (DFT) calculations, and X-ray absorption of fine structure (XAFS) studies), thus revealing the electrochemical mechanism of heavy metals' ions interaction on various faces of tin dioxide nanocrystal. Researches showed that the face {110} has the lowest potential barrier for ions' superficial diffusion, hence showing the best electrochemical characteristics for detection of heavy metals ions.

Group of researchers from India and Saudi Arabia [29] showed that the SnO_2 Quantum Dots (QDs) are effective electron mediators for efficient electrochemical sensor fabrication for the easy, cost effective and accurate on-site detection of cadmium ions in a solution. Tin Dioxide QDs were obtained by low-temperature hydrothermal process and were pure SnO_2 with an average size of 2–4 nm and having perfect crystallinity with tetragonal crystal structure. The manufactured sensor exhibited a high sensitivity of $\sim 77.5 \times 10^2$ nA ppm^{-1} cm^{-2}, a low detection limit of ~ 0.5 ppm and response time of <2 s.

In [30] it is reported the preliminary research of the electrochemical sensor for an organophosphate (OP) on the basis of nanoscale SnO_2 as matrix material. Researchers used both pure tin dioxide, and the one, doped by Ni and Cr for the process of chemical synthesis in solution with the subsequent calcinations at 1000 °C. The matrix electrode material consisting of the agglomerated nanoscale crystal nature particles was immobilized by acetylcholinesterase (AChE). Sensitivity to organophosphate (from 0.01 to 100 ppm in PBS) was recorded in cyclic current-voltage (CV) measurement in the interval –1.5–1.5 V. The greatest sensitivity at the level of 10 ppb was shown by Ni-SnO_2 composite. The mechanism of sensitivity is explained by oxidation-reduction reactions on the surface of a sensitive material involving different forms of pre-adsorbed oxygen.

9.2.3 SPR-Sensor Applications for Tin Dioxide

Presence of free electrons in a tin dioxide thin-film, as in a degenerate semiconductor makes it possible to observe a surface plasmon resonance (SPR) in it [31]. This phenomenon appeared to be enough sensitive to changes of an environment. In a number of recent works, tin dioxide has been used as a material for sensors of biological media composition based on the SPR phenomenon [32, 33]. The measured sensitive value in such sensors is the refraction index of a material contacting with the controlled environment.

Using tin dioxide [33] as a sensing element in a fiber optic in SPR technique makes it possible to obtain a miniature probe for online monitoring, remote sensing of ammonia with high sensitivity, selectivity, stability, durability and low cost. Since the SPR characteristics depend significantly on the thickness of the sensitive material layer, then the probes with different thicknesses of SnO_2 have been investigated. Electromagnetic field distribution for the multilayer structure of the probe reveals the enhancement of evanescent field at the tin oxide-ammonia gas interface which in its turn manifests the highest shift in resonance wavelength at a definite thickness.

A theoretical analysis of a thin-film SnO_2 sensor for detection of volumetric media refraction index was presented in [34]. The analysis was carried out on the basis of Fresnel's equations for reflection coefficients. It showed that tin dioxide usage as the dielectric matching layer between a prism and the waveguide layer increases angular sensitivity. The offered configuration of a refraction index sensor using SnO_2 is perspective for effective monitoring of biomolecular interactions and application in a biosensorics.

9.3 Tin Dioxide as a Material for Electrodes

At the first sight, tin dioxide – as an electrode material does not belong to CBRN area. However, such electrodes are applicable in ecologically safe solar power, and tin dioxide application as a part of electrodes in lithium-ion batteries (LiBs) significantly extends time of their operation, and allows to save resources considerably and to reduce environmental pollution.

Electric characteristics of tin dioxide as degenerate semiconductor in combination with its wide forbidden zone and, hence, with optical transparency allow using it actively as transparent electrodes in the most different optoelectronic devices for a long time.

9.3.1 For Li-Ion Batteries

Traditionally graphite is used as an electrode in Li-ion batteries (LiBs). However, tin dioxide has better operational characteristics with its theoretical capacity (782 mAh g^{-1}) which is twice more than graphite capacity, for example, (372 mAh g^{-1}) [35]. Usage of SnO_2 was limited due to a significant change of oxide volume (more than 300%) during the battery charge/discharge procedure that led to destruction of an electrode. Many researchers succeeded to overcome this defect using various nanoporous structures based on tin dioxide or with the embedded tin dioxide nanoparticles, also nanotubes, shell/core nanostructures, etc. [36–39]. In the majority of such structures the size of the embedded SnO_2 particles does not exceed 4–5 nm, since a larger size can not secure the reversibility of Sn/SnO_2 conversion process [40, 41]. The usage of tin dioxide in complex nanocomposites made it possible to increase the operational capacity of the electrodes to a level approaching 2000 mAh/g. In addition to high throughput, SnO_2-containing electrodes have another qualitive advantage for lithium-ion batteries – a long-term multiple cyclicity. As it was well summarized in [41], the nanostructured SnO_2 with a high reversibility of Sn/SnO_2, suppressed aggregation of Sn and stable passivating solid-electrolyte interphase (SEI) layer during cycling would be an ideal material for realization of high-performance LiBs.

After A. Geim and K. Novoselov reported in 2004 [42] on the production of graphene fixed on silicon oxide layers, such a material became widely used in electronics. In 2009 [35] it was reported on the creation of a composite material of graphene nanosheets and SnO_2 nanoparticles used as an anode material for LiBs. The Japanese researchers created nanoporous mobile 3-D structure as SnO_2 nanoparticles (3.3–7.5 nm) composite which are uniformly located between graphene sheets. Nanopores between SnO_2 and nanosheets can play a role of buffer space in the charging/discharging process that significantly increases efficiency, capacitive properties and strength of such material (810 mAh/g) compared with usual SnO_2. In subsequent years several more groups of scientists used tin dioxide nanoparticles in graphene nanosheets for creation of highly effective electrodes for LiBs [36–38], at that, reversible capacities of the created electrodes exceed the theoretical capacity of SnO_2 almost by 1.5–2 times.

Enough complex three-dimensional graphene-like network structures with uniformly built-in SnO2@Sn nanoparticles doped by nitrogen, and encapsulated by carbon (N-C) were created and studied in work [39, 43] as anodes of high-performance LiBs. Structures, denoted as N-CSnO_2Sn/3D-GNs, has been fabricated by means of a low-cost and scalable method: an in situ hydrolysis of Sn salts and immobilization of SnO_2 nanoparticles on the surface of 3D-GNs, followed by an in situ polymerization of dopamine on the surface of the SnO_2/3D-GNs, and finalized by carbonization. The created composites promoted highly efficient insertion/extraction of Li(+). It is provided by three-layer structure of composite nanoparticles. The outermost N-C layer with graphene-like structure of the N-CSnO_2Sn nanoparticles can effectively buffer the large volume changes, enhance electronic conductivity, and prevent SnO_2/Sn aggregation and pulverization during discharge/charge. The middle layer function (namely tin dioxide) was to facilitate the cyclic charge/discharge process by participating in the reaction SnO_2 + Li(+) → Sn + Li_2O. The inner Sn layer with large theoretical capacity can guarantee high lithium storage in the composite. This novel hybrid anode exhibits highly stable capacity of up to 901 mAh/g, with ~89.3% of capacity retention after 200 cycles at 0.1 A/g and superior high performance rate, as well as a long lifetime of 500 cycles with of 84.0% retention at 1.0 A/g. Such high operational rates are combined with structural integrity of the whole electrode in which tin dioxide plays an important role.

However, electrodes with good performance may be obtained not only of a graphene. Asymmetric membrane structure for stabilization of the LiBs anode based on SnO_2 with excellent electrochemical characteristics was created in [40] using a combined sol-gel technique followed by carbonization. The reached specific capacity of an asymmetric membrane electrode on tin dioxide in 500 mAh/g managed to be kept at the level of 96% after 400 cycles at a current density of 280 mA/g (~0.5 C). The current density doubling leads to a decrease in the general power only by 36%. The membrane has mesh porous structure which provides high conductivity, multiple channels for diffusion and free volumes for an electrode expansion. The main technological step influencing properties of an electrode is the carbonization temperature. The membrane electrode shows the best characteristics in case of carbonization at 500 °C. At the same time the size of SnO_2 nanoparticles is ~3.9 nm. At temperature increase to 800 °C electrode loses 51% of the capacity for 100 cycles due transformation of nanoparticles into large tin spheres (~40 nm).

In February 2017 Wei and colleagues [44] reported the inexpensive way of synthesizing of uniform anode material for LiBs from ultrasmall (~3 nm) particles of SnO_2 which are uniformly distributed in polymer of a styrylpyridinium (SbQ).

Formation of an electrode was carried out by UF hardening method. Ability of SbQ polymer to a photocrosslinking in-situ allowed synthesizing uniform ultramassive material. Styrylpyridinium being strong binding for SnO_2 nanoparticles provides effective change of SnO_2 anodes' volume during cyclic charge- discharge process. Steady specific capacity 572.5 mAh/g of the created SnO_2 electrode is obtained at the current density of 0.2 C (156.2 mA/g) after 150 cycles. Even at high current density 5 C (3905 mA/g) the specific capacity is 440.2 mAh/g.

The priority requirements to anode materials of LiBs were realized by the Korean researchers [45] having synthesized anodes on SnO_2/NiO nanotube (m-SNT decorated by mesoporous Ag nanoparticles). The material was synthesized by an electrospinning treatment followed by fast calcination and subsequent chemical reduction. Such a problem as considerable change of volume at cycling is effectively reduced in one-dimensional porous hollow structure. This structure also provides a short lithium-ion diffusion length. The reversible capacity of the m-SNT anodes was significantly improved using metallic nickel (Ni) nanoparticles converted from NiO nanograins during the lithiation process of Li_2O reversible decomposition. Ag nanoparticles uniformly decorated on the m-SNT via a simple chemical reduction process significantly improve rate capability and also contribute to long-term cyclability. The m-SNTAg anodes exhibited excellent cycling stability without capacity fading after 500 cycles with a high capacity of 826 mAh/g at a high current density of 1000 mA/g. Furthermore, even at a very high current density of 5000 mA/g, charge-specific capacity remained as high as 721 mAh/g, corresponding to 60% of its initial capacity at a current density of 100 mA/g. The Fig. 9.6 shows schematic illustration for the synthetic route of the m-SNT@Ag and long-term cycling stability of the m-SNTAg at a current density of 500 mA/g [45].

Thus, the nanostructured SnO_2 with high reversibility of Sn/SnO_2 obtained due to the suppressed aggregation of Sn and to stable layer of SEI at cycling would be an ideal material for high-performance LiBs.

Problems of tin-dioxide's volume change in electrodes of LiBs, and at the same time the satisfaction of the requirement to their charging efficiency and cyclic stability were solved in [41] by creation of polydophamine (PDA) – coated SnO_2

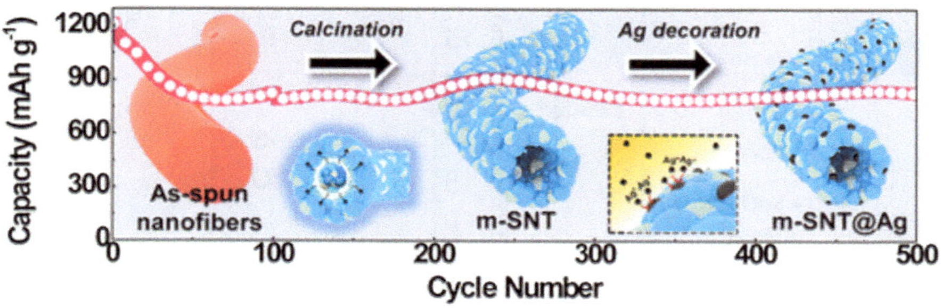

Fig. 9.6 Schematic illustration for the synthetic route of the m-SNTAg and long-term cycling stability of the m-SNTAg at a current density of 500 mA/g [45]

nanocrystals, composed of hundreds of PDA-coated "corn-like" SnO_2 nanoparticles (diameter ~5 nm) decorated along a "cob". In a combination, the corn-like nanostructure and PDA protection provided excellent electrochemical characteristics of SnO_2 electrode together with excellent long-term cyclic stability throughout more than 300 cycles, high reversibility of Sn/SnO_2 and excellent speed.

9.3.2 *For Solar Cells*

Unique combination of electrophysical properties of tin dioxide as degenerate semiconductor (small work function, high density of electrons in the forbidden zone) and considerable optical width of its forbidden zone (~4 eV), a possibility of obtaining thin layer coatings makes this material to be very convenient as electron transporting layer (ETL) for solar elements, especially new perovskits solar elements – perspective cheap silicon substitutes [46–48]. Many researchers note positive impact of low-temperature technologies of SnO_2 nanolayers production for electron selective layers [46–51].

Thin films of amorphous tin dioxide were offered in [47] as electron transporting layer (ETL) in planar heterojunction n-i-p organohalide lead perovskite and organic bulk heterojunction solar cells. Films were prepared by chemical bath deposition from a non-toxic aqueous bath of tin chloride at very low temperatures (55 °C) and do not require post-annealing treatment. Low mobility of electrons, specific for SnO_2 film samples, is successfully compensated by low work function with ideal alignment of zones on the border of SnO_2/methylammonium lead iodide ($MAPbI_3$) and strong blocking of holes due to the deep SnO_2 valence zone. The schematic picture of an amorphous SnO_2 film deposition method and the volt-current characteristics of a solar cell with a-SnO_2 ETL are given at the Fig. 9.7 from

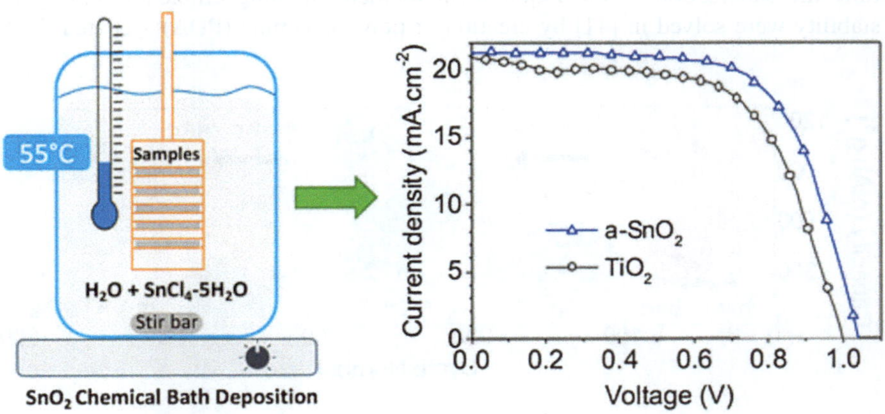

Fig. 9.7 The schematic picture of SnO_2 amorphous film deposition method and the volt-current characteristic of a solar cell with a-SnO_2 ETL [47]

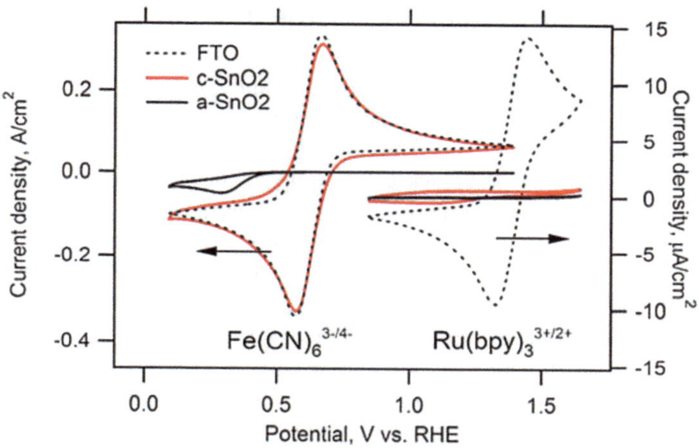

Fig. 9.8 CV characteristics of SCs with different electron selective layers [52]

[47]. It is seen that usage of a-SnO_2 ETL prepared by chemical bath deposition provides a solar cell with the better operational characteristics in comparison with TiO_2 ETL.

Group of the Swiss and Czech scientists led by M. Grätzel (the famous creator of the same name electrochemical solar cell) [52] presented in 2017 an ultrathin electron selective layers (ESL) of SnO_2 fabricated by atomic layer deposition (ALD) for application in planar perovskite solar cells as powerful alternatives to other oxides such as TiO_2. The obtained layers besides being used for photoelectrodes in perovskite and dye-sensitized solar cells also may be applied in photoelectrochemical water splitting. The various studies of the layers' physical properties showed that not calcinated low-temperature ALD-grown SnO_2 layers are amorphous and perfectly pinhole-free for thicknesses down to 2 nm. Such a thickness and electric properties of layers allows designing a photoelectrode with thinner electron selective layers that potentially minimizes resistance losses. Figure 9.8 shows the advantages of amorphous low-temperature ALD-grown SnO_2 layers application as ESL. These layers have the minimal hardly noticeable hysteresis of CV characteristics of SCs which is one of their negative issues.

The reduction of CV hysteresis researchers [49] tried to achieve using the yttrium-doped tin dioxide (Y-SnO_2) electron selective layer. This material was synthesized by an in situ hydrothermal growth process at 95 °C. Two main effects of yttrium doping which supplemented and strengthened the tin dioxide characteristics of were revealed in the studies. The first, it improves electrons transfer from perovskite to ESL due to formation of well leveled and more uniform distribution of SnO_2 nanosheet massifs, thus providing the best contacts of perovskite with SnO_2 nanosheet, secondly, it secures the charges recombination decrease at NSA-perovskite interfaces due to the forbidden zone growth and to the switching of the band energy levels up, and their alignment by perovskite energy levels.

9.4 Catalytic Applications of Tin Dioxide

Tin dioxide in catalysis, including photo-electrocatalytic applications, is also useful for environmental safety. It promotes the decomposition of many harmful chemicals polluting the aquatic environment. Therefore the material is widely and successfully used also in such CBRN processes as cleaning of the polluted environments, for example, sewage. For these aims they use the catalytic properties of the material stimulating reactions of pollutants decomposition, which at their turn solves two ecologically important problems simultaneously: a cleaning problem and a problem of utilization of harmful compounds. As a rule, tin dioxide in this case is used as a part of an electrode in the electrochemical reactor [53–56].

In work [53] tin dioxide was used as a part of Ti/SnO_2-RuO_2 electrode for which the electrocatalytic degradation of bromkrezol green (BCG, growth mediums for microorganisms) was investigated in details. The carried-out electrode material characterization by means of the scanning electron microscopy, X-ray diffractometry and X-ray fluorescence spectrometry showed "cracking dirt" structure and excellent specific surface area of Ti/SnO_2-RuO_2 electrode. Efficiency of BCG removal from the Ti/SnO_2-RuO_2 electrode was determined basing on chemical consumption of oxygen and on ultra-violet-visible absorbing spectrometry (UV-Vis spectroscopy changes are shown in the Fig. 9.9).

The major factors defining the BCG removal efficiency were in descending order: initial pH_0, reaction temperature, and current density and electrolysis duration time. The BCG removal efficiency reached 91% under optimal experimental conditions (initial concentration – 100 mg/L, initial pH_0 – 7, temperature of reaction – 30 °C, current density – 12 m/cm^2 and the electrolysis time – 150 min).

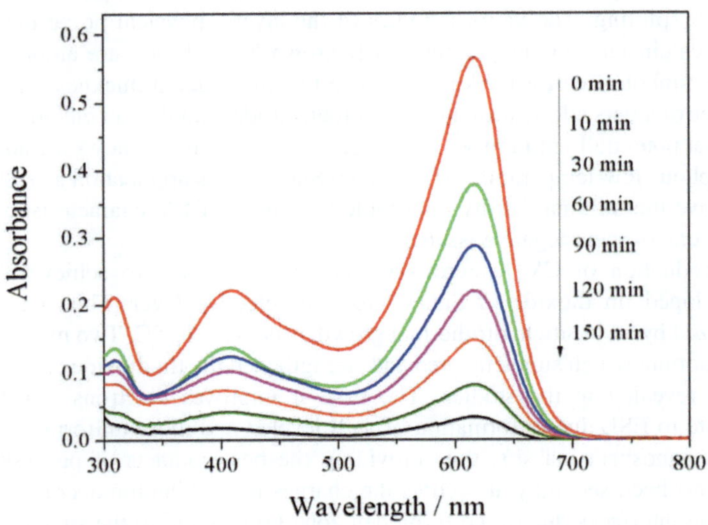

Fig. 9.9 UV-Vis spectroscopy changes of BCG electrocatalytic degradation [53] with electrolysis time

Iranian researchers [54] also used SnO_2 together with Sb_2O_4 on a titanic plate for obtaining Ti/SnO_2-Sb_2O_4 electrode for electrocatalytic phenol removal from sewage. The phenol decomposition mechanism on a surface of Ti/SnO_2-Sb_2O_4 anode depends on the solution's. The phenol chemical degradation on an electrode was studied also under the influence of current density, the supporting electrolyte and the phenol initial concentration. Various strong radicals formation in electrolysis with Ti/SnO_2-Sb_2O_4 anode led to quick (during 1 h) oxidation of phenol up to benzoquinon from 100 to <1 mg/L under optimal conditions. Speed of TOS degradation on the anode surface was 49% during 2 h.

In [55] tin dioxide was used as a part of Ti/SnO_2-Sb tubular porous electrode (anode) for decomposition of pyridine of sewage. Tubular porous Ti was covered by SnO_2-Sb by Pechini's method. The performance data of the reactor were studied under the influence of flow on a static pattern, initial pyridine concentration, supporting electrolyte concentration, current density and pH. The pyridine removal coefficient reaches a maximum (98%) under optimal operating conditions: 100 mg/L of pyridine initial concentration, 10 g/L, the supporting concentration of electrolyte, 30 m/cm^2 density of current and pH 3.

Chinese researchers [56] applied tin dioxide for creation of the nanostructured high-porous 3D electrode of the Ti/Sb-SnO_2-GR based on three-dimensional porous graphene hydrogel. The electrode was formed by layer-by-layer sedimentation. The 3D porous electrode has the high potential for oxygen release (2.40 V), smaller resistance to charge transfer (29.40 Ohms·cm^2), higher porosity (0.90), the elevated coefficient of a roughness (181) and higher charge value (57.4 mC/cm^2) in comparison with flat Ti/Sb-SnO_2 electrode. Electrocatalytic oxidation of rhodamine B (RhB) as a test electrochemical reaction also showed better characteristics, stability and low specific power consumption.

The SnO_2 nanoparticles microspheres decorated by Ag, obtained by a simple hydrothermal (one-pot hydrothermal method) [57] were successfully applied in catalytic reduction of 4-nitrophenol to 4-aminophenol by potassium borohydride (KBH_4) as a model reaction. The known toxic pollutant 4 nitrophenol (4-NP), is widely present in industrial drains and agricultural sewage, and its chemical reduction to 4-aminophenol (used for production of analgetics, cosmetic and anticorrosive materials) is the most preferable. The catalytic characteristics of the obtained composite structures were enough high, so that normalized rate constant (κ_{nor}) was 6.20 L/(min g). Besides the mentioned, they demonstrate good renewability after the first five cycles.

Tin dioxide usage in various structures improves the production characteristics of gaseous hydrogen [58] and ozone [59] which is useful both as for power, and ecological safety.

In [58] tin dioxide was applied as a part of one-dimensional nanowires with core-cover type structure of the CeO_2/SnO_2 grown up on a three-dimensional porous disk in the form of the nano-wood. The structure was effectively used for obtaining H_2 by thermochemical separation of water at temperatures up to 800 °C. Three-dimensional a core- cover type "nanowire wood" increased the production of H_2 by 45.5% at 800 °C in comparison with usual thin-film CeO_2 covered disks as standards.

Theoretical calculations using a theory of the density functional were applied by Gibson and colleagues [59] to analyze electrochemical mechanism of ozone production by decomposition of water on the SnO_2 catalyst doped by Ni and Sb. The SnO_2 face (110) appeared to be the most stable according to calculations. This surface was used for water decomposition modeling in the presence of doping by Sb and Ni, paying special attention to stages of O_2 and O_3 formation. The ozone formation takes place according to the Langmuir–Hinshelwood's mechanism, i.e. on the surface of the catalyst. The adsorption energy calculation (EADS), the Gibbs' free energy (ΔG_{rxn}) and barriers of activation (E_{act}) obtained for two final stages of ozone formation serves as a basis for new materials development with higher catalytic efficiency.

9.5 The Nanostructured Thin Films of Tin Dioxide

Authors have conducted a series of studies of tin dioxide thin films obtained using polyvinil acetate (PVA) polymers in sol-gel method to improve structuring.

In the said studies the content of the precursor (Bisacetylacetonato dichlorotin (BADCT)) in the initial solution varied from 1 to 10%. The polymer (PVA) content in the solution was 1%. After deposition on the glass substrate, the samples were annealed until the organic components were removed and a transparent tin dioxide layer was formed. AFM three dimensional tin dioxide film's surface picture is shown at Fig. 9.10.

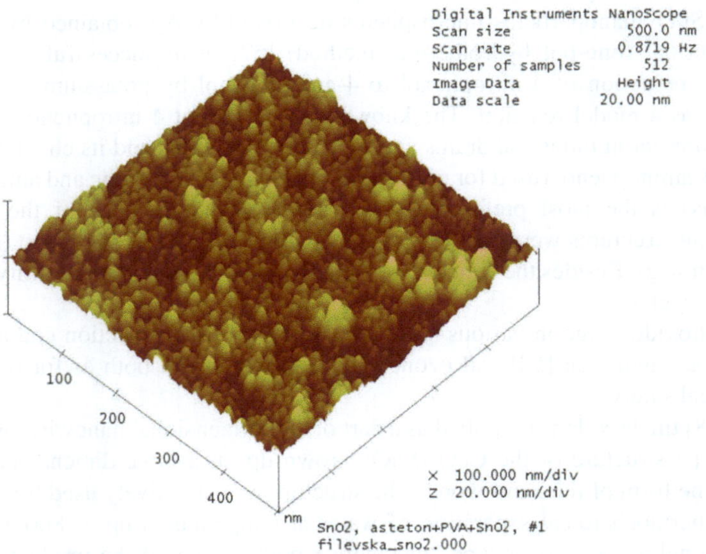

Fig. 9.10 AFM three dimensional tin dioxide film's surface picture

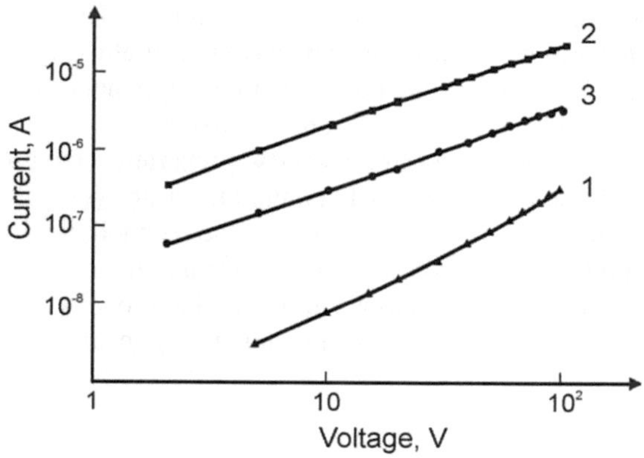

Fig. 9.11 CV characteristic of the SnO_2 film ($T = 290\,K$): *1* – in the air, *2* – after heating to 410 K at the pressure of about 10^{-3} mm Hg and again, cooling to room temperature, *3* – in 15 min after the atmospheric air was let into the chamber [60]

Preliminary researches showed that these films may be essentially applicable as CBRN material, in a gas sensorics, at least.

The conductivity of the studied films in vacuum and in the air differs by more than an order of magnitude, which indicates the considerable influence of adsorption interaction with oxygen in the air [60]. Figure 9.11 shows CV characteristics of the SnO_2 film at room temperature.

The films' conductivity changes during heating and subsequent cooling at vacuum up to the initial temperature are reversible and repeatable, which shows the electrical characteristics of the SnO_2 films stability and allows using them as adsorptive-sensitive elements for gas sensors.

The obtained films of the nanostructured tin dioxide demonstrate the photoluminescence ($\sim$1.9 and $\sim$2.2) eV at a room temperature. This makes them perspective as for their application as PhL sensors for non contact mediums content control.

The SPR parameters were studied on the films by means of polarization- modular spectroscopy (PMS) in the work [61]. The SPR phenomena established in the films of tin dioxide makes it possible to use them as a sensor material in SPR transducers of gases and biological mediums.

9.6 Conclusion

The results of papers for the latest two years show the effectiveness of tin dioxide usage to ensure safety as a convenient, cheap, common material with suitable physicochemical properties. The use of nanoscale forms of SnO_2 as a material for gas sensors of conductometric type, electrochemical sensors, sensors on the SPR effect, the material for electrodes of lithium-ion batteries and solar cells, catalysts for the pollutant decomposition is considered.

Dependence of its properties on manufacturing technology, significant volume change when used in LiBs electrodes, low selectivity to different hydrocarbons, are successfully overcome by researchers. The transition to the creation of various material nanoforms, the extensive use of alloying, the formation of composites and heterostructures, including organic components, are just some of the methods used by developers of devices on tin dioxide.

The mentioned in the review preparation methods of nanoscale forms of SnO_2 and composite materials containing it are generally simple, inexpensive and predominantly safe. In these cases, the technological variability of tin dioxide from a shortcoming becomes an advantage and interesting for further study.

The main applications of tin dioxide show its promising as a material for ensuring environmental safety, health and energy security.

References

1. Jarzebski ZM, Marton JP (1976) Physical properties of SnO_2 materials. I. Preparation and defect structure. J Electrochem Soc 123(7):199–205. https://doi.org/10.1149/1.2133010
2. Comini E (2006) Metal oxide nano-crystals for gas sensing. Anal Chim Acta 568:28–40
3. Hwang I-S, Lee J-H (2011) Gas sensors using oxide nanowire networks: an overview. J Nanoeng Nanomanuf 1:4–17. https://doi.org/10.1166/jnan.2011.1002
4. Sun Y-F, Liu S-B, Meng F-L, Liu J-Y, Jin Z, Kong L-T, Liu J-H (2012) Metal oxide nanostructures and their gas sensing properties: a review. Sensors 12:2610–2631. https://doi.org/10.3390/s120302610
5. Woo HS, Na CW, JH Lee (2016) Design of highly selective gas sensors via physicochemical modification of oxide nanowires: overview. Sensors 16:1531–23. https://doi.org/10.3390/s16091531, www.mdpi.com/journal/sensors.
6. Granqvist CG (2007) Transparent conductors as solar energy materials: a panoramic review. Sol Energy Mat Sol Cells 91:1529–1598.
7. Ginley D, Hosono H, Paine DC (eds) (2011) Handbook of transparent conductors. Springer, New York
8. Papargyri S, Tsipas DN, Papargyris DA, Botis AI, Papargyris AD (2005) Review on the production and synthesis of nanosized SnO_2. Solid State Phenom 106:57–62
9. Batzill M, Diebold U (2005) Review: the surface and materials science of tin oxide. Prog Surf Sci 79:47–154
10. Pan J, Shen H, Mathur S (2012) One-dimensional SnO_2 nanostructures: synthesis and applications. J Nanotechnol 2012. Article ID 917320. https://doi.org/10.1155/2012/917320, www.hindawi.com/journals/jnt/2012/917320/
11. Das S, Jayaraman V (2014) SnO_2: a comprehensive review on structures and gas sensors. Prog Mat Sci 66:112–255. https://doi.org/10.1016/j.pmatsci.2014.06.003
12. Nazarov DV, Bobrysheva NP, Osmolovskaya OM, Osmolovsky MG, Smirnov VM (2015) Atomic layer deposition of tin dioxide nanofilms: a review. Rev Adv Mater Sci 40:262–275
13. Mohanta D, Ahmaruzzaman M (2016) Tin oxide nanostructured materials: an overview of recent developments in synthesis, modifications and potential applications. RSC Adv 6:110996–111015. https://doi.org/10.1039/C6RA21444D
14. Dastkhoon M, Ghaedi M, Asfaram A, Arabi M, Ostovan A, Goudarzi A (2017) CuSnS/SnO_2 nanoparticles as novel sorbent for dispersive micro solid phase extraction of atorvastatin in human plasma and urine samples by high-performance liquid chromatography with UV detection: application of central composite design (CCD). Ultrason Sonochem 36:42–49. https://doi.org/10.1016/j.ultsonch.2016.10.030

15. Kwon YJ, Kang SY, Wu P, Peng Y, Kim SS, Kim HW (2016) Selective improvement of NO_2 gas sensing behavior in SnO_2 nanowires by ion-beam irradiation. ACS Appl Mater Interfaces 8(21):13646–13658. https://doi.org/10.1021/acsami.6b01619
16. Hong SN, Kye YH, Yu C-J, Jong UG, Ri G-C, Choe C-S, Kim KH, Han JM (2016) Ab initio thermodynamic study of the $SnO_2(110)$ surface in an O_2 and NO environment: a fundamental understanding of the gas sensing mechanism for NO and NO_2. Phys Chem Chem Phys 18:31566–31578. https://doi.org/10.1039/C6CP05433A
17. Dang TV, Hoa ND, Duy NV, Hieu NV (2016) Chlorine gas sensing performance of on-chip grown ZnO, WO_3, and SnO_2 nanowire sensors. ACS Appl Mater Interfaces 8(7):4828–4837. https://doi.org/10.1021/acsami.5b08638
18. Li YX, Guo Z, Su Y, Jin XB, Tang XH, Huang JR, Huang XJ, Li MQ, Liu JH (2017) Hierarchical morphology-dependent gas-sensing performances of three-dimensional SnO_2 nanostructures. ACS Sens 2(1):102–110. https://doi.org/10.1021/acssensors.6b00597
19. Johari A, Johari A, Bhatnagar MC, Sharma M (2014) Structural, optical and sensing properties of pure and Cu-doped SnO_2 nanowires. J Nanosci Nanotechnol 14(7):5288–922014
20. Bhardwaj N, Pandey A, Satpati B, Tomar M, Gupta V, Mohapatra S (2016) Enhanced CO gas sensing properties of Cu doped SnO_2 nanostructures prepared by a facile wet chemical method. Phys Chem Chem Phys 18(28):18846–54. https://doi.org/10.1039/c6cp01758d
21. Yang F, Guo Z (2015) Comparison of the enhanced gas sensing properties of tin dioxide samples doped with different catalytic transition elements. J Colloid Interface Sci 448:265–74. https://doi.org/10.1016/j.jcis.2015.02.045
22. Degler D, Rank S, Müller S, de Carvalho HWP, Grunwaldt JD, Weimar U, Barsan N (2016) Gold-loaded tin dioxide gas sensing materials: mechanistic insights and the role of gold dispersion. ACS Sens 1(11):1322–1329. https://doi.org/10.1021/acssensors.6b00477
23. Teterycz H, Licznerski BW (2006) Properties of selective gas-sensitive SnO_2/RuO_2/Pt composition and detection mechanism. J Electrochem Soc 153(5):H94–H104
24. Lyson-Sypien B, Kusior A, Rekas M, Zukrowski J, Gajewska M, Michalow-Mauke K, Graule T, Radecka M, Zakrzewska K (2017) Nanocrystalline TiO_2/SnO_2 heterostructures for gas sensing. Beilstein J Nanotechnol 8:108–122. https://doi.org/10.3762/bjnano.8.12
25. Chen W, Li Q, Xu L, Zeng W (2015) Gas sensing properties of ZnO-SnO_2 nanostructures. J Nanosci Nanotechnol 15(2):1245–1252
26. Chesler P, Hornoiu C, Mihaiu S, Vladut C, Calderon Moreno JM, Anastasescu M, Moldovan C, Firtat B, Brasoveanu C, Muscalu G, Stan I, Gartner M (2016) Nanostructured SnO_2-ZnO composite gas sensors for selective detection of carbon monoxide. Beilstein J Nanotechnol 7:2045–2056. https://doi.org/10.3762/bjnano.7.195
27. Wang L, Li J, Wang Y, Yu K, Tang X, Zhang Y, Wang S, Wei C (2016) Construction of 1D SnO_2-coated ZnO nanowire heterojunction for their improved n-butylamine sensing performances. Sci Rep 6:35079. https://doi.org/10.1038/srep35079
28. Jin Z, Yang M, Chen SH, Liu JH, Li QX, Huang XJ (2017) Tin oxide crystals exposed by low-energy {110} facets for enhanced electrochemical heavy metal ions sensing: X-ray absorption fine structure experimental combined with density-functional theory evidence. Anal Chem 89(4):2613–2621. https://doi.org/10.1021/acs.analchem.6b04977
29. Bhanjana G, Dilbaghi N, Kumar R, Umar A, Kumar S (2015) SnO_2 quantum dots as novel platform for electrochemical sensing of cadmium. Electrochim Acta 169:97–102. https://doi.org/10.1016/j.electacta.2015.04.045
30. Khan N, Athar T, Fouad H, Umar A, Ansari ZA, Ansari SG (2017) Application of pristine and doped SnO_2 nanoparticles as a matrix for agro-hazardous material (organophosphate) detection. Sci Rep 7:42510. https://doi.org/10.1038/srep42510
31. Grinevich VS, Filevska LM, Matyash IE, Maximenko LS, Mischuk ON, Rudenko SP, Serdega BK, Smyntyna VA, Ulug B (2012) Surface plasmon resonance investigation procedure as a structure sensitive method for SnO_2 nanofilms. Thin Solid Films 522:452–456
32. Yang D, Lu H-H, Chen B, Lin C-W (2010) Surface plasmon resonance of SnO_2/Au bi-layer films for gas sensing applications. Sensors and Actuators B Chem 145(2):832–838

33. Pathak A, Mishra SK, Gupta BD (2015) Fiber-optic ammonia sensor using Ag/SnO_2 thin films: optimization of thickness of SnO_2 film using electric field distribution and reaction factor. Appl Opt 54(29):8712–8721. https://doi.org/10.1364/AO.54.008712
34. Kaur D, Madaan D, Sharma VK, Kapoor A (2016) High angular sensitivity thin film tin oxide sensor. In: Proceeding of international conference on condensed matter and applied physics (ICC 2015). AIP conference proceedings, vol 1728, p 020210. http://dx.doi.org/10.1063/1.4946261
35. Paek SM, Yoo EJ, Honma I (2009) Enhanced cyclic performance and lithium storage capacity of SnO_2/graphene nanoporous electrodes with three-dimensionally delaminated flexible, structure. Nano Lett 9(1):72–75. https://doi.org/10.1021/nl802484w
36. Lin J, Peng Z, Xiang C, Ruan G, Yan Z, Natelson D, Tour JM (2013) Graphene nanoribbon and nanostructured SnO_2 composite anodes for lithium ion batteries. ACS Nano 7(7):6001–6006. https://doi.org/10.1021/nn4016899
37. Xia G, Li N, Li D, Liu R, Wang C, Li Q, Lü X, Spendelow JS, Wu J, Zhang G (2013) Graphene/Fe_2O_3/SnO_2 ternary nanocomposites as a high-performance anode for lithium ion batteries. ACS Appl Mater Interfaces 5(17):8607–8614. https://doi.org/10.1021/am402124r
38. Zhou D, Song WL, Li X, Fan LZ Confined porous graphene/SnO_x frameworks within polyaniline-derived carbon as highly stable lithium-ion battery anodes. ACS Appl Mater Interfaces 8(21):13410–72016. https://doi.org/10.1021/acsami.6b01875
39. Li Y, Zhang H, Chen Y, Shi Z, Cao X, Guo Z, Shen PK (2016) Nitrogen-doped carbon-encapsulated SnO^2Sn nanoparticles uniformly grafted on three-dimensional graphene-like networks as anode for high-performance lithium-ion batteries. ACS Appl Mater Interfaces 8(1):197–207. https://doi.org/10.1021/acsami.5b08340
40. Wu J, Chen H, Byrd I, Lovelace S, Jin C (2016) Fabrication of SnO_2 asymmetric membranes for high performance lithium battery anode. ACS Appl. Mater. Interfaces 8(22):13946–13956. https://doi.org/10.1021/acsami.6b03310
41. Jiang B, He Y, Li B, Zhao S, Wang S, He Y, Lin Z (2017) Polymer-templated formation of polydopamine-coated SnO_2 nanocrystals: anodes for cyclable lithium-ion batteries. Angew Chem Int Ed Engl 56(7):1869–1872. https://doi.org/10.1002/anie.201611160
42. Novoselov KS, Geim AK, Morozov SV, Jiang D, Zhang Y, Dubonos SV, Grigorieva IV, Firsov AA (2004) Electric field effect in atomically thin carbon films. Science 306:666–669. https://doi.org/10.1126/science.1102896
43. Repetsky SP, Vyshyvana IG, Kruchinin SP, Molodkin VB, Lizunov VV (2017) Influence of the adsorbed atoms of potassium on an energy spectrum of grapheme. Metallofiz Noveishie Tekhnol 39:1017–1022
44. Wei H, Xia Z, Xia D (2017) One step synthesis of uniform SnO_2 electrode by UV curing technology toward enhanced lithium-ion storage. ACS Appl Mater Interfaces 9(8):7169–7176. https://doi.org/10.1021/acsami.6b15820
45. Kim C, Jung JW, Yoon KR, Youn DY, Park S, Kim ID (2016) A high-capacity and long-cycle-life lithium-ion battery anode architecture: silver nanoparticle-decorated SnO_2/NiO nanotubes. ACS Nano 10(12):11317–11326. https://doi.org/10.1021/acsnano.6b06512
46. Singh T, Singh J, Miyasaka T (2016) Role of metal oxide electron-transport layer modification on the stability of high performing perovskite solar cells. Chem Sus Chem 9(18):2559–2566. https://doi.org/10.1002/cssc.201601004
47. Barbe J, Tietze ML, Neophytou M, Murali B, Alarousu E, Labban A, Abulikemu M, Yue W, Mohammed OF, McCulloch I, Amassian A, Del Gobbo S (2017) Amorphous tin oxide as low temperature-processed electron transport layer for organic and hybrid perovskite solar cell. ACS Appl Mater Interfaces 9(13):11828–11836. https://doi.org/10.1021/acsami.6b13675
48. Lee Y, Paek S, Cho KT, Oveisi E, Gao P, Lee S, Park J-S, Zhang Y, Humphry-Baker R, Asiri AM, Nazeeruddin MK (2017) Enhanced charge collection with passivation of the tin oxide layer in planar perovskite solar cells. J Mater Chem A 5(25):12729–12734

49. Yang G, Lei H, Tao H, Zheng X, Ma J, Liu Q, Ke W, Chen Z, Xiong L, Qin P, Chen Z, Qin M, Lu X, Yan Y, Fang G (2017) Reducing hysteresis and enhancing performance of perovskite solar cells using low-temperature processed Y-doped SnO_2 nanosheets as electron selective layers. Small 13(2):1601769–1601769. https://doi.org/10.1002/smll.201601769
50. Rodionov VE, Shnidko IN, Zolotovsky A, Kruchinin SP (2013) Electroluminescence of Y_2O_3:Eu and Y_2O_3:Sm films. Mater Sci 31:232–239
51. Ermakov V, Kruchinin S, Fujiwara A (2008) Electronic nanosensors based on nanotransistor with bistability behaviour. In: Bonca J, Kruchinin S (eds) Proceedings of NATO ARW "Electron transport in nanosystems". Springer, pp 341–349
52. Kavan L, Steier L, Grätzel M (2017) Ultrathin buffer layers of SnO_2 by atomic layer deposition: perfect blocking function and thermal stability. J Phys Chem C 121(1):342–350. https://doi.org/10.1021/acs.jpcc.6b09965
53. Bai H, He P, Chen J, Liu K, Lei H, Zhang X, Dong F, Li H (2017) Electrocatalytic degradation of bromocresol green wastewater on Ti/SnO_2-RuO_2 electrode. Water Sci Technol 75(1):220–227. https://doi.org/10.2166/wst.2016.509
54. Loloi M, Rezaee A, Aliofkhazraei M, Rouhaghdam AS (2016) Electrocatalytic oxidation of phenol from wastewater using Ti/SnO_2-Sb_2O_4 electrode: chemical reaction pathway study. Environ Sci Pollut Res Int 23(19):19735–43. https://doi.org/10.1007/s11356-016-7110-6
55. Li D, Tang J, Zhou X, Li J, Sun X, Shen J, Wang L, Han W (2016) Electrochemical degradation of pyridine by Ti/SnO_2-Sb tubular porous electrode. Chemosphere 149:49–56. https://doi.org/10.1016/j.chemosphere.2016.01.078
56. Asim S, Zhu Y, Rana M, Yin J, Shah MW, Li Y, Wang C (2017) Nanostructured 3D-porous graphene hydrogel based Ti/Sb-SnO_2-Gr electrode with enhanced electrocatalytic activity. Chemosphere 169:651–659. https://doi.org/10.1016/j.chemosphere.2016.11.119
57. Hu M, Zhang Z, Luo C, Qiao X (2017) One-pot green synthesis of Ag-decorated SnO_2 microsphere: an efficient and reusable catalyst for reduction of 4-nitrophenol. Nanoscale Res Lett 12:435. https://doi.org/10.1186/s11671-017-2204-8
58. Seo K, Lim T, Mills EM, Kim S, Ju S (2016) Hydrogen generation enhanced by nano-forest structures. RSC Adv 6:12953–12958. https://doi.org/10.1039/C5RA26226G
59. Gibson G, Wang Z, Hardacre C, Lin WF (2017) Insights into the mechanism of electrochemical ozone production via water splitting on the Ni and Sb doped SnO_2 catalyst. Phys Chem Chem Phys 19(5):3800–3806. https://doi.org/10.1039/c6cp06906a
60. Filevska LM, Chebanenko AP, Grinevych VS, Simanovych NS (2016) The electrical characteristics of nanoscale SnO_2 films, structured by polymers. Photoelectronics 25:62–67
61. Grinevich VS, Filevska LM, Rudenko SP, Stetsenko MA, Maximenko LS, Serdega BK, Smyntyna VA (2016) Radiation modes of surface plasmons in SnO_2 thin films (in Ukrainian). In: 7th international scientific and technical conference on "Sensor Electronics and Microsystem Technologies" (SEMST-7), Book of Abstract, 30 May–3 June 2016, Odessa (Astroprint Odessa), p 181

Chapter 10
New Experimental Approach to Measure the Photocatalytic Activity of the TiO_2 Nanosamples

I. Donchev, V. Smatko, J. Briancin, D. Kupka, and E. Kovacova

Abstract Titanium dioxide (TiO_2) samples are important industrial nano-objects with wide applications. For example nowadays TiO_2 is the most abundantly used white pigment. The pigment is still in use, both in the production of contemporary art and for the conservation of older artwork as a retouching pigment. But next to its positive characteristics, the pigment has a major potential drawback: its photo-catalytic activity that can cause degradation of artworks in which it is used. So it is important to create methods to test the photo-catalytic activity of different quality of titanium dioxide samples. In this paper a new experimental set-up to measure the photo-catalytic activity of TiO_2 samples is described.

Keywords TiO_2 · Photo-catalytic activity measurements

10.1 Introduction: Photocatalytic Activity of Titanium Dioxide

Several methods exist to assess photocatalytic activity of photocatalytic semiconductors such as titanium dioxide. Test reactions such as the conversion of isopropanol to acetone [1] and dye degradation reactions [2, 3] are commonly described to characterize catalyst powders. However, they require a certain level of expertise and equipment. The same limitation holds for other methods that have been proposed such as photoconductivity measurements [4], the evaluation of CO_2

I. Donchev (✉)
South-Ukrainian National Pedagogical University after K.D. Ushinskij, Odessa, Ukraine
e-mail: donchev@pdpu.edu.ua

V. Smatko · E. Kovacova
Institute of Electrical Engineering, Slovak Academy of Sciences, Bratislava, Slovakia

J. Briancin · D. Kupka
Institute of Geo-technics, Slovak Academy of Sciences, Kosice, Slovakia

J. Bonča, S. Kruchinin (eds.), *Nanostructured Materials for the Detection of CBRN*, NATO Science for Peace and Security Series A: Chemistry and Biology,
https://doi.org/10.1007/978-94-024-1304-5_10

from an enclosed paint film [5], ESR analysis [6] and full characterization of the pigment. Finally, accelerated aging tests are a valuable tool in the field of paint development. These tests can assess the stability of paint over time, which, in a paint containing titanium dioxide, can be related to the photocatalytic activity of the pigment. However, these tests are time consuming and cannot be performed on original material [7]. Thus, there are currently no routine methodologies to determine the photocatalytic activity of TiO_2 pigments prior to their use by artists or conservators.

10.2 Experimental Part: Gas Analysis

Gas analysis in continuous flow-through or intermittent (stop-flow) mode was applied to measure the volatile organic carbon oxidation caused by photo-catalytic activity of the TiO_2 samples. The experimental set-up has the following features.

Incurrent air pumped from outdoor was filtered and dried in two steps using Nafion® tubing (MD Series, Perma Pure) followed by chemical drying with magnesium perchlorate. Further, the dry air was directed through mass flow controller (Cole Parmer) to maintain constant STP-corrected flow rate of the incurrent air in the system. In our set-up incurrent gas stream passed through purge vessels in which volatile organic compounds from the liquid diffused into carrier gas. In this study toluene was sparged by carrier gas using standard gas washer at room temperature.

Eight-channel multiplexer (RM-8, Sable Systems International) was used for automatic switching air flow between individual reaction chambers. For continuous flow-through mode, the solenoid valves of the multiplexer were adjusted to allow permanent constant flow of the air through the chambers. Otherwise, the channels were air-tight sealed during specific time intervals. The incurrent O_2, $CO_2$and VOCs vapour concentrations are referred to as the baseline values. The excurrent gas was water vapour scrubber again by combined Nafion membrane/anhydride drying before analysis.

Paramagnetic oxygen analyser (PA-10a) and infrared carbon dioxide analyser (CA-10 Sable Systems) were calibrated for zero air and ambient air O_2 and CO_2 concentrations span. Measuring both gas species gives the most accurate results and allowed mathematical compensation for the enrichment of CO_2 caused by O_2 consumption and the O_2 enrichment/dilution caused by CO_2 consumption/production. Air tight photoreaction chambers were made from standard GL-45 threaded glass bottles equipped with three-port PTFE caps (VICI AG International). Two ports were connected to incurrent and excurrent air delivery tubing for sampling the bottle gas.

In continuous mode, permanent constant flow of the air through the chambers is allowed. Individual mass flow control valves regulate a fixed flow rate into each chamber. The analysers continuously measure the concentrations of O_2 and CO_2 of the gas streams coming from individual chambers relative to background (baseline) O_2 and CO_2 concentration. Continuous mode was not suitable for our purposes because of very small concentration differences between baseline O_2 and CO_2 concentrations and respective values of gas coming from reaction chambers.

During the stop-flow mode the chamber is flushed with fresh baseline air, and then sealed for a period of time (discontinuous gas exchange cycles). During enclosure period the air surrounding the catalyst becomes sufficiently depleted of O_2 and enriched in CO_2 to be measured, after which it is swept through the analysers. The air sampling procedure is automated, using gas stream multiplexer with multiple solenoid valve pairs operated under computer control. The system then sequentially opens and flushes each chamber with a return to baseline levels between chambers. Figure 10.1 shows O_2 and CO_2 concentrations recorded from analysis of air samples from successive chambers. The system periodically analyses reference air to obtain stable baseline signal between peeks from other chambers. This sequence of sampling can last a variable length of time with minimum maintenance requirements.

The data acquisition was performed by a specific software application developed in Lab – VIEW ® (National Instruments). The software operated on a PC with an attached serial port hub connected to all measurement devices in the test equipment via serial link (RS 232). The raw data from the measurement devices were displayed on-line on the screen and recorded in a spread sheet file for further calculations.

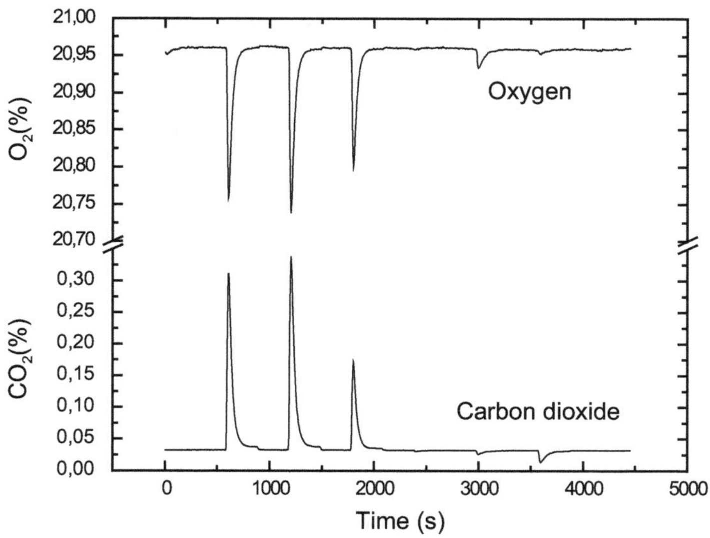

Fig. 10.1 Oxygen and carbon dioxide concentrations in air samples taken from successive reaction chambers in the course of stop-flow mode analysis

10.3 Results

Before the start of experiment, the photocatalytic reactor was flushed with the baseline gas mixture containing VOC, until stable baseline O_2 and CO_2 concentrations were achieved. When UV-radiation was applied, the concentration changes of the gases were monitored. Thus, any CO_2 produced in the system yielded an upward-going signal. Oxygen, in contrast, showed the valleys that trend downward from the baseline value, corresponding to oxygen consumption during photocatalytic reaction.

During the photocatalytic oxidation of toluene, the concentration changes were observed for carbon dioxide and oxygen respectively (Fig. 10.1).

References

1. Irick G (1972) Determination of the photo-catalytic activities of titanium dioxides and other white pigments. J Appl Polym Sci 16(9):2387–2395
2. Rauf MA, Ashraf SS (2009) Fundamental principles and application of heterogeneous photocatalytic degradation of dyes in solution. Chem Eng J 151(1–3):10–18. https://doi.org/10.1016/j.cej.2009.02.026
3. Doushita K, Kawahara T (2001) Evaluation of photocatalytic activity by dye decomposition. J Sol Gel Sci Technol 22(1–2):91–98
4. Chin J, Scierka S, Kim T, Forster A (2003) Photoconductivity technique for the assessment of pigment photo-reactivity. In: Proceedings of the 81st annual meeting of the FSCT, Nov 2003
5. Jin C, Christensen PA, Egerton TA, Lawson EJ, White JR (2006) Rapid measurement of polymer photo-degradation by FTIR spectrometry of evolved carbon dioxide. Polym Degrad Stab 91(5):1086–1096
6. Ceresa E, Burlamacchi L, Visca M (1983) An ESR study on the photo-reactivity of TiO_2 pigments. J Mater Sci 18(1):289–294. https://doi.org/10.1007/BF00543837
7. Feller RL (1994). In: Borland D Accelerated aging: photochemical and thermal aspects. Getty Conservation Institute. ISBN:0-89236-125-5.1

Chapter 11
Boron-Containing Nanostructured Materials for Neutron-Shields

L. Chkhartishvili

Abstract Chapter aims to overview the studies on development of boron-containing nanostructured materials, which can serve as working body of the neutron shields designed to protect first of all people – both the general population and staff of nuclear plants.

Keywords Boron-containing nanostructured materials · Neutron shields

11.1 Introduction

One of the major challenges of our time is the creating an environment in which people would be protected from the harmful effects of ionizing radiation. The above has been confirmed once again by the strongest effects of the earthquake and subsequent tsunami that occurred in Japan recently. As a result, there were small leaks of radioactive substances and the danger of severe leakage.

In the light of these developments, the IAEA (International Atomic Energy Agency) has revised approaches to the problem of nuclear safety and emergency responses to nuclear accidents. Taking into account the existing risk, more attention should be paid to the detection of sources of different types of radiation, the measurement of fluxes of emitted elementary particles, as well as the development of facilities for radiation protection of the environment, population, and special personnel working at nuclear plants.

From this perspective, a particularly important task is the protection from the neutron irradiation, which is caused by two factors:

L. Chkhartishvili (✉)
Department of Engineering Physics, Georgian Technical University, Tbilisi, Georgia

Laboratory for Boron-Containing and Composite Materials, Ferdinand Tavadze Institute of Metallurgy and Materials Science, Tbilisi, Georgia

J. Bonča, S. Kruchinin (eds.), *Nanostructured Materials for the Detection of CBRN*, NATO Science for Peace and Security Series A: Chemistry and Biology, https://doi.org/10.1007/978-94-024-1304-5_11

- First, in nuclear weapons and nuclear energy production there are used neutron reactions and correspondingly wastes of nuclear plants also are neutron-radioactive.
- Second, the neutron irradiation is one of the most dangerous to human health and life because of the large mass of neutrons combined with the ability to penetrate deeply into the living tissues.

Let us briefly characterize types of existing neutron sources and the consequences of the neutron exposure for human health and life. The information given below mainly is based on IAEA data [1] and text-book [2].

There are several fundamentally different neutron sources.

- Neutron radiation of a very high intensity occurs during a nuclear weapon testing or its use in hostilities. The most dangerous neutron flux would be formed by the neutron bomb explosion because it is a weapon that is not intended for significant infrastructural damages, but primarily to kill people and other living organisms. It should be noted that work on neutron weapons have long ceased.

As for the issue of exploring the possibilities of protection against nuclear weapons in general, it is beyond the scope of our consideration here.

- Among the sources of the neutron fluxes related to the peaceful, i.e., non-military, uses, the most important are nuclear reactors intended for the electrical power production or energy supply of special-engines from the heat released during a nuclear reaction.

Similarly, nuclear reactors are operated using the neutron chain-reaction. But reaction in the reactor is controllable, the system does not reach the explosion-state and, therefore, the neutron flux through the reactor protective shell is relatively weak.

As heavy nuclei contain neutrons in relatively high proportion when they decay there are formed several free neutrons and nuclei-fragments supersaturated with neutrons. The decay process can be initiated by the nuclear–neutron collision: because of electrical neutrality even very low-energy neutron is capable to achieve the nucleus vicinity, where it can be captured by the short-acting potential. The result is a creation of the compound nucleus which shortly decays with emission of more neutrons. The example of reactions of this type which occurs in nuclear reactors is a decay of the uranium nucleus with formation of rubidium and cesium nuclei and neutrons emission:

$$ {}^{235}_{92}\mathrm{U} + {}^{1}_{0}n \rightarrow {}^{92}_{37}\mathrm{Rb} + {}^{140}_{55}\mathrm{Cs} + 4{}^{1}_{0}n. \tag{11.1} $$

So instead of a single neutron it will be arisen 4 neutrons. Additional neutrons cause the disintegration of other uranium nuclei and thus number of neutrons increases avalanchely: the chain-reaction grows on. In every act of such collisions, the certain amount of useful heat releases which finally is transferred to the user.

- The emergence of one more artificial source of neutron radiation is closely related to the nuclear weapons production and the nuclear reactors operation. It is a spent nuclear fuel i.e. a neutron radioactive waste.
- Weak artificial neutron sources are neutron generators, whose appearance is due to the fact that currently neutron fluxes are increasingly utilized in various areas of human life and industry. These areas include: science (e.g. neutron radiography of materials, radiobiological analysis in anthropology, zoology and botany), industry (e.g. oil wells logging, radioactivation analysis of rocks and deposits in the ore-processing enterprises, control of the nuclear reactor functioning, express analysis of the products under the processing, nondestructive analysis of the closed areas, namely, during the customs inspection), agriculture (e.g. determination of the proteus content in the grain crops and the like), environmental protection (e.g. detection of radioactive wastes, soil water pollutions, sediment toxins, some other hazardous and toxic substances), and medicine (e.g. diagnosis of fibrous degradation, radiation therapy of cancer).

Depending on the purpose, neutron generators are of different types. For example, in so-called deuterium and deuterium–tritium generators, where nuclear reactions involve hydrogen heavy isotopes deuterium ${}^{2}_{1}\mathrm{H}$ (D) and tritium ${}^{3}_{1}\mathrm{H}$ (T), energies of the emitted neutrons are ~2.5 and ~14.3 MeV, respectively. In this case, the maximal value of the integrated intensity is of about 10^{12} n/s.

In the generator of other type that uses the spontaneous decay of the heavy nucleus of californium Cf, emitted neutrons are with energies above 7.5 MeV. For scientific purposes commonly there are used neutron fluxes produced in proton accelerators with so-called photoneutron reaction in which high-energy (accelerated) protons detach neutrons from the nuclei of heavy atoms constituting the target.

- For weak neutron sources serve the natural ones in the form of uranium deposits. For example, it takes place the reaction of a spontaneous decay of uranium with formation of lanthanum and bromine and emission of 3 neutrons:

$$ {}^{235}_{92}\mathrm{U} \rightarrow {}^{87}_{35}\mathrm{Br} + {}^{145}_{57}\mathrm{La} + 3{}^{1}_{0}n. \quad (11.2) $$

- At the Earth surface, extremely weak natural source of neutrons is the cosmic radiation.

Biological effects of neutron-irradiation of living tissues are very complicated. For this reason, the consequences of the neutron exposure for human health and life depend, on the one hand, on the level of exposure, and, on the other hand, on the type of radiation.

For relatively low levels of radiation energy impact on living cells is the only reason of radiation damage. In such cases, the living matter can self-regenerate and therefore weak irradiation does not lead to lasting harm.

However, at high doses the rapid destruction of cells takes place. In addition, for survived but heavily damaged cells exposure can have serious and long-term harmful effects. Namely, the radiation damage of the cell nucleus can lead to the undesirable biological effects that may occur after several years or even a decade.

The neutron fluxes tend to provide high values of the LET (linear-energy-transfer) to living tissue. Here, the linearity of the energy transfer means that within the corresponding energy range the parameter characterizing the irradiation effect is proportional to the irradiated energy. Quantitatively, this characteristic is determined by the energy released by radiation beam per unit length of its trajectory into the tissue. For this reason, the primary physical damage caused by neutron irradiation is more dangerous to humans than other types of damages from exposure at the same level of flux.

Secondary radiation damage of the tissue from neutrons, heavy particles, is associated with the ionization of the medium and the subsequent formation of active toxic chemicals. Because a living tissue mainly is composed of water H_2O in the liquid state ($\sim$80%) containing some dissolved gases incident radiation causes ionization of many simple molecules containing light chemical elements, like hydrogen H and oxygen O, transforming them into pairs of positive ion H_2O^+ and negatively charged free electron e^-. Subsequently, these free electrons are very likely to be captured by neutral water molecules, which they turn into negative ions H_2O^-. Pairs of molecular ions of water with charges of both signs are unstable. Finally it leads to their dissociation into simple free radicals OH• and H•. Due to the presence of unpaired electrons, these are very active chemically. They together with other molecules form some new free radicals R•, which are able to initiate undesirable processes in a biological system.

It should be also noted that because of the high oxygen content in tissues the neutron exposure can lead to the different set of harmful chemical reactions called as oxygen effect.

Typically, the flow of heavy particles is ionizing because they are charged. These include, for example, protons and α-particles respectively, nuclei of hydrogen ${}_1^1H$ and helium ${}_2^4He$. Neutron irradiation is the exception in this regard: because of the lack of electric charge neutrons are unable to cause a primary ionization of the tissue detaching electrons from water molecules. But, due to the same electrical neutrality, they penetrate quite deeply into tissue and come in the short-range strong interaction with the charged nuclei of hydrogen atoms – protons. Neutrons detaching protons cause secondary ionization because protons are charged heavy particles. For this reason, neutron radiation is more dangerous than other kinds of ionizing radiation – it penetrates the living tissue deeply and causes there strong ionization.

The basic dosimetric quantity characterizing the degree of exposure – averaged absorbed dose or total energy of absorbed radiation per unit of body weight, Gray (Gy = J/kg), cannot adequately describe the biological effects of radiation, since the nature of damage to living tissue largely depends on the type of radiation. By introducing suitable dimensionless weight factors, the concept of equivalent dose, Sievert (Sv = J/kg), takes into account relative shares of different types of irradiation damage (the concept of effective dose similarly takes into account the sensitivity of different organs of the body to the given type of exposure). For neutrons the dependence of radiation weighting factor w on their average energy E has a complicated form and can be approximated as

$$w \approx 5 + 17\exp(-(\ln 2E\,[\mathrm{MeV}])^2/6). \tag{11.3}$$

Permissible dose limits are introduced for the Reference Man, an adult of Caucasian race specially described for this purpose. For the general population the total annual dose should not exceed ~1, in particular for the skin 50 mSv/y. For the nuclear plants staff same doses are, respectively, ~50 and 500 mSv/y.

Currently, there are available a number of reports on the use of boron-based materials, in particular, low-boron steels or Fe–B alloys (see e.g. [3]), in containers and armors designed for neutron shielding of devices, equipment and vehicles, but not directly people. These issues will not be considered here: the present chapter aims to overview the studies on development of boron-containing nanostructured materials which can serve as working body of the neutron shields designed to protect first of all people – both the general population and staff of nuclear plants. Present review mainly is based on recent works by two Georgian teams affiliated with Department of Engineering Physics of the Georgian Technical University and Laboratory for Boron-Containing & Composite Materials of the Ferdinand Tavadze Institute of Metallurgy & Materials Science.

11.2 Why Boron-Containing Materials?

Thus, there is a need in reliable neutron shields usable for various purposes. The working body of any of these facilities must be made from a material containing isotopes of chemical elements that are strong absorbers of neutrons. In the Table 11.1, there are listed the nuclides with highest neutrons-capturing cross sections (in barns).

One can see that neutrons-capturing by the nuclei is accompanied by γ-radiation, detaching of protons or α-particles, and fragmentation (f) reactions as well. Some of these elements are highly radioactive (RA), some of them are not able to form structural materials or are too expensive. One more important characteristic for neutron-shielding materials components is monotonic. i.e. non-resonance type of the cross-section σ dependence on the incident neutrons mean energy E. Taking into account all the requested conditions, boron isotope ^{10}B seems to be especially suitable for mentioned purposes.

The ^{10}B nuclei have especially high capture cross section for so-called slow neutrons, which are primarily used in heavy nuclei decay chain reactions in nuclear reactors. Suffice to say that the capture cross section of a thermal neutron with a standard velocity of ≈2200 m/s is $\sigma \approx 3835$ b. According to the temperature scale of energy, thermal neutron kinetic energy of $E \approx 0.025$ eV corresponds to the neutron beam with "room temperature" of 20 °C, whereas for another stable isotope of boron ^{11}B, it consists of only 0.0055 b [4]. Natural isotopic composition of boron is $^{\mathrm{nat}}\mathrm{B} = 19.9\%\ ^{10}\mathrm{B} + 80.1\%\ ^{11}\mathrm{B}$, which leads to a weighted value of the thermal

Table 11.1 Cross section of thermal neutrons capturing by some nuclides

Nuclide	Interaction	Cross section, b
$^{135}_{54}$Xe (RA)	(n, γ)	2.7×10^6
$^{157}_{64}$Ga	(n, γ)	2.6×10^5
$^{155}_{54}$Ga	(n, γ)	6.1×10^4
$^{149}_{62}$Sm	(n, γ)	4.2×10^4
$^{113}_{48}$Cd	(n, γ)	2.1×10^4
$^{151}_{63}$Eu	(n, γ)	9.2×10^3
$^{242}_{95}$Am (RA)	(n, f)	8.0×10^3
$^{3}_{2}$He	(n, p)	5.3×10^3
$^{10}_{5}$B	(n, α)	3.8×10^3
$^{199}_{80}$Hg	(n, γ)	2.2×10^3
$^{241}_{94}$Pu (RA)	(n, f)	1.4×10^3
$^{6}_{3}$Li	(n, α)	9.2×10^2
$^{235}_{92}$U(RA)	(n, f)	6.8×10^2
$^{157}_{72}$Hf	(n, γ)	5.6×10^2

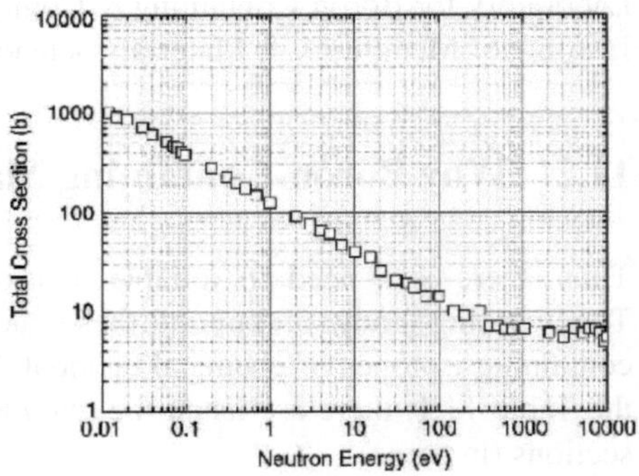

Fig. 11.1 Variation of total cross section of elemental boron of natural isotopic composition as a function of neutron energy

neutrons capture cross-section of 767 b. It should be noted that the ^{10}B nuclei retain the ability of strong absorption for high-energy neutrons as well.

In general, interactions between boron stable nuclei and neutron include elastic, inelastic, and non-elastic scatterings. According to the JENDAL (Japanese Evaluated Nuclear Data Library) [5], the last, which means capturing of the neutron, mainly determines the total cross-section of the ^{10}B–n reaction. And since the capture cross-section of a particle is inversely proportional to its momentum, it turns out that total cross section σ of the neutron interaction with ^{10}B nucleus is nearly inversely proportional to the neutron momentum, i.e. square-root of the neutron energy E:

$$\sigma/\sigma_0 \approx (E_0/E)^{1/2}. \quad (11.4)$$

Same also is almost true for elemental boron of natural isotopic composition (Fig. 11.1).

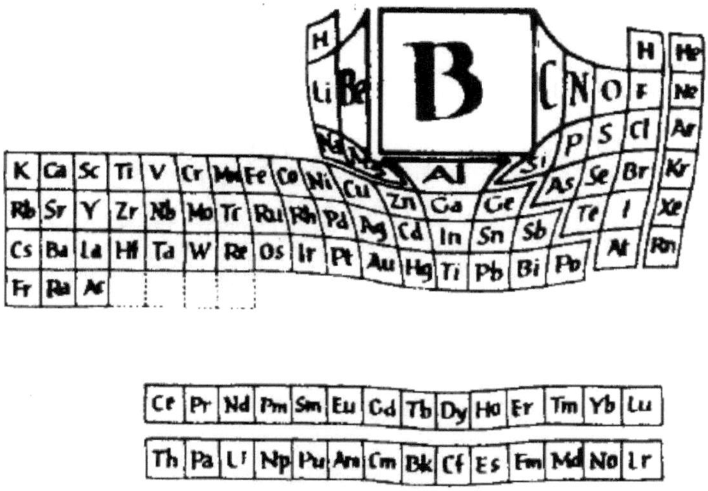

Fig. 11.2 Relative importance of boron to other chemical elements by W.N. Lipscomb

The ^{10}B–n reaction has two channels with and without accompanying γ-radiation with probabilities of 93.9 and 6.1%, respectively:

$$ {}^{10}_{5}\mathrm{B} + {}^{1}_{0}n \rightarrow {}^{4}_{2}\mathrm{He} \uparrow + {}^{7}_{3}\mathrm{Li} + 2.31\,\mathrm{MeV} + {}^{0}_{0}\gamma + 0.48\,\mathrm{MeV}, \tag{11.5} $$

$$ {}^{10}_{5}\mathrm{B} + {}^{1}_{0}n \rightarrow {}^{4}_{2}\mathrm{He} \uparrow + {}^{7}_{3}\mathrm{Li} + 2.79\,\mathrm{MeV}. \tag{11.6} $$

Together with strong absorption of neutrons, advantage of boron is related to its very rich structural chemistry. Relative importance of boron to other chemical elements by W.N. Lipscomb [6] is presented in Fig. 11.2. Why is the role of boron in forming of the various solid-state structures so great? In the final analysis, understanding of the diversity of crystalline lattices containing boron atoms reduces to their electronic structure in isolated state. Configuration of valence electrons peculiar to the free boron atom is $2s^2 2p$. It is an energetically unstable configuration, which in lattice tends to more stable ones, first to $2s^2 2p^2$ and then to $2s^2 2p^3$. Thus, boron is strongly distinct electron-acceptor.

Electron-deficiency of elemental boron structures (their descriptions can be seen e.g. in monograph [7]) should be a reason, why all-boron crystalline forms, as well as amorphous boron, exhibit very complex, clustered structures. Slightly distorted regular boron icosahedron B_{12} (Fig. 11.3) and its fragments serve for the main structural motifs of crystalline and amorphous modifications of boron and boron-rich solids.

Crystalline and amorphous modifications of boron and boron-rich solids are considered as structures stabilized by intrinsic point defects and/or certain impurities in high concentrations precisely compensating electron-deficiency inherent in corre-

Fig. 11.3 Boron icosahedron B_{12} – main structural motif of crystalline and amorphous modifications of boron and boron-rich solids

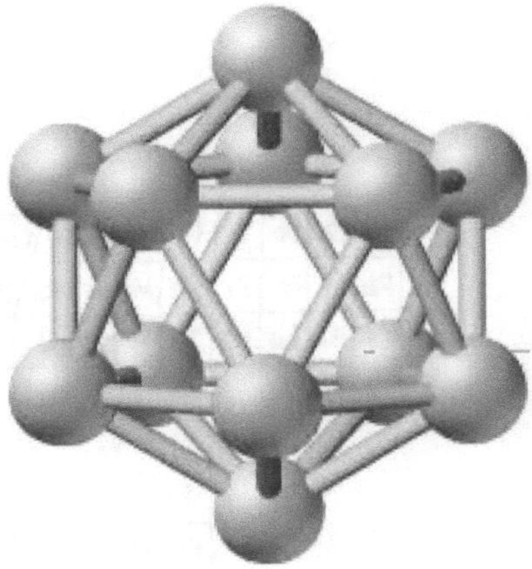

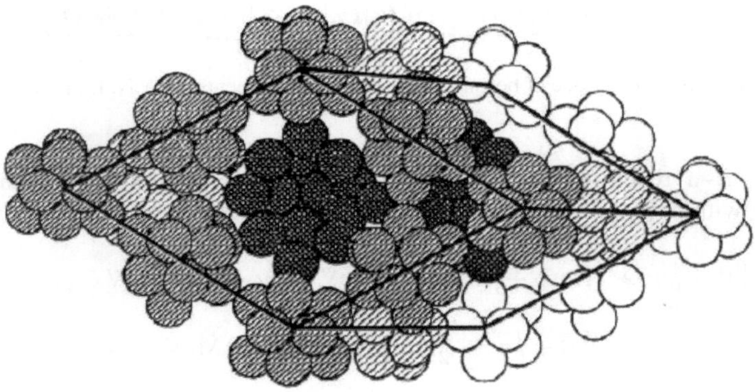

Fig. 11.4 Idealized unit cell of β-rhombohedral boron

sponding ideal structures. For instance, unit cell of β-rhombohedral boron, which is believed to be ground-state structural modification of boron, consists of 105 regular atomic sites (Fig. 11.4), but in real crystals with partially occupied regular sites and interstitials there are ~106.5 boron atoms per cell.

Because all the boron 3D phases are constructed from the interconnected icosahedra, as a rule most of boron atoms are members of the almost regular atomic triangles and surrounded by 6 nearest neighbors having 5 intra- and 1 inter-icosahedral bonds. This circumstance leads to the possibility to synthesize 2D boron phases: boron sheets with triangular lattices, as well as boron nanotubes and borophenes – cylindrically or spherically rolled boron surfaces. Detailed atomic geometries of actually synthesized and also theoretically predicted boron nanostructures have been presented in the encyclopedia [8].

Acceptor behavior of boron atoms and their clusters also favors formation of the huge number of borides (see e.g. monograph [9]), i.e. compounds of boron B with metals Me, which usually are characterized by the donor behavior. Only B/Me binary crystalline compounds with chemical formulas from Me_5B to MeB_{66} can be counted of ~250 with ~40 different crystal sublattices. Like the pure boron, higher borides are characterized by the clustered structures based on icosahedron and/or other (smaller or bigger) boron-cages.

Hexaborides MeB_6 consist of two simple cubic sublattices with Me atoms and boron octahedra B_6 in sites, respectively; while diborides MeB_2 can be imagined as a layered structures alternating B and Me plane atomic networks. As for the lower borides, among them one can find wide variety of one- and two-dimensional structural motifs constructed from boron atoms: isolated and coupled atoms; zigzag, straight, branched, paired and three-throw chains; corrugated plane- and plane stacked-networks. Great variety of the possible combinations of B–B, B–Me, and Me–Me bonds allows tuning of the boride properties by controlling their composition and structure.

Among higher borides, one can find all kinds of electronic structures: they can be insulating, semiconducting, semimetallic or metallic. As for the lower borides combining strength and hardness with deal of plasticity usually they are "metal-like" what according to Samsonov means specific kind of electronic structure revealed in high conductivity, in some cases even exceeding that of the pure metal.

Compounds of boron with non-metals characterized by the higher electron affinity, i.e. boron hydrides, carbides, nitrides, oxides, silicides, and arsenides have lesser complicated, common structures. For example, boron nitride BN can be found in layered hexagonal and rhombohedral or denser cubic and wurtzite-like structures, as well as in form of nanotubes, fullerenes and other nanostructures [10]. They differ from metal-like borides by physical properties as well being semiconductors or wide-gap insulators.

Strong (mainly covalent) B–B bonds make boron-based solids refractory and resistive to the aggressive environments, while diversity of atomic and electronic structures and, consequently, physical and chemical properties yields extremely large sphere of technological applications for boron compounds and composites. At the beginning of the Century, boron-containing materials altogether more than ~200 were used in ~250 fields: from glass to detergent industry, in metallurgical, nuclear, tooling, vacuum and solid-state electronic applications etc. [11].

In addition, it should be emphasized lightness and other technologically attractive properties of structural modifications of elemental boron and boron-rich solids. This set of properties makes them promising for making neutron shields with high technical characteristics from the materials containing high concentrations of boron isotope ^{10}B.

11.3 Boron-Containing Bulk Materials as Neutron Shields

Results of our investigation of boron-containing materials' neutron-shielding properties were reported in articles [12–15], presented at conferences [16–19], and summarized in review [20].

As known, protective properties of a material is determined by the particles penetration depth d_0. This parameter depends, on the one hand, on the concentration n of centers interacting with incident particles and, on the other hand, on the cross-section σ of an elemental act of interaction:

$$d_0 = 1/n\,\sigma. \tag{11.7}$$

The physical meaning of the penetration depth is as follows. Let the intensity of the flux incident on the sample surface is equal to J_0. Then the flux J at the depth of d will be

$$J = J_0 \exp(-d/d_0). \tag{11.8}$$

Thus, d_0 is the distance which must to pass particles in the material to decrease flux value in the direction of the initial propagation in $e \approx 2.72$ times. So, the neutron shield's penetration depth is nothing else as effective thickness of the layer protecting against neutrons.

It is clear that to maximize the neutron shielding properties of boron compounds and composites it should be used the ^{10}B-monoisotopic material and not a material with the same chemical composition, but with a natural isotopic composition of boron. For this reason, we can assess the maximal concentration of the ^{10}B isotope atoms as the total concentration of boron atoms in a crystalline structure according to the formula

$$n = N/V, \tag{11.9}$$

where N is the number of B-atoms in the unit cell and V is the unit cell volume.

In view only of neutron-protective properties, structural modifications of elemental boron seem to be preferable, because their n is rather high. However, the neutron shield materials have to meet some other requirements as well, such as adaptability of the material obtaining process, its mechanical strength, good adhesion to the substrate, etc. For this reason, a neutron-shield designer have to make compromises and instead of elemental boron use boron compounds, in which the concentration of neutron-absorbing ^{10}B-atoms are reduced by the presence of other chemical elements.

One more important requirement is related to the fact that the absorption of a neutron by the ^{10}B nucleus yields the release of helium into a gaseous state. For this reason, if the crystalline structure does not contain large voids enough to accumulate and wide channels enough to smooth gas-release, the internal pressure of the helium released during neutron shield exploitation can damage it – promote fracturing and cracking in the working part.

The crystal structures of elemental boron as well as boron-rich compounds being characterized by the presence of large crystallographic voids meet the last requirement. For example, in the β-rhombohedral boron unit cell only ~36.5% of the space is filled by the regular atoms. We have calculated the maximal concentration of boron atoms in 4 main crystalline (α- and β-rhombohedral and α- and β-tetragonal) and also amorphous modifications of boron.

The α-rhombohedral boron is interesting primarily because its structure serves as basis for such important compounds as boron carbide and also oxide, phosphide and arsenide. Suffice it to say that boron carbide is formed by adding about 3 carbon atoms in each unit cell of the α-rhombohedral boron crystal. It is clear that these compounds are characterized approximately by the same concentration of boron atoms.

Boron forms β-rhombohedral lattice-based solid solutions with number of metals accommodating their atoms within large crystallographic voids that are inherent in this structure. For this reason, these materials must be characterized by approximately the same concentration of boron atoms.

Crystals of α- and β-tetragonal boron are believed to be boron-rich structures stabilized by special impurities in high concentrations.

When amorphous boron is considered as solidified boron melt, its structure is represented as a chaotic network with icosahedral clusters in its sites instead of individual atoms. According to this model, concentration of boron atoms in amorphous phase is comparable with that in the boron crystalline phases or even slightly exceed them due to shifting the free icosahedra in voids and, therefore, material compressing during its melting. But, actually in the amorphous boron real structure there are found also icosahedra jointed by edges or faces, as well as isolated atoms and ultra-small quasi-planar clusters. As we pursue the goal of only to estimate the concentration of boron atoms, we can adopt above simplified structural model of amorphous boron imagining it as a system of randomly close-packed perfect icosahedra. At average the neglecting of presence of other structural units should be compensated by the chaotic distribution of icosahedra in the space.

The maximal possible concentration of boron atoms in the boron solid structures has estimated based on the value 1.78 Å for the equilibrium distance between a pair of interacting boron atoms according to the quasi-classical B–B pair interatomic potential [21].

Calculations of maximal concentration n of boron atoms were restricted to those boron crystalline compounds, which like the structural modifications of elemental boron are characterized by the large crystallographic voids.

Namely, layered (hexagonal h-BN and rhombohedral r-BN) boron nitride crystals examined. In these structures, because of weakness of van der Waals forces the interlayer distance is too large to facilitate the free migration of gaseous helium to the sample's surface.

Parameter n was also calculated for metal diborides MeB_2, which possess layered-like (not truly layered because of strong B–Me bonds between neighboring layers) structure.

Finally, we assessed the maximal concentration of boron atoms in layered crystals of boron oxide B_2O_3 and boric acid H_3BO_3.

Of course, because of poor mechanical properties above listed materials are not directly suitable for the neutron shields manufacture. However, they are interesting as the most common precursor-substances or components for a variety of boron-containing composites. Note that porosity of the composition leads to further reducing in n.

Table 11.2 Maximal possible concentration of boron atoms in some 3D solid materials

Material	n, cm^{-3}
α-Rhombohedral boron/Boron carbides, etc.	1.4×10^{23}
β-Rhombohedral boron/Solutions of metals in boron	1.3×10^{23}
α-Tetragonal boron	1.3×10^{23}
β-Tetragonal boron	1.3×10^{23}
Amorphous boron	1.2×10^{23}
Layered boron nitrides	1.7×10^{23}
Zirconium diboride	6.5×10^{22}
Boron oxide	1.3×10^{23}
Boron acid	1.4×1^{22}
"Close-packed" boron	2.5×10^{23}

In principle, it is possible to limit the concentration of boron atoms by their concentration in the "close-packed" crystal built from the boron atoms. Let emphasize that it is a purely hypothetical object, not confirmed experimentally.

Obtained concentrations are summarized in Table 11.2.

Results of calculation have revealed that:

- The maximal possible concentration of atoms in all the 3D solid structural modifications of boron are almost the same: $n = (1.2\text{–}1.4) \times 10^{23}$ cm^{-3}.
- The concentration of boron atoms in some crystalline compounds of boron can be comparable (e.g. B_2O_3) to that in the structures of elemental boron or even exceed it (e.g. layered BN). This result once again underlines the effect of the large voids in the 3D all-boron structures.
- Upper limit of B-atoms concentration achievable in hypothetical "close-packed" boron is almost the twice than that in real crystals.

In the logarithmic scale, the dependence of the neutrons penetration depth as a function of their energy calculated for the material containing the maximal possible concentration of ^{10}B atoms is the line

$$\lg d_0\,[\text{cm}] \approx 1/2 \lg E\,[\text{eV}] - 2.182. \tag{11.10}$$

For ^{10}B-isotope concentration actual values, penetration depths of neutrons of different energies can be easily recalculated if above formula is properly adjusted.

Note that for low-energy neutrons penetration depth is rather small: for "room temperature" neutron flux $d_0 \sim 10\,\mu$m. This means that in the neutron shields made from the boron-containing material the role of the working part plays only the thin surface layer. Thus, expensive ^{10}B-isotopically enriched material should be used only for growth of the shield's surface layer, whereas a much thicker substrate can be made from the same material of natural isotopic composition.

In general, the powder plasma spraying has several advantages if compared with other methods of obtaining boron coatings: this method is technically simple, allows to vary boron content in powder compositions, if necessary presents opportunities of varying in the substrate temperature, coating precursor-components for a short time are in high-temperature plasma flow, remote coating process can be automated applying so-called plasma pistol, etc.

In particular, for boron powder delivery into the plasma jet, it can be used a powder-carrier which may be a component of certain composition [22]. We established the possibility of obtaining B–Al and B–Fe composite powders coatings with particles (of sizes of 10–20 μm) having satisfactory fluidity, which allows their direct use for coatings. However, the production of the B–Fe powder particles is possible only after the chemical treatment of the plasma-treatment products to remove the carrier-powder and resulting phases. In to some extent, this makes it difficult to obtain boron powders suitable for use in plasma technology. To simplify the process of plasma coating using boron powders, we propose [23] an alternative technological scheme of dispersion of the melted crystalline boron to obtain crystalline boron powders in the range of dispersion ensuring the powder fluidity and its use for the plasma production of coatings.

At the same time, it is necessary to pay attention to the development of chemical methods for the preparation of highly ultrafine powders [24, 25], including nanocrystalline powders of boron and boron-containing materials which may also have good prospects for the use in boron-containing coatings grown by the plasma technology.

11.4 Boron-Containing Nanomaterials as Neutron Shields

The achievability of the highest limit of concentration of boron atoms in hypothetical "close-packed" 3D boron crystals yields that effective solving of the neutron-shield problem actually should be based on recently discovered elemental boron 2D nanostructures: sheets, nanotubular and fullerene-like surfaces. These materials high concentration of the ^{10}B isotope atoms may combine with unique physical properties.

Calculations for boron nanotubular bundles were performed within frames of geometric models developed for their atomic structures in articles [26–32], conference presentations [33–37], and reviews [38, 39].

Obtained concentrations are summarized in Table 11.3. Like the 3D structural modification of boron, calculations for nanotubular boron were conducted for B–B bond length of 1.78 Å.

As for the bundles of multi-walled boron nanotubes of modest sizes, according to our single-parameter geometric model typically the concentrations of boron atoms are expected in the narrow range (1.6–1.8) $\times$ 10^{23} cm^{-3}.

Among bulk all-boron structures and boron compounds useful for neutron-shield applications layered boron nitrides demonstrate maximal concentration of the neutron-capturing centers. Then, one could expect advantages of materials based on nano-BN.

Table 11.3 Concentration of atoms in bundles of ultra-small boron nanotubes

Nanotube	n, cm^{-3}
(1,0)	2.2×10^{23}
(1,1)	2.4×10^{23}
(2,0)	1.6×10^{23}
(2,1)	2.7×10^{22}
(2,2)	2.1×10^{23}

Especially, bundles of multi-walled BN nanotubes can contain very high concentrations of ^{10}B nuclei, the effective neutron-capturing centers. So, nanotubular BN can serve for effective neutron-shielding material. The key objectives of the recent NIAC Program phase-I report [40] are to investigate, both computationally and experimentally, which forms, compositions, and layerings of hydrogen, boron, and nitrogen containing materials will offer the greatest shielding in the most structurally robust combination against GCR (galactic cosmic radiation), secondary neutrons, and SEP (solar energetic particles). The objectives and expected significance of this research are to develop a space radiation shielding materials system that has high efficacy for shielding radiation and that also has high strength for load bearing primary structures. The boron nitride nanotubes can theoretically be processed into structural material and used for load bearing structures. Furthermore, the nanotubular BN can be incorporated into high hydrogen polymers and the combination used as matrix reinforcement for structural composites. In this regard, it should be indicated BN nanotubes structure is attractive for hydrogen storage and hydrogenation [41].

11.5 Problem of Accompanying Gamma-Radiation

Usually, neutron radiation is accompanied by the electromagnetic radiation. Manufacturing of shields from boron nanomaterials can help to solve the problem of neutron-shielding in conjunction with the electromagnetic shielding [42]. The point is that boron sheets as well as nanotubes of any chirality are expected to be metallic, in contrast, on the one hand, the bulk semiconducting modifications of elemental boron and, on the other hand, the carbon nanotubes, some of which (depending on chirality) are also semiconducting.

"Metallic" boron nitride [43] can serve for alternative of metallic nanostructures of elemental boron. Boron nitride shell structures of chemical composition BN_x with boron excess ($x \ll 1$) contaminated with carbon and oxygen were synthesized [44] in process of melting of a boron-rich material in a boron nitride crucible with the nitrogen source in the form of high-purity pressed boron nitride rods, which held up the crucible. Obtained material (Fig. 11.5) is found to be conductive despite the fact that all the boron nitrides of stoichiometric chemical composition BN are insulators.

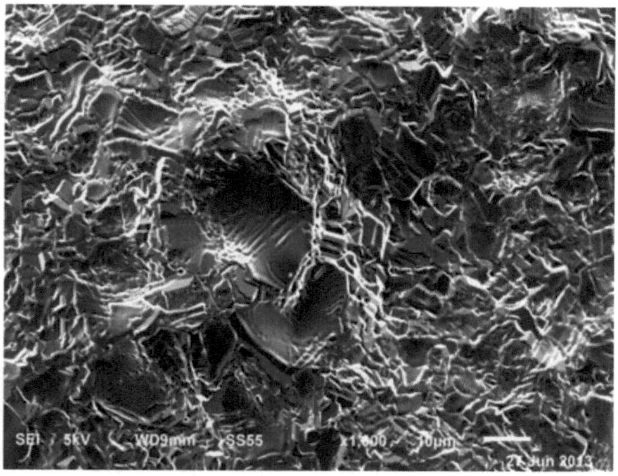

Fig. 11.5 Outer surface microstructure of BN_x shell – metallic-colored deposit entirely covering a BN rod

Table 11.4 Sizes of simulated cylindrical nuclear reactors

Geometry	Base diameter, m	Height, m
1	3.84	4.50
2	2.88	3.20
3	12.00	8.80

"Metallic" boron nitride is modeled as a mixture of structural modifications of semiconducting boron and boron carbide heavily doped with nitrogen. First principle calculations performed within the quasi-classical approximation prove that nitrogen impurities, accommodated in large crystallographic voids characteristic of all-boron and boron-rich crystalline lattices, create donor electron-states inside the conduction band and then lead to the local "metallization" of chemical bonds in these initially covalently bonded structures.

As mentioned, most of ^{10}B–n elementary acts is accompanied by radiation of a high-energy (0.48 MeV in the limit of zero-velocity neutrons incident on a stationary boron nucleus) γ-quantum. For this reason, one of the main problems associated with the use of boron-containing neutron-shields is the need to provide additional protection against the concomitant γ-radiation.

It can be provided assessments of radiation properties of B–Me layered composites as neutron/gamma shields [45]. Numerical calculations were performed [46] using the computing code MicroShield for a boron–tungsten system. There were tested three geometries of the neutron source, which corresponded to the parameters of standard cylindrical nuclear reactors (Table 11.4). For each of these geometries, there are considered two cases: lead Pb as standard material often used for protection against γ-radiation and tungsten W as a heavy metal forming a number of compounds with boron. It was assumed that these metallic layers are placed on the outer surface of the boron-containing layer, so that the γ-rays concomitant neutron-capture in boron must pass through the γ-absorbing metal layer.

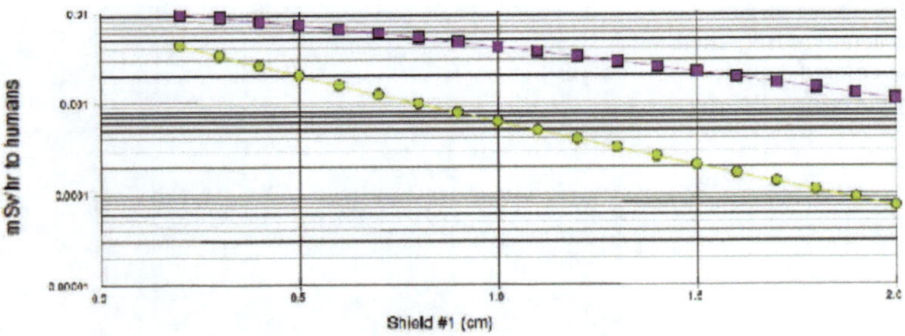

(a)

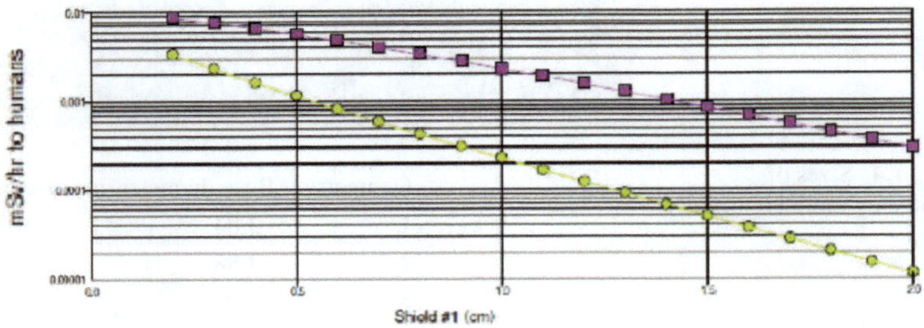

(b)

Fig. 11.6 Attenuation of γ-radiation concomitant the neutrons-capture in boron-containing shield of cylindrical nuclear reactor 1 versus thickness of (**a**) lead and (**b**) tungsten

In each of the six analyzed cases, the thickness of the protective layer was varied in the range of 0.2–2.0 cm, and the calculations were performed in 19 points of the interval. It was obtained the pairs of curves (Figs. 11.6, 11.7, and 11.8) of exposure dose, mR/h with (upper curves) and without (lower curves) taking into account the accumulation factor. The use of this factor meant the account in the main beam of already scattered rays. Flux densities were calculated at the same form. Conventionally, the γ-ray flux incident on the protective layer was assumed to be of 10^9 ph/s, i.e., the equivalent of 1 GBq (1 Becquerel is the source of radioactivity, which is 1 decay/s). Since the curves define the degree of attenuation of gamma radiation, the absolute value of output flow from the protective layer is easily counted for each part of the incident flux.

For geometries 1 and 2 dose rate is calculated for a distance of 200 cm from the center, and for the geometry 3 – 410 cm. These distances approximately correspond to regions near the surfaces of cylinders and, therefore, allow comparison of the results. Instead of the exposure dose, curves are presented for effectively equivalent (or effective) dose in mSv/h. As we see, initial dose rates are too low and tungsten layer further reduces them (remember that natural background is in the range of 10–20 mSv).

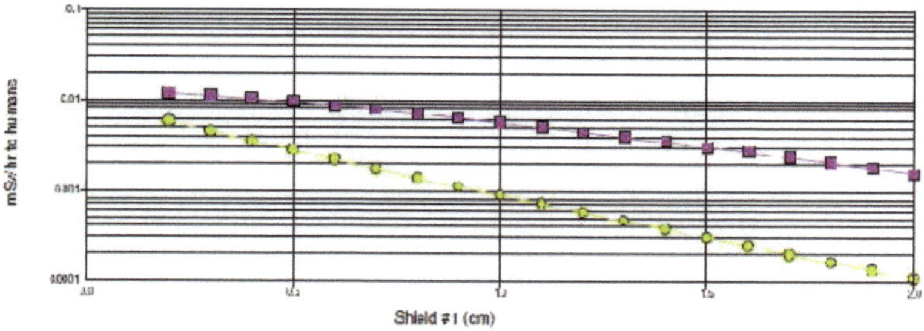

(a)

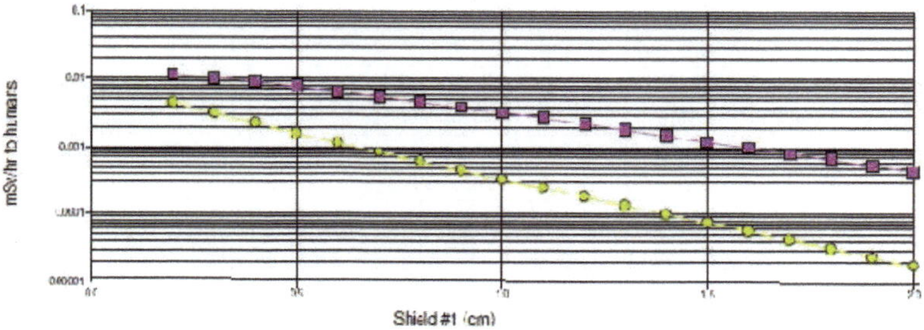

(b)

Fig. 11.7 Attenuation of γ-radiation concomitant the neutrons-capture in boron-containing shield of cylindrical nuclear reactor 2 versus thickness of (**a**) lead and (**b**) tungsten

These results lead to the conclusion that monoisotopic ^{10}B-layered compositions of boron and tungsten can be used successfully to protect against the different neutron flux intensities. In this regard it should be emphasized that the state diagram of the W–B system is among the most studied phase diagrams of transition metal–boron systems. This indicates the existence of a series of binary W/B phases which differ in crystal structure and value of the W:B ratio. This allows thinking the adhesion in such layered structures will be good due to the formation of a transition layer, the mixture of tungsten borides, as it is in the case of well-known technique of growing boron fiber coating of tungsten wire.

11.6 Neutron-Detectors Based on Boron-Containing Materials

Detection/measuring of neutron fluxes, together with neutron-shielding, are the important constituents of the protection against neutron-irradiation.

Existing neutron detectors based on boron-containing nanomaterials have been described in [47–49]. Their physical-technical parameters were introduced as well.

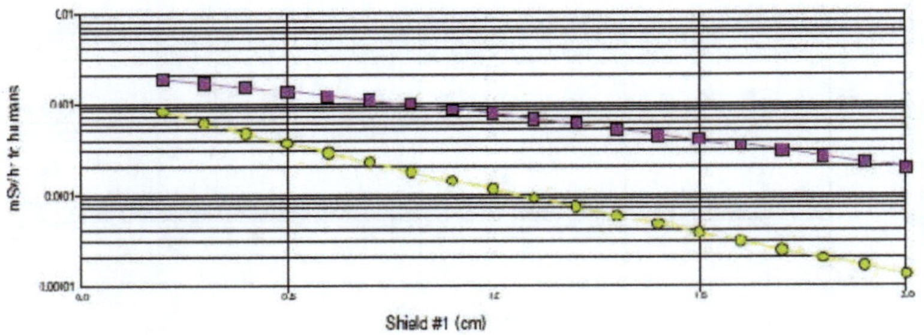

(a)

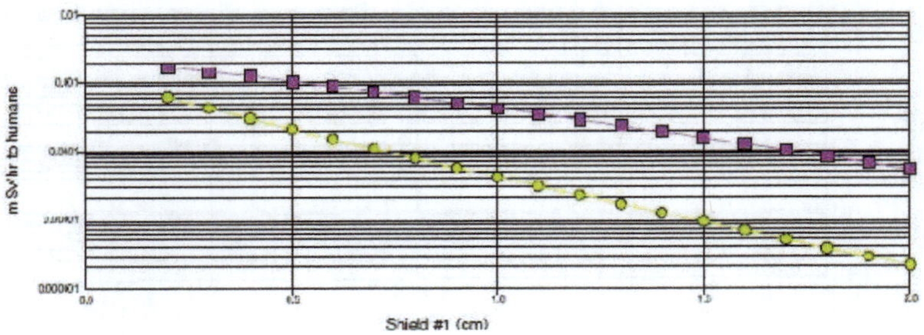

(b)

Fig. 11.8 Attenuation of γ-radiation concomitant the neutrons-capture in boron-containing shield of cylindrical nuclear reactor 3 versus thickness of (**a**) lead and (**b**) tungsten

Estimated ranges of values of these parameters are:

- Thickness of the effective working layer in a semiconductor device $\sim 10\ \mu$m.
- Releasing rate of the ^{10}B decay products, lithium and helium, induced by neutrons absorption $\sim 10^{15}/\mathrm{cm}^3$ s.
- Generating rate of electron–hole pairs in process of neutron absorption $\sim 10^{22}/\mathrm{cm}^3$ s.
- Rate of rise in the temperature in process of neutron absorption ~ 10 K/s.
- Device mean operating time $\sim 10^{-4}$ s.

Using obtained results and relevant multipliers, engineers and designers of solid-state detectors of neutron radiation can easily recalculate parameters for devices of different designs and different values of the incident neutron flux.

In addition, certain recommendations can be made about improvements in physical-technical characteristics of neutron-radiation flux detecting and measuring devices made from ^{10}B-containing materials. The general conclusion is that ^{10}B-isotopically enriched semiconducting modifications of elemental boron, semiconducting boron compounds, and boron-doped common semiconductor materials can serve as working bodies for thin-film, high-reliable, high-sensitive, and fast-acting robust solid-state electronic neutron-detectors of various types.

11.7 Conclusions

In conclusion we can note that the application of thin coatings made from the ^{10}B-isotopically modified materials (e.g., powders) can not only greatly simplify the problem of protection against neutron irradiation of materials, appliances and products (this may allow the use of available construction materials instead of expensive, for example, specially alloyed steels and alloys), but also ensure the safety of personnel of plants using neutron sources and the effective protection of the environment from radioactive contamination.

Such thin layers will be useful coatings not only for radiation disks, but also for consumables such as protective clothing – suits, gloves, and shoes designed for personal defense against neutron radioactive materials. From nanotubular bundles it can be, for example, woven an elastic material suitable for the manufacture of neutron-protective clothing instead of hard neutron shields.

Acknowledgements Author gratefully acknowledges the Science and Technology Center in Ukraine for partial support within the Grant STCU 6204.

References

1. IAEA (1996) International basic standards for protection against ionization radiation and for safety of radiations. Safety Ser 115:1–354
2. Martin BR (2008) Nuclear and particles physics. An introduction. Wiley, Chichester
3. Nakaidze S, Gharibashvili V, Antadze M, Tsagareishvili O (2004) Neutron-absorption deformable boron-rich ferrous alloys. J Solid State Chem 177(2):592–595
4. Sears VF (1992) Neutron scattering lengths and cross sections. Neutron News 3(1):26–37
5. JENDL – 4.0 (2010) Sublibrary: neutron reactions
6. Lipscomb WN (1981) Borides and boranes. J Less-Comm Met 82(11–12):1–20
7. Tsagareishvili GV, Antadze ME, Tavadze FN (1991) Production and structure of boron. Metsniereba, Tbilisi
8. Chkhartishvili L (2016) All-boron nanostructures. In: Kharisov BI, Kharissova OV, Ortiz-Mendez U (eds) CRC concise encyclopedia of nanotechnology, chap 7. CRC Press, Boca Raton, pp 53–69
9. Samsonov GV, Serebryakova TI, Neronov VA (1975) Borides. Atomizdat, Moscow
10. Chkhartishvili L (2016) Boron nitride nanostructures. In: Kharisov BI, Kharissova OV, Ortiz-Mendez U (eds) CRC concise encyclopedia of nanotechnology, chap 8. CRC Press, Boca Raton, pp 70–99
11. Acarkan N (2002) Boron products and their uses. In: Erarslan K (ed) Proceedings of the 1st international symposium on boron. Dumlupinar University, Kutahya, pp 1–5
12. Tsagareishvili O, Chkhartishvili L, Gabunia D, Gachechiladze A, Shalamberidze S, Gabunia L, Tushishvili M (2009) Some aspects of use of boron-containing materials for neutron shield. In: Konuk A, Kurama H, Ak H, Iphar M (eds) Proceedings of the 4th international symposium on boron. Osmangazi University, Eskisehir, pp 257–266
13. Chkhartishvili L, Tsagareishvili O, Gabunia D (2012) ^{10}B-based materials for neutron-shielding. In: Khantadze J, Chkhartishvili L, Gabunia D, Ramazashvili D (eds) Proceedings of the 1st international conference on "Modern technologies & methods of inorganic materials science". Meridian, Tbilisi, pp 188–202

14. Pulariani IN, Chkhartishvili LS, Gabunia DL, Mirijanashvili ZM (2014) Estimation of neutron-absorbing characteristics of CrB_2–TiB_2–Al_2O_3 composite coatings. In: Sinyaev DN (ed) Materials of the 2nd All-Russia scientific internet-conference with international participation "Nanotechnology in theory & practice", Kazan', IE Sinyaev DN, pp 131–136
15. Mirijanashvili ZM, Pulariani IN, Chkhartishvili LS, Gabunia DL, Gharibashvili VI, Kandelaki AZ, Kupreishvili IL (2015) Wear-resistant (Cr,Ti)B_2-based composites containing Al_2O_3 nanoinclusions. In: Khantadze J, Ramazashvili D, Tavadze L, Chkhartishvili L (eds) Proceedings of the 2nd international conference on "Modern technologies & methods of inorganic materials science". Sachino, Tbilisi, pp 202–208
16. Chkhartishvili LS, Tsagareishvili OA, Gachechiladze AA, Gabunia DL, Mirijanashvili ZS, Tavadze GF, Peradze TA, Eristavi AG (2010) On possibility of application of boron-containing coatings as neutron shield. In: Proceedings of the 6th international conference on MEE. IPMS, Kyiv, pp 73–73
17. Chkhartishvili L, Gabunia D, Tsagareishvili O (2011) ^{10}B powder-based neutron shield coating materials. In: Abstracts of the 17th international symposium on boron, borides and related materials. Istanbul Technical University, Istanbul, pp 294–294
18. Chkhartishvili LS, Tsagareishvili OA, Gabunia DL, Jmukhadze TT (2011) Nano-structural ^{10}B as an effective shielding material against neutron irradiation. In: Proceedings of the 3rd international conference on high mat tech. IPMS, Kyiv, pp 270–270
19. Gabunia DL, Chkhartishvili LS, Tsagareishvili OA, Gachechiladze AA, Mikeladze AG, Gharibashvili VI, Mirijanashvili ZM (2012) Boron-containing materials – effective tools of protection from neutron irradiation. In: Proceedings of the 7th international conference on MEE. IPMS, Kyiv, pp 49–49
20. Chkhartishvili L (2013) Interaction between neutron-radiation and boron-containing materials. In: Kharisov BI, Kharissova OV, Mendez UO (eds) Radiation synthesis of materials and compounds, chap 3. CRC Press/Taylor & Francis, Boca Raton, pp 43–80
21. Chkhartishvili L, Lezhava D, Tsagareishvili O (2000) Quasi-classical determination of electronic energies and vibration frequencies in boron compounds. J Solid State Chem 154(1):148–152
22. Gabunia DL, Tsagareishvili OA, Darchiashvili MD, Gabunia LD, Darsavelidze GS (2004) Gas-thermal treatment of the boron powders and obtaining of coatings. In: Proceedings of the 3rd international conference on MEE. IPMS, Kyiv, pp 368–369
23. Gabunia DL, Tsagareishvili OA, Gachechiladze AA, Tavadze GF (2008) Obtaining of the β-rhombohedral boron powders. In: Proceedings of the 5th international conference on MEE. IPMS, Kyiv, pp 115–115
24. Gabunia D, Gachechiladze A, Mikeladze A, Tsagareishvili O, Chkhartishvili L (2009) Obtaining opportunities of boron carbide powders based hetero-modulus ceramics. In: Fiore S, Belviso C, Giannossi ML (eds) Micro et Nano – Sciantiæ Mare Magnum, 1. SA Publishers, Bari, pp 86–86
25. Gachechiladze A, Gabunia D, Mikeladze A, Tsagareishvili O, Chkhartishvili L (2009) On technological process of synthesis of boron carbide nanodispersive powders. In: Konuk A, Kurama H, Ak H, Iphar M (eds) Proceedings of the 4th international boron symposium. Osmangazi University, Eskisehir, pp 67–71
26. Chkhartishvili L (2009) On quasi-classical estimations of boron nanotubes ground-state parameters. J Phys Conf Ser 176:012013
27. Chkhartishvili L (2009) Molar binding energy of the boron nanosystems. In: Konuk A, Kurama H, Ak H, Iphar M (eds) Proceedings of the 4th international boron symposium. Osmangazi University, Eskisehir, pp 153–160
28. Chkhartishvili L (2011) Geometrical models for bare boron nanotubes. In: Borisenko VE, Gaponenko SV, Gurin VS, Kam CH (eds) Physics, chemistry and applications of nanostructures. World Scientific, Singapore, pp 118–121

29. Chkhartishvili LS (2011) Geometrical structure of chiral and degenerated nanotubular forms of elemental boron. In: Vityaz' PA, Labunov VA, Orlovich VA, Dobrego KV, Zhornik VI, Shpilevskij EM (eds) Nanostructures in condensed matter. Belarusian State University Publishers Center, Minsk, pp 324–329
30. Chkhartishvili L (2011) Nanotubular boron: ground-state estimates, chap 8. In: Chikoidze E, Tchelidze T (eds) New developments in materials science. Nova Science Publishers, New York, pp 67–80
31. Chkhartishvili L (2012) Geometry of all-boron nanotubes with reduced numbers of bonds. In: Gerasimov A, Chkhartishvili L (eds) Conference on 2nd international conference on "Nanotechnologies". Nekeri, Tbilisi, pp 81–85
32. Chkhartishvili L, Mamisashvili N, Maisuradze N (2015) Single-parameter model for multi-walled geometry of nanotubular boron. Solid State Sci 2015(47):61–67
33. Chkhartishvili L (2008) Quasi-classical estimations of the boron nanotubes ground-state parameters. In: Abstracts of the 16th international symposium on boron, borides and related materials. Tokyo University, Matsue, pp 126–126
34. Chkhartishvili L (2009) Ground-state parameters of the stable tubular structures of nanoboron. In: Chikoidze E, Trapaidze L, Tchelidze T (eds) Materials science days. Universal, Tbilisi, pp 11–12
35. Chkhartishvili L (2012) Geometry of all-boron nanotubes with reduced numbers of bonds. In: Trans. Int. Sci. Conf. Ded. 90th Anniv. GTU "Basic paradigms in science and technology development for the 21st century". Technical University Press, Tbilisi, pp 383–386
36. Chkhartishvili L, Mamisashvili N, Maisuradze N (2014) Single-parameter model for multi-walled geometry of nanotubular boron. In: Abstracts of the 18th international symposium on boron, borides and related materials. University of Hawaii, Honolulu, pp 118–118
37. Chkhartishvili LS (2015) Modeling atomic geometry of fibers and bundles of nanotubular boron. In: Proceedings of the 5th international conference on high mat tech. IPMS, Kyiv, pp 47–47
38. Chkhartishvili L (2011) Micro- and nano-structured boron, chap 6. In: Perkins GL (ed) Boron. Compounds, production and application. Nova Science Publishers, New York, pp 221–294
39. Chkhartishvili L (2011) Nanoboron (An overview). Nano Stud 3:227–314
40. Thibeault SA, Fay CC, Lowther SE, Earle KD, Sauti G, Kang JH, Park C, McMullen AM (2012) Radiation shielding materials containing hydrogen, boron, and nitrogen: systematic computational and experimental study – phase I (NASA innovative advanced concepts final report). NASA, Hampton, pp 1–31
41. Chkhartishvili L (2014) Hydrogen nanoreservoirs made of boron nitride, chap 27. In: Kharisov BI, Kharissova OV, Rasika Dias HV (eds) Nanomaterials for environmental protection. Wiley, Hoboken, pp 453–467
42. Chkhartishvili L, Murusidze I (2012) Band structure of all-boron 2D metallic crystals as a prospective electromagnetic shielding material. In: Proceedings of the international conference on fund. & appl. nano electro magn. Belarusian State University, Minsk, pp 11–11
43. Becker R, Chkhartishvili L (2014) On possible nature of metallic conductance of boron–nitrogen compounds. In: Abstracts of the 3rd international conference on "Nanotechnologies". Georgian Technical University, Tbilisi, pp 13–14
44. Becker R, Chkhartishvili L, Avci R, Murusidze I, Tsagareishvili O, Maisuradze N (2015) "Metallic" boron nitride. Eur Chem Bull 4(1–3):8
45. Chkhartishvili LS, Nabakhtiani GN, Gigineishvili AV, Tsagareishvili OA, Gabunia DL, Rostomashvili ZI (2012) Assessment of attenuation of accompanying gamma-radiation in boron–metallic neutron shields. In: Proceedings of the 7th international conference on MEE. IPMS, Kyiv, pp 79–79
46. Nabakhtiani G, Chkhartishvili L, Gigineishvili A, Tsagareishvili O, Gabunia D, Rostomashvili Z, Dekanosidze S (2013) Attenuation of gamma-radiation concomitant neutron-absorption in boron–tungsten composite shields. Nano Stud 8:259–266

47. Chkhartishvili L, Tsagareishvili O, Tavadze G (2014) Neutron detectors based on ^{10}B-containing nanomaterials. In: Nuclear radiation nanosensors and nanosensory systems. Georgian Technical University, Tbilisi, pp 81–83
48. Chkhartishvili L (2014) Neutron-fluence nanosensors based on boron-containing materials, chap 26. In: Kharisov BI, Kharissova OV, Rasika Dias HV (eds) Nanomaterials for environmental protection. Wiley, Hoboken, pp 445–449
49. Chkhartishvili L, Tsagareishvili O, Tavadze G (2016) Neutron detectors based on ^{10}B-containing nanomaterials, NATO science for peace and security series sub-series B. Physics and biophysics, chap 12. In: Kervalishvili PJ, Yannakopoulos PH (eds) Nuclear radiation nanosensors and nanosensory systems. Springer, Dordrecht, pp 187–196

Part II
Nanosensors

Chapter 12
Zeolite-Based Interfaces for CB Sensors

A. S. Fiorillo, J. D. Vinko, F. Accattato, M. G. Bianco, C. D. Critello, and S. A. Pullano

Abstract Nano-porous synthetic zeolite can be used to interface bio-systems to microelectronic circuits, which are integrated onto Si wafers. Similar micro-devices have the potentiality not only of acquiring and processing biochemical signals in situ, but also of measuring physico-chemical parameters involved in matter conversion, joining with full rights the emerging technology of "laboratory-on-a-chip". Wearable, miniaturized devices capable of fast responses which are also small in size, lightweight and low power consuming have rendered them helpful detecting-to-warn support in case of CBRN threats. Thin layers of Zeolite were deposited on both glass and silicon wafers using a technique fully compatible with standard integrated circuit manufacture. The experimental results of various tests carried out in order to confirm their trapping capabilities of the toxic agents acetone, chlorine, and methanol are reported in this article.

Keywords Nanoporous materials · Zeolite · Sensors · Toxic agents · Wearable devices · Embedded systems

12.1 Introduction

Iono-electronic interaction plays a fundamental role in many biochemical and biological investigations, mainly when process-monitoring is carried on by means of miniaturized devices the principle of operation of which is based on biochemical- or biophysical-to-electrical energy conversion. The actual trend in many fields (from medical to pharmaceutical, from industrial to civil as well as military) is focused

A. S. Fiorillo (✉) · F. Accattato · M. G. Bianco · C. D. Critello · S. A. Pullano
Laboratory of Biomedical Applications, Technologies and Sensors – BATS, Department of Health Sciences, University Magna Græcia of Catanzaro, Catanzaro, Italy
e-mail: nino@unicz.it

J. D. Vinko
ERA Physics, Velletri, Italy

J. Bonča, S. Kruchinin (eds.), *Nanostructured Materials for the Detection of CBRN*, NATO Science for Peace and Security Series A: Chemistry and Biology, https://doi.org/10.1007/978-94-024-1304-5_12

on the emerging technology of laboratory-on-a-chip (LOC) and there are already excellent, tangible examples reported in specific literature [1]. Although they have been primarily employed in medicine (their use in cancer research is very common), their small size, lightness and low power consumption have made this class of devices a helpful support for detecting-to-warn in case of Chemical, Biological, Radiological, and Nuclear (CBRN) threats. It is clear that wearable, miniaturized devices capable of a fast response are desirable in case of air contamination due to highly toxic substances [2]. Undoubtedly, efficiency combined with a simple fabrication process are two essential characteristics of iono-electronic devices for the manufacture of LOC.

In this chapter we explain how thin layers of nano-porous synthetic zeolite can be used to interface bio-systems to microelectronic circuits, preferably of the MOSFET type, which are integrated on Si wafers.

Next, one of the most important properties of zeolite is its capacity to exchange ions and trap molecules, compatibly with pore dimensions. Consequently, when it is deposited on the gate of an N channel metal–oxide–semiconductor (NMOS) circuit [3], we can expect that the induced changes in its electrical state will modify the electric field, which in turn will modulate the Field Effect Transistor (FET) channel. The FET channel is capacitively coupled with the Gate electrode in intimate contact with the zeolite which in some way acts as an additional capacitive layer. For this reason the capacitance of the nanoporous layer was characterized before and after exposure to toxic agents.

However, the proposed micro-device has the potential not only to acquire and process biochemical signals in situ, but also to measure physico-chemical parameters involved in matter conversion. So, finally, various tests were carried out in order to confirm its trapping capabilities with respect to water, urea, miRNA, and Rhodamine B molecules, from a medical standpoint, and acetone, chlorine, and methanol as toxic agents in CBRN threats.

12.2 Materials and Methods

In a previous work, we investigated the feasibility of fabricating low-noise iono-electronic microsystems to directly transduce biological signals into electrical signals by using a layer of synthetic zeolite deposited on the gate of a NMOS, in order to interface bio-systems to microelectronic circuits integrated on Si wafers. In similar devices the capacity of zeolite to change its electrical state, as result of ion exchange and molecule trapping, can be exploited to modify the electric field, which modulates the FET channel, when deposited on a MOSFET gate [4, 5].

Zeolite is a nanoporous crystalline aluminosilicate material with an open, regular, three-dimensional structure that consists of rigid tetrahedral units composed of $(SiO_2)_n$ and $(AlO_4)_n$ which are linked by oxygen atoms, generating regular intra-crystalline pores and channels with molecular dimensions ranging between 0.3 and 1.3 nm, and characterized by the following properties [6, 7]:

- good heat conductivity,
- variable hydrophilicity/hydrophobicity,
- ion-exchange and molecule-trapping and sieving capabilities,
- modulus of elasticity: 35 GPa,
- relatively low dielectric constant: $\varepsilon_r = 2.7$.

The chemical composition of zeolites is expressed as:

$$Ay/m^{m+}[(SiO_2)_x \cdot (AlO_2^-)_y] \cdot zH_2O, \tag{12.1}$$

where A is a positive charge-compensating cation with m valence, required to balance the excess negative charge in the structure; $(x + y)$ is the number of tetrahedra per crystallographic unit cell; x/y is the silicon/aluminum ratio and z is the number of water molecules associated with the framework. The z value depends not only on the nature of the zeolite, but also on the temperature and the partial pressure of water vapor in the environment. The minimum value of the x/y ratio is 1 (Si/Al $\geqslant$ 1) and usually varies from 1 to 5, as provided by the Loewenstein rule, but there is no maximum limit [8, 9]. The extra-framework cations are not covalently bonded to the zeolite structure, which confers ion-exchange properties.

According to a low-temperature technique compatible with standard Integrated Circuit (IC) technology described by Fiorillo, type 3A, 4A, and 13X zeolite powders with different degrees of porosity were spun onto silicon wafers and activated both in an O_2 plasma reactor, and under UV rays [10–14].

12.2.1 Zeolite Type 3A, 4A, and 13X

Zeolite 3A is a potassium form of aluminosilicate having a porous dimension of 3 Å. Its chemical formula is:

$$K_{12}[AlO_2 \cdot SiO_2]_{12} \cdot 27H_2O. \tag{12.2}$$

Table 12.1 Physico-chemical parameters of vegetable oils

Oil	Density g/cm^3 20 °C	Viscosity cP 20 °C	Smoke point °C	Flash point °C	Iodine value	Ref.
Sunflower	0.917–0.923	63.7	209	316	136	[14]
Soybean	0.919–0.925	60.7	242	330	132	[14]

Zeolite 13X (courteously furnished, as well as zeolite 3A, by the UOP-Honeywell plant in Reggio Calabria, Italy) has a porous dimension of more than 7 Å and its chemical formula is:

$$Na_{86}[(AlO_2)_{86} \cdot (SiO_2)_{106}] \cdot H_2O. \quad (12.3)$$

Zeolite 4A (furnished by the Chinese Luoyang Jianlong Micro-nano New Materials Co., Ltd.) has a porous dimension of about 4 Å and its chemical formula is:

$$Na_{12}[AlO_2 \cdot SiO_2]_{12} \cdot 27H_2O. \quad (12.4)$$

The various zeolite powders were deposited on glass substrates by means of a spin-coating process. They were mixed at 50% w/w with different vegetable oils (10 g of powder plus 10 g of vegetable oil), with a preference for those with high iodine value, the physico-chemical parameters of which are reported in Table 12.1.

Zeolite 3A and zeolite 13X were tested after deposition on the extended gate of a NMOS integrated circuit [3]. The absorption of different biomolecules such as urea and miRNA and their interaction with the zeolite framework were investigated using different techniques, based mainly on FT-IR, a spectroscopic absorption technique for the qualitative analysis of a sample [15–19]. Several studies underline the fact that circulating small nucleic acids such as miRNAs are directly involved in many physiological and pathological processes, particularly in processes of cancer initiation and progression but also in cardiovascular and metabolic disorders, such as diabetes [20–24]. Several papers have described the interactions of nucleic acid bases with minerals and zeolite, particularly those involving Van der Waals forces [25–29]. Zeolite 13X was also used for Rhodamine B entrapment. Rhodamine molecules were loaded into zeolite pores, following an experimental procedure present in literature [20]. Rhodamine B is a fluorescent material that emits electromagnetic waves in the wavelength of 544 nm.

The experimental investigations presented in this work were aimed at electrically characterizing the absorbing properties of the Zeolite modified electrodes (ZMEs).

12.2.2 Zeolite Modified Electrodes

Zeolite, due to its morpho-structural, chemical-physical and mechanical strength properties, is widely used in the fabrication of electrochemical biosensors called zeolite-modified electrodes in which the traditional metal electrode is modified with the nanoporous biomediator in order to selectively transduce the biosignal of interest, improving the response of the sensor.

The main advantage of this is the possibility of combining the specificity of charge transfer reactions with molecular sieving and the ion-exchange properties of aluminosilicates in a single device. In addition, the hydrophilic character of zeolite renders them suitable for the co-immobilization of enzymes and mediators in the preparation of biosensors.

Electrodes are fabricated on an insulating substrate (a thin glass layer in the present work) by means of e-beam evaporation of the metallic layer followed by deposition of a thin film containing zeolite particles using the spin-coating method. This technique results in the deposition of a very thin (between 5 and 7 μm), uniform (roughness <500 nm) zeolite layer. The masks used for metallization on the glass substrates were designed by CAD Rhinoceros 5 software and realized through 3D printing (see Fig. 12.1). The glass substrates were shaken with isopropyl alcohol for 15 s, then dried under a stream of N_2 gas, rinsed in deionized water, and, again, dried under N_2 gas. The metallic coating was carried out in two successive steps in order to obtain two distinct nickel layers, each with a thickness of 50 nm. The resulting electrode has dimensions of $5 \times 5\,mm^2$ with a 7 mm contact section (see Fig. 12.1).

In the second step of the process, the mixtures of zeolite and the various oils (a mole ratio of 40–60 zeolite and 60–40 oil, respectively) were spun onto the nickel electrode deposited on the glass substrate for 60 s and at different speeds (3000–5000 rpm/min), and then heated for 16 h in a furnace at 120 °C.

Fig. 12.1 Mask and electrode deposited on glass

During the curing process of the zeolite mixture which is performed at low temperatures, the oil blend also makes calcination unnecessary in which the product must be exposed to higher temperatures, resulting in the risk of damage when ZMEs are coupled with silicon-integrated circuits. Finally, the samples were then cooled to room temperature.

The best results were obtained for oils having high iodine values and low viscosity, such as sunflower oil and soybean oil which gave comparable results in terms of adhesion, roughness, resistance to scratches, and mechanical strength. Finally, ZMEs were activated at 120 °C and a pressure of 350 Pa, in an O_2-plasma reactor (STS Multiplex).

12.3 Iono-Interface Characterization

Morphological and structural characterization was performed by Scanning Electron Microscopy (SEM) as shown in Fig. 12.2, optical profilometry and Fourier Transform Infrared (FT-IR) spectroscopy and the evaluation of the dielectric properties of the film was carried out through capacitance measurements. By means of this technique it was possible to evaluate the characteristics of the iono-selective layer and analyze the shapes and sizes of the zeolite grains.

Fig. 12.2 Cross section of the ZME deposited on glass

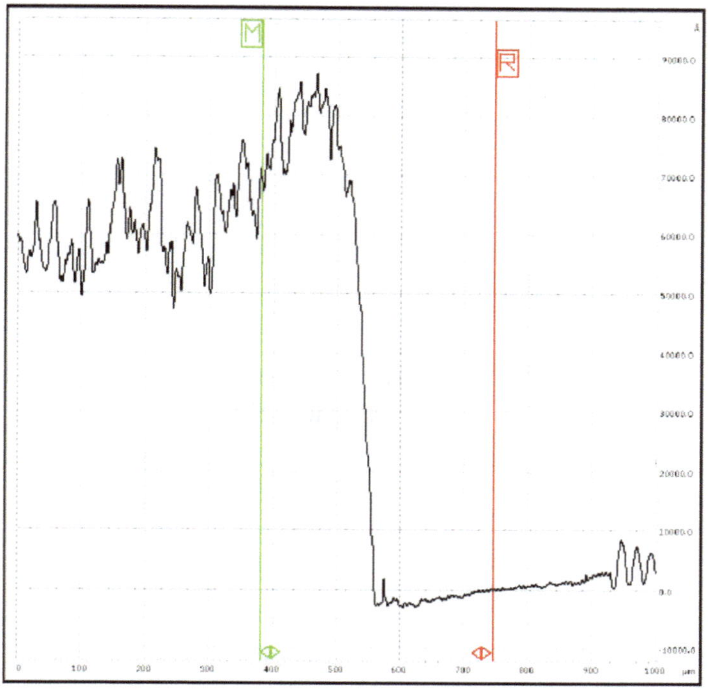

Fig. 12.3 Profilometric analysis of layer obtained at 5000 rpm

Surface roughness analysis such as the one seen in Fig. 12.3, performed with a profilometer on both kind of samples, shows the average thickness of the 4A sample in which the spin coating carried out at 5000 rpm was 5 µm, whereas that of the sample obtained at 3000 rpm was 7 µm.

The fabrication of thin films with low relative dielectric constants (ε_r) and capacitance (C) and therefore the evaluation of the dielectric behavior of the iono-selective film is crucial. Capacitance measurements at different frequencies (from 20 Hz to 1 kHz) were performed by means of a Keysight E4980AL precision LCR meter. The standard deviation of the capacitance values for our samples was higher at lower frequencies and decreased as the frequency increased. The sample capacitance versus frequency is reported in Fig. 12.4, showing the minimum value of about 0.4 and 0.2 nF for the 5 and 7 µm layers, respectively.

In order to quantify the interference of humidity, taking into account the peculiarity of zeolite of retaining water, the capacitance of the individual samples were measured at regular intervals for 30 min after the application of a pressure reduction, obtained by means of a vacuum pump. From the capacitance analysis of our electrodes at 1 kHz under vacuum conditions, it emerges that the relative

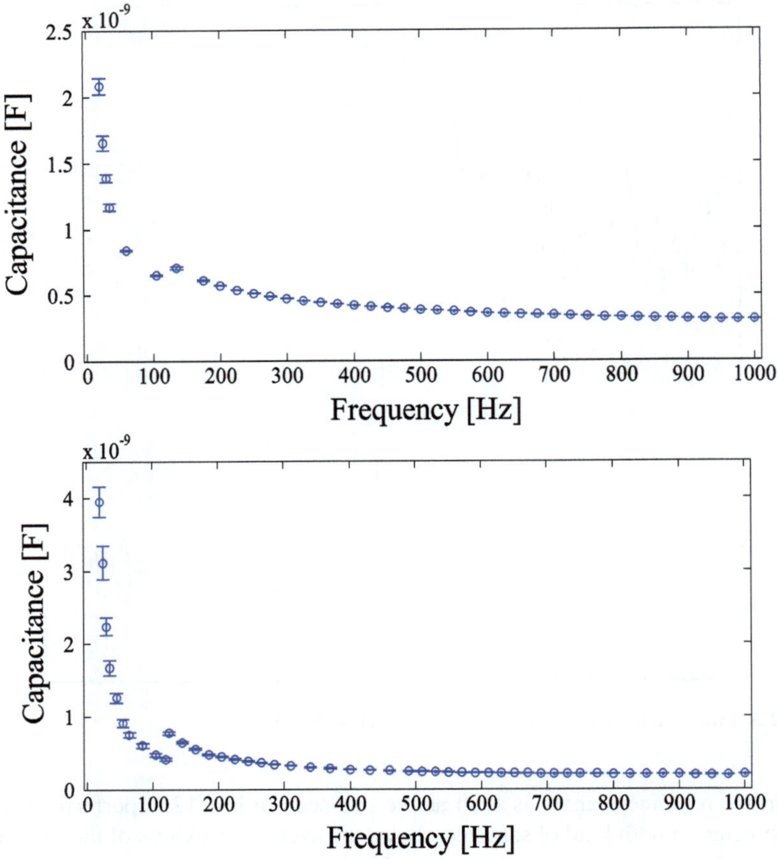

Fig. 12.4 Capacitance vs frequency for the 5 μm sample (upper) and the 7 μm sample (lower)

dielectric constant shows a minimum value of 3.06 for the 5 μm sample and 3.13 for the 7 μm sample (the capacitive trend is shown in Fig. 12.5). The dielectric constant of the films is therefore generally doubled due to the exposure to air.

The gas-selective properties of zeolite were determined by alternating its exposure to nitrogen and chlorine vapors. When chlorine was introduced into the bell, the capacitance value for both electrodes remained almost unchanged, while exposure to nitrogen slowly restored the starting values (see Fig. 12.6).

Based on this result it is possible to demonstrate the selective properties of the film:

- chlorine, due to its kinetic diameter (320 pm), is absorbed by the zeolite framework;

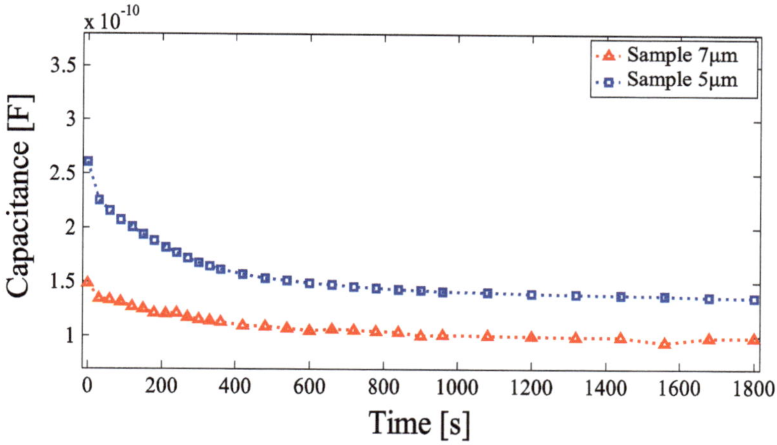

Fig. 12.5 Capacitance behavior in vacuum at 1 kHz

- during the exposure of zeolite to nitrogen, mainly due to its kinetic diameter (364 pm) the gas is not absorbed at all or is minimally absorbed by the zeolite layer, resulting in a constant capacitance as shown in Fig. 12.6.

Analysis spectra are shown below, performed both on the sample and the individual components in order to evaluate the contribution of each to the overall spectroscopic profile (Fig. 12.7).

Typical peaks of Si-O bond-stretching are present in the spectra of zeolite, glass and sample at around 1060 cm^{-1}.

The area between 950 and 1200 cm^{-1} is in fact representative of the stretching of the -O and Al-OH bonds (Si, Al). In the sample and zeolite 4A, the peak at around 2700 cm^{-1} is due to the water present in the sample; another characteristic peak of zeolite is present at around 550 cm^{-1} (not visible in this analysis), characterizing the vibration of the external tetrahedron. Generally, the samples are most affected by the influence of oil in the area between 2100 and 2600 cm^{-1}, and zeolite both from 3500 to 4000 cm^{-1} and between 800 and 1200 cm^{-1}. The reflectance value in the latter area of the spectrum also emerged as being most affected by the presence of glass as a substrate.

12.4 Conclusion

The absorption properties of a zeolite thin layer with respect to some toxic agents typical of CBRN threats were investigated. Absorption analyses were performed on zeolite-modified electrodes through capacitive measurements for the fabrication of a microelectronic, embedded, wearable system for CBRN detection. Based on the

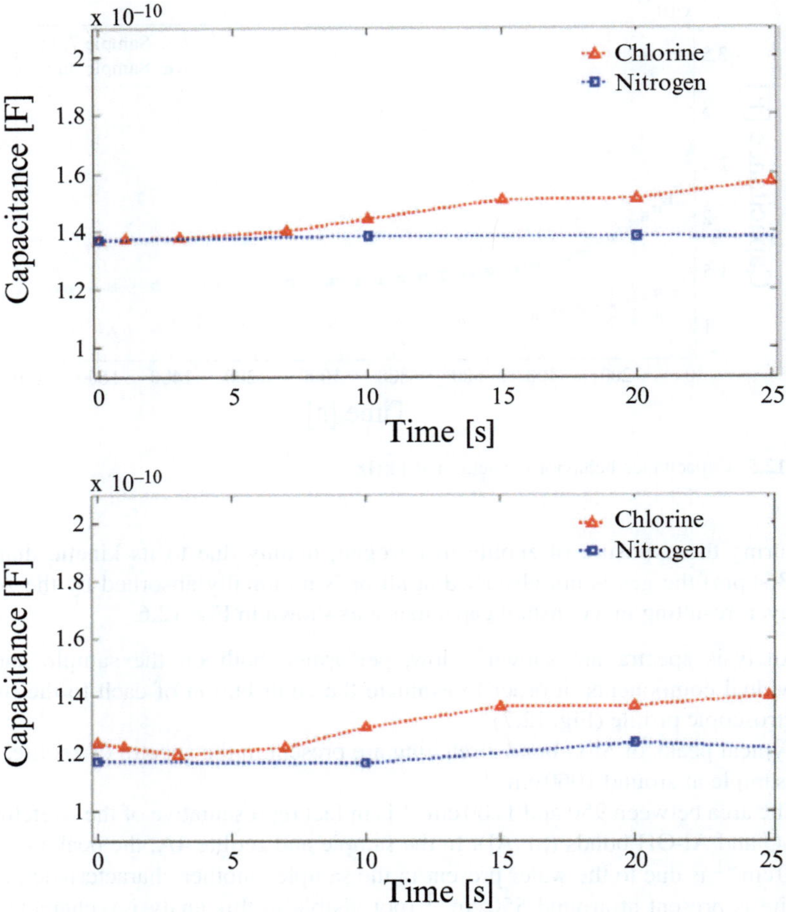

Fig. 12.6 Chlorine/Nitrogen adsorption at 1 kHz in 5 μm (upper) and 7 μm (lower) samples

capacitive measurements it is possible to demonstrate the selective properties of the film and in particular the fact that in the case of exposure to toxic agents with a kinetic radius smaller than the pores of zeolite, the agent is absorbed and causes an increase in the capacitance.

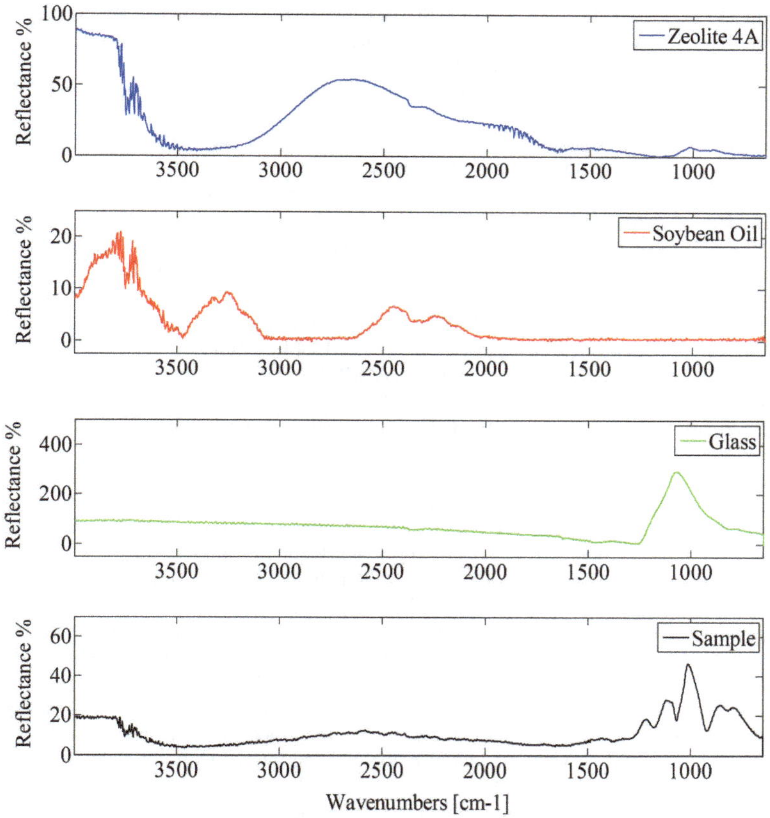

Fig. 12.7 FT-IR Analysis of the ZME electrode

References

1. Ghallab YH, El-Hamid HA, Ismail Y (2015) Lab on a chip based on CMOS technology: system architectures, microfluidic packaging, and challenges. IEEE Des Test 32(6):20–31
2. Li H, Liu X, Li L, Genov R, Mason AJ (2017) CMOS electrochemical instrumentation for biosensor microsystems: a review. Sensors (Basel) 17(1):74
3. Fiorillo AS, Pullano SA, Tiriolo R, Vinko JD (2016) Iono-electronic interface based on innovative low temperature zeolite coated NMOS (Circuits) for bio-nanosensor manufacture. In: Bonča J, Kruchinin S (eds) Nanomaterials for security. NATO science for peace and security series A: chemistry and biology. Springer, Dordrecht
4. Barrer RM (1982) Hydrothermal chemistry of zeolites. Academic, London, pp 152
5. Breck DW (1984) Zeolite molecular sieve. Malabar, Krieger
6. Rhodes CJ (2010) Properties and applications of zeolites. Sci Prog 93:223–284
7. Universal Oil Products (2010) An introduction to zeolite molecular sieves. UOP, Des Plaines. www.eltrex.pl/pdf/karty/adsorbenty/ENG-Introduction%20to%20Zeolite%20Molecular %20Sieves.pdf

8. Larin AV (2013) The Loewenstein rule: the increase in electron kinetic energy as the reason for instability of Al–O–Al linkage in aluminosilicate zeolites. Phys Chem Miner 40(10):771–780
9. Kim SA, Kamala-Kannan S, Lee KJ, Park YJ, Shea PJ, Lee WH, Kim HM, Oh BT (2013) Removal of Pb(II) from aqueous solution by a zeolite-nanoscale zero-valent iron composite. Chem Eng J 217:54–60
10. Fiorillo AS (2009) Deposition of layers of porous materials, layers thus obtained and devices containing them, EP2132768 A2, 16 Dec 2009
11. Fiorillo AS (2012) Deposition of zeolite thin layers onto silicon wafers for biomedical use. IEEE Trans Nanotechnol 11:654–656
12. El Roz M, Lakiss L, Valtchev V, Mintova S, Thibault-Starzyk F (2012) Cold plasma as environmentally benign approach for activation of zeolite nanocrystals. Microporous Mesoporous Mater 158:148–154
13. Wang Z, Mitra A, Wang H, Huang L, Yan Y (2001) Pure silica zeolite films as low-k dielectrics by spin-on of nanoparticle suspension. Adv Mater 13:1463–1466
14. Bockisch M (1998) Fats and oils handbook. AOCS Press, Champaign
15. Bottoni U, Tiriolo R, Pullano SA, Dastoli S, Amoruso GF, Nistico SP, Fiorillo AS (2016) Infrared saliva analysis of psoriatic and diabetic patients: similarities in protein components. IEEE Trans Biom Eng 63(2):379–384
16. Fiorillo AS, Tiriolo R, Pullano SA (2015) Absorption of urea into zeolite layer integrated with microelectronic circuits. IEEE Trans Nanotechnol 14(2):214–217
17. Larrubia MA, Ramis G, Busca G (2000) An FT-IR study of the absorption of urea and ammonia over V_2O_5-MoO_3-TiO_2 SCR catalysts. Appl Catal B Environ 27:145–151
18. Rodionov VE, Shnidko IN, Zolotovsky A, Kruchinin SP (2013) Electroluminescence of Y_2O_3:Eu and Y_2O_3:Sm films. Mater Sci 31:232–239
19. Ermakov V, Kruchinin S, Fujiwara A (2008) Electronic nanosensors based on nanotransistor with bistability behaviour. In: Bonca J, Kruchinin S (eds) Proceedings of the NATO ARW "Electron transport in nanosystems". Springer, pp 341–349
20. Vilaca N, Amorim R, Machado AF, Parpot P, Pereira MF, Sardo M, Rocha J, Fonseca AM, Neves IC, Baltazar F (2013) Potentiation of 5-fluorouracil encapsulated in zeolites as drug delivery systems for in vitro models of colorectal carcinoma. Colloids Surf B Biointerfaces 112:237–244
21. Bartel DP (2004) MicroRNAs: genomics, biogenesis, mechanism, and function. Cell 116(2):281–297
22. Creemers EE, Tijsen AJ, Pinto YM (2012) Circulating microRNAs: novel biomarkers and extracellular communicators in cardiovascular disease? Circ Res 110(3):483–495
23. Pescador N, Pérez-Barba M, Ibarra JM, Corbaón A, Martínez-Larrad MT, Serrano-Ríos M (2013) Serum circulating microRNA profiling for identification of potential type 2 diabetes and obesity biomarkers. PLoS One 8(10):e77251
24. Guay C, Regazzi R (2013) Circulating micro RNAs as novel biomarkers for diabetes mellitus. Nat Rev Endocrinol 9(9):513–521
25. Hashizume H, Theng BKG (2007) Adenine, adenosine, ribose and 5′-amp adsorption to allophane. Clay Clay Miner 55(6):599–605
26. Benetoli LOB, Santana H, Zaia CTBV, Zaia DAM (2008) Adsorption of nucleic acid bases on clays: an investigation using Langmuir and Freundlich isotherms and FT-IR spectroscopy. Monatshefte für Chemie 139(7):753–761
27. Negron-Mendoza A, Ramos-Bernal S, Gamboa de Buen I (2010) A thermoluminescence study of bio-organic compounds adsorbed in a clay mineral. IEEE Trans Nucl Sci 57(3):1223–1227
28. Cleaves II HJ, Jonsson CM, Jonsson CL, Sverjensky DA, Hazen RM (2010) Adsorption of nucleic acid components on rutile (TiO_2) surfaces. Astrobiology 10(3):311–323
29. Baú JPT, Carneiro CEA, de Souza IG Jr, de Souza CMD, da Costa ACS, di Mauro E, Zaia CTBV, Coronas J, Casado C, de Santana H, Zaia DAM (2012) Adsorption of adenine and thymine on zeolites: FT-IR and EPR spectroscopy and X-ray diffractometry and SEM studies. Orig Life Evol Biosph 42(1):19–29

Chapter 13
Nanosensor Devices for CBRN-Agents Detection: Theory and Design

Yu. Shunin, S. Bellucci, V. Gopeynko, T. Lobanova-Shunina, A. Kiv, D. Fink, A. Mansharipova, R. Mukhamediyev, and Yu. Zhukovskii

Abstract Pressing challenges of recent decades, associated with agents that are aggressive towards humans – substances and radiation of chemical, biological, radiological, and nuclear (CBRN) agents – require scientific and technological responses. These responses lie in the areas of agent detection and protection from them. The mentioned bio destructive agents can be divided into 2 groups: (1) chemical and biochemical, and (2) radiative (leading to chemical destruction of biomass). In this study, we consider models of universal track nanosensors that are capable of producing a correlated electrical response to the flow of active agents.

Y. Shunin (✉)
Institute of Solid State Physics of the University of Latvia, Riga, Latvia

Ventspils University College, Ventspils, Latvia
e-mail: yu_shunin@inbox.lv; jurjs.sunins@venta.lv

S. Bellucci
INFN-Laboratori Nazionali di Frascati, Frascati-Rome, Italy

V. Gopeynko
ISMA University, Riga, Latvia

Ventspils University College, Ventspils, Latvia

T. Lobanova-Shunina
Faculty of Mechanical Engineering, Transport and Aeronautics, Riga Technical University, Riga, Latvia

A. Kiv
Department of Materials Engineering, Ben-Gurion University of the Negev, Beer-Sheva, Israel

D. Fink
Departamento de Fisica, Universidad Autónoma Metropolitana-Iztapalapa, México, DF, México

A. Mansharipova · R. Mukhamediyev
Kazakh-Russian Medical University, Almaty, Kazakhstan

Y. Zhukovskii
Institute of Solid State Physics of the University of Latvia, Riga, Latvia

J. Bonča, S. Kruchinin (eds.), *Nanostructured Materials for the Detection of CBRN*, NATO Science for Peace and Security Series A: Chemistry and Biology,
https://doi.org/10.1007/978-94-024-1304-5_13

Keywords CBRN-agents · Agents detection · Nanotrack based bionanosensor · Nanotrack based devices for CBRN-agents detection · Enzymatic detection tools

13.1 Introduction: CBRN-Agents Nature

Contemporary extremist groups have a wide variety of potential agents and delivery means to choose from *chemical, biological, radiological, or nuclear* (CBRN) attacks. Identification, protection and decontamination are the main scientific and technological responses for the modern challenges of CBRN events [1].

Decontamination, in general, is defined as the removal of hazardous materials from the areas where it is not wanted. Decontamination is utilized to reduce the dose that humans or biological organisms may receive from a component or surface, to reduce the potential for airborne CBRN-agents, or to reduce the disposal cost associated with components or the materials [1].

An expanded abbreviation can be used for destructive agents, particularly, the additional symbol 'E' in the acronym CBRNE means '**E**xplosive' for **Chemical, Biological, Radiological, Nuclear, and Explosive** materials (so called 'SEE-burn') which are considered very dangerous for people, animals, and the environment. Usually CBRNE materials are used for weapons of mass destruction. However, the focus of our attention is on CBRN-agents, which can be detected, controlled, and managed. Typical CBRN-agents are presented in Table 13.1.

The second group of CBNR-agents presented in Table 13.2 characterizes the influences and effects of aggressive materials included in particular warfare devices and tools. For example, we should mention Radiological Dispersion Devices, which can be of any size, determined only by the amount of radioactive material and explosives. A variety of radioactive materials are commonly available and could be used in any RDD, including ^{137}Cs, ^{90}Sr, ^{60}Co. Moreover, these agents can also be constituent parts of the devices and tools used in hospitals, universities, factories, construction companies, and laboratories, being possible sources for such radioactive materials. A passive RDD is a system in which unshielded radioactive material is dispersed or placed manually at the target.

An explosive variant of RDD (*a dirty bomb*) is any system that uses the explosive force of detonation to disperse radioactive material. An atmospheric RDD is any system in which radioactive material is converted into a form that is easily transported by air currents.

Another type of aggressive action is the radiation, which outruns any direct contact with an aggressive material. Subsequent destructive biochemical reactions are additional destructive post-factors that can be identified biochemically.

Radiations induces the radiation sickness, resulting from the effects of various types of ionizing radiation and characterized by symptoms, depending on the type of damage, its dose, the location of the radiation source, the distribution of the dose in time and the body of a living being (for example, a person). The most important types of ionizing radiation: X-rays, γ-rays, $\beta^{+,-}$-particles, α-particles, n-neutrons, p-protons (see also Table 13.2).

Table 13.1 Typical chemical and biochemical CBRN-agents

Agent groups and types	Identification features
Chemical	
Cyanides	
Sodium or potassium cyanides	Sodium or potassium cyanides are white-to-pale yellow salts that can be easily used to poison food or drinks
Hydrogen cyanide and cyanogen chloride	Hydrogen cyanide (HCN) and cyanogen chloride (ClCN) are colorless-to-pale yellow liquids that will turn into a gas near room temperature. HCN has a characteristic odor of bitter almonds, and ClCN has an acrid choking odor and causes burning pain in the victim's eyes
Mustard agent	Mustard is a blister agent that poses a contact and vapor hazard. Its color ranges from clear to dark brown depending on purity, and it has a characteristic garliclike odor. Mustard is a viscous liquid at room temperature
Nerve agents	Sarin, tabun, and VX are highly toxic military agents that disrupt a victim's nervous system by blocking the transmission of nerve signals
Toxic industrial chemicals	Chlorine and phosgene are industrial chemicals that are transported in multiton shipments by road and rail. The effects of chlorine and phosgene are similar to those of mustard agent. Organophosphate pesticides such as parathion are in the same chemical class as nerve agents
Biochemical	
Anthrax	*Bacillus anthracis*, the bacterium that causes anthrax, is capable of causing mass casualties. Symptoms usually appear within one to six days after exposure and include fever, malaise, fatigue, and shortness of breath
Botulinum toxin	*Botulinum toxin* is produced by the bacterium *Clostridium botulinum*, which occurs naturally in the soil. Crude but viable methods to produce small quantities of this lethal toxin have been found in terrorist training manuals
Ricin	*Ricin* is a plant toxin that is 30 times more potent than the nerve agent VX by weight and is readily obtainable by extraction from common castor beans
Viruses biological warfare (BW)	*Germ warfare* is the use of biological toxins or infectious agents such as bacteria, viruses, and fungi with the intent to kill or incapacitate humans, animals or plants as an act of war. *Biological weapons* ('bio-agents') are living organisms or replicating entities (viruses, which are not universally considered "alive") that reproduce or replicate within their host victims. *Entomological (insect) warfare* is also considered a type of biological weapon. This type of warfare is distinct from nuclear warfare and chemical warfare, which together with biological warfare make up NBC, the military acronym for nuclear, biological, and chemical warfare using weapons of mass destruction (WMDs)

Different types of ionizing radiation have a different destructive effect and different ways of influencing biological tissues. Accordingly, the same biological dose of radiation corresponds to the same absorbed dose. Therefore, to describe the effect of radiation on living organisms, the concept relative biological effectiveness of RBE of radiation is introduced [2].

Table 13.2 Typical radiological and nuclear CBRN-agents

Agent groups and types	Identification features
Radiological and nuclear device	An RDD (Radiological Dispersion Device) is a conventional bomb not a yield producing nuclear device. RDDs are designed to disperse radioactive material to cause destruction, contamination, and injury from the radiation produced by the material (^{137}Cs, ^{90}Sr, ^{60}Co)
Improvised Nuclear Device (IND)	An IND is intended to cause a yield-producing nuclear explosion. An IND could consist of diverted nuclear weapon components, a modified nuclear weapon, or indigenous-designed device
Radiological	
^{131}I	β-, γ decays, *half-life*-appr. 8 days. β-decay mutations and cell death, as well as – surrounding tissues to a depth of several millimeters. It concentrates mainly in the thyroid gland
^{90}Sr	*half-life* – appr. 28.8 years. ^{90}Sr falls into the environment mainly with emissions from nuclear power plants and nuclear explosions. It is extremely dangerous. It is deposited, mainly, in bone tissues
^{137}Cs	*half-life* – 30 years. One of the main components of radioactive contamination of the biosphere. The release of ^{137}Cs into the environment occurs mainly as a result of accidents at nuclear power plants and nuclear weapons tests
^{60}Co	*half-life* appr. 5.3 years
^{241}Am	*half-life* appr. 433 years
Note	The contribution of these radioactive components at *Chernobyl event 1986, April* was (appr.): ^{131}I – 1.8×10^{18} Bq, ^{137}Cs – 8.5×10^{16} Bq, ^{90}Sr – 1×10^{16} Bq. The total activity of substances released into the environment was, according to various estimates, up to 14×10^{18} Bq
Nuclear atomic warfare, thermonuclear warfare	
α-particle	Basic radiations:
β-particle	${}^{241}_{95}\mathrm{Am} \rightarrow {}^{237}_{93}\mathrm{Np} + {}^{4}_{2}\alpha$, ${}^{90}_{38}\mathrm{Sr} \rightarrow {}^{90}_{39}\mathrm{Y} + \beta$, γ-rays, X-rays – Nuclear atomic warfare,
γ-rays	X-rays – thermonuclear warfare hydrogen bomb,
n-neutrons	n-neutrons – *Neutron bomb*,
	p-ptotons – the sources of intense proton radiation are accelerators of charged particles (till 10 GeV). Even greater energy are found in outer space. Proton radiation is the main component of galactic and solar cosmic radiation

RBE of ionizing radiation is a dimensionless coefficient characterizing the efficiency biological effect of various ionizing radiations, is defined as the ratio of the dose of some of the reference radiation D_0 to the dose of this radiation D_x: RBE $= D_0/D_x$ [2]. As a quality of standard take X-rays with a certain spectrum, D_0 and D_x.

RBE is measured by Quality factor (Q). Coefficient (Q) to account for the biological effectiveness of different types of ionizing radiation in determining the equivalent dose. To obtain an equivalent dose, the absorbed dose of the radiation in question must be multiplied by the quality factor. For X-ray, beta and gamma radiation, the coefficient $Q = 1$, proton and neutron radiation (fast neutrons) $Q = 10$, alpha radiation $Q = 20$.

For metrology and control of aggressive and biologically active agents it is essential to observe the evolution of the decontamination level. The effectiveness of the decontamination can be expressed as the decontamination factor (*DF*). It is the ratio of the contamination level of a material before decontamination to the contamination level of a material after decontamination:

$$DF = LB/LA,$$

where *LB* is the contamination *L*evel of a material or *C*omponent *B*efore the decontamination application, *LA* is the contamination *L*evel measured immediately *A*fter decontamination application.

A decontamination process that removes material will result in a *DF* greater than 1. The percentage of contamination removed from the surface can be given by the Percent contamination removed = $(1 - 1/DF) \times 100$; If $DF = 10$, the percent contamination removed = 90% If $DF = 100$, the percent contamination removed = 99% [1].

13.2 Bionanosensors: Polymer Nanoporous Model Structures

Since the 1960s, it has been known that energetic (with tens of MeV or more) heavy (with atomic masses being usually larger than that of Ar) ion irradiation ('swift heavy ions', SHI) introduces very narrow (~some nm) but long (typically 10–100 μm) parallel trails of damage in irradiated polymer foils, the so-called latent ion tracks. The damage shows up primarily by the formation of radiochemical reaction products. Whereas the smaller ones readily escape from the irradiated zone, thus leaving behind themselves nanoscopic voids, the larger ones tend to aggregate towards carbonaceous clusters. Thus, emerging structural disorder along the tracks modifies their electronic behavior.

The newly created intrinsic free volume enables electrolytes to penetrate into the polymer, thus forming parallel liquid nanowires. In case of tracks penetration through all the foil, the conducting connections emerge between the front and back sides of the foil. The ion track technology is particularly intended to biosensing applications. In this case, the ion tracks are functionalized directly by attaching organic or bioactive compounds (such as enzymes) to their walls.

Consider the scheme of the glucose detection (see Fig. 13.1) and also in [3].

The description of the sensing reaction of glucose with the enzyme GO_x looks as follows:

(a) the overall net reaction is: Glucose ($C_6H_{12}O_6$) + O_2 (due to enzyme-induced oxidation) gluconic acid ($C_6H_{12}O_7$) + O;

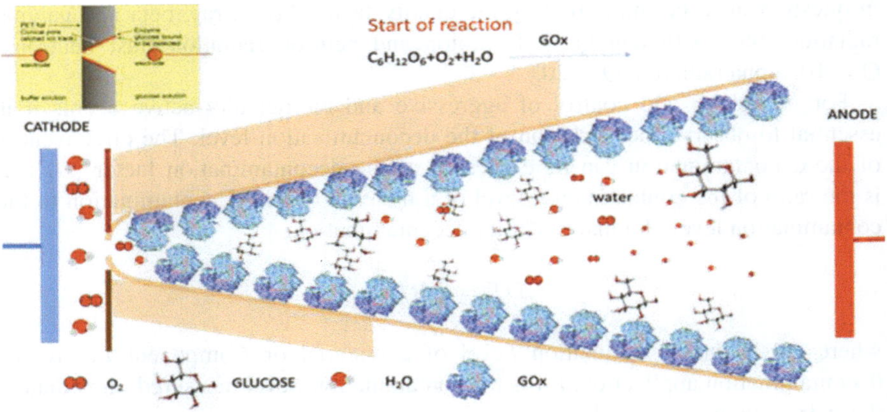

Fig. 13.1 General scheme describing the detection scheme and modified polymer. Principle arrangement of experimental setup to study voltage-current dependences in ion track-containing foils embedded in electrolytes

(b) this remaining O attaches to some H_2O to form peroxide H_2O_2;
(c) the product: gluconic acid dissociates around pH = 7. Thus, the conductivity of the liquid changes (essentially if the product is enriched in the track's confinement); this is what is measured by the sensor [3, 4].

In particular, a complicated biochemical kinetics of basic reaction of glucose detection depends on track qualities (e.g., track creation mechanism, foil material properties), enzyme (GO_x) distribution on the track surface, geometry of the etched track etc. All these factors are the subject for the nearest special research. Moreover, the detailed kinetics of reaction is the object of 3D-modelling to design the optimal geometry of nanosensor active space. This allows creating optimal nanosensors with the increased efficiency.

Experimental and theoretical calibration dependences demonstrate similar trends. The proposed device can serve to detect physiologically relevant glucose concentrations. The catalytic sensor can be made re-usable due to the formation of diffusible products from the oxidative biomolecular recognition event. Moreover, we can develop a multi-agent packet nanosensor, suitable for application as a human breathing analyzer in relation to cancer detection, hepatitis, and so on [4–10].

The main detection problems of various agents in the proposed schemes of nanotrack devices are connected with the exploration and creation of effective chemical reactions capable of producing stable positive and negative ion fluxes within the working zone (as an electrolyte, Fig. 13.1). These ion fluxes should be strongly correlated with the tested agent concentrations [3–8]. The detection of glucose concentration in the human blood on the polymer nanotrack based technology has well-tested and stable experimental results [11, 12].

The recent advancements in the field of nanosensor design allow monitoring and tracking biomolecules in such areas as the environment, food quality and healthcare.

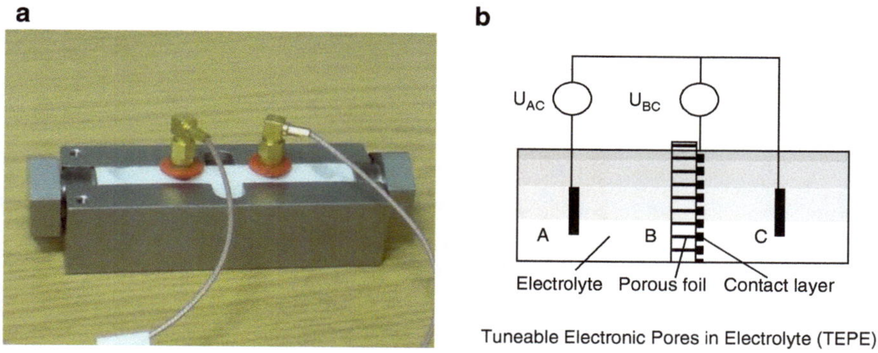

Fig. 13.2 Favourable platform for measuring enrichment of chemical reaction products via electrical current transmitted through the nanopores: (**a**) nanosensor system prototype [8]; (**b**) principal electrical scheme of the sensor current [8]

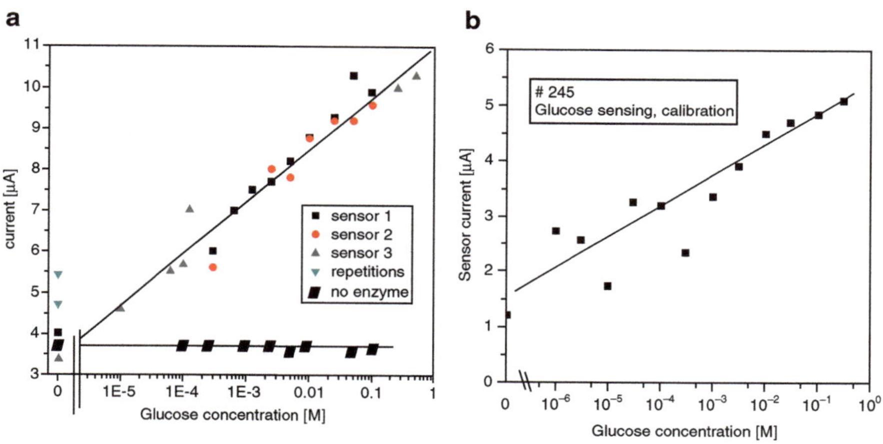

Fig. 13.3 (**a**) Experimental calibration curves on 2007 year glucose testing series [3], Experimental calibration dependence performance comparison of three identically produced track-based glucose detectors against a calibration curve I (+5 V) vs. glucose concentration; (**b**) Similar experimental calibration curve on 2016 year glucose testing series [12]

The presently developed ion track-based nanosensors (see Fig. 13.2) provide high sensitivity, reliable calibration (see Fig. 13.3), small power and low cost.

Using simulation of chemical kinetics for glucose oxidation with glucose oxidase (see Fig. 13.4), we have obtained theoretical calibration dependences, when the concentration of H^+ is proportional to the concentration of the detected glucose.

Simulation schemes are directed on concentrations of H^+ which are equal to glucose concentrations. In cases of stable glucose inputs we can write typical simulation equations which leads to the saturated values of H^+ ions concentrations (see Fig. 13.5):

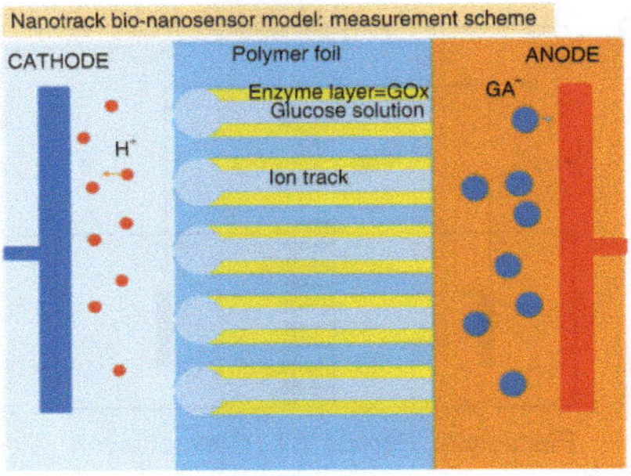

Basic chemical reactions of Glucose detection: and simulation equations are:

$$C_6H_{12}O_6 + O_2 \xrightarrow{GOx} C_6H_{12}O_7 + O$$

$$O + H_2O \rightarrow H_2O_2, H_2O_2 \xrightarrow{catalaza} \frac{1}{2}O_2 + H_2O$$

$$C_6H_{12}O_7 \rightarrow C_6H_{12}O_7^- + H^+$$

$$Anode: H_2O \rightarrow \frac{1}{2}O_2 + 2H^+ + 2e^-$$

$$Cathode: 2H^+ + 2e^- \rightarrow H_2$$

$$Ultimate\ \ reaction: C_6H_{12}O_6 + H_2O \rightarrow C_6H_{12}O_7 + H_2$$

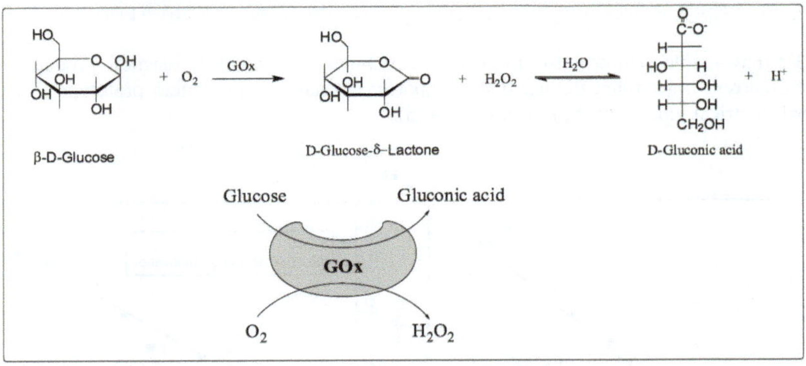

Fig. 13.4 General model of glucose detection process on the ion track-containing foils embedded in electrolyte and basic set of biochemical reaction [3, 5, 8]

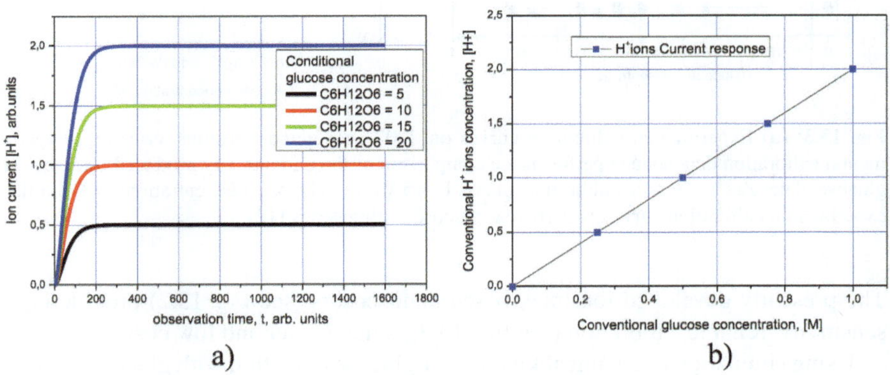

Fig. 13.5 (**a**) Simulation of H^+ ion current via observation time in case of saturation; (**b**) Theoretical model of typical calibration dependence based on chemical kinetics results: simulation of induced H^+ ion current via glucose concentration

$$\begin{cases} \dfrac{d}{dt}[C_6H_{12}O_6] = -w_1, \\ \dfrac{d}{dt}[H_2O_2] = w_1 - w_2, \\ \dfrac{d}{dt}[O_2] = -w_1 + w_2, \\ \dfrac{d}{dt}[H^+] = 2w_1, \end{cases} \tag{13.1}$$

where $w_1 = k_1[C_6H_{12}O_6][O_2][GO_x]$, $w_2 = k_2[H_2O_2]$ are the reactions rates, k_1 and k_2 are simulation parameters.

We can point out that theoretical simulation results demonstrate similar linear trend in comparison to experimental data (see, Figs. 13.3 and 13.5b).

Taking into account the probable process of H^+ ions recombination we can modify simulation equations (13.1) as:

$$\begin{cases} \dfrac{d}{dt}[C_6H_{12}O_6] = -w_1, \\ \dfrac{d}{dt}[H_2O_2] = w_1 - w_2, \\ \dfrac{d}{dt}[O_2] = -w_1 + w_2, \\ \dfrac{d}{dt}[H^+] = 2w_2 - 2w_3, \end{cases} \tag{13.2}$$

where $w_1 = k_1[C_6H_{12}O_6][O_2][GO_x]$, $w_2 = k_2[H_2O_2]$, $w_3 = k_3[2H^+][2e^-]$ are the reactions rates, and k_1, k_2, k_3 are again simulation parameters. Corresponding simulation results are presented in Fig. 13.6.

13.3 Novel Bio-agents Devices: Technological Solutions

The main idea of the modification of the considered bionanosensor model requires finding effective detection reactions of the controlled agents that will provide a stable and correlated electrical response. An equally fundamental problem is the creation of a multi-packet nanosensor tuned to a set of agents. The market presents various technological solutions for the denoted problems. Consider some of them.

13.3.1 The Breath Analyzer Mint

The company Breathometer (USA) began selling a mobile oral health analyser. The device is called Mint. However, this devise has very limited applications. Mint is a compact device with built-in sensors to analyse the condition of the mouth. A person

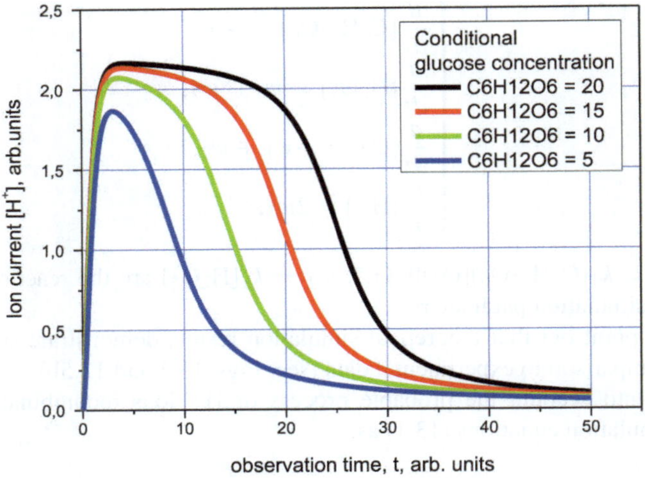

Fig. 13.6 Simulation of H^+ ion current via observation time in case of H^+ ions recombination

needs to keep the mouth closed for about 30 s and then put the mouthpiece of the device into the mouth. The gadget sucks in the air, passes it through special analysers and produces a result: from 'A' – 'everything is fine' up to 'F' – 'it's time to see a doctor'. Mint sensors detect volatile sulphur compounds. They are the cause of bad breath from the mouth when, for example, periodontal disease occurs. The manufacturer notes that the device is designed for a primary diagnosis, not for the treatment. All data appears in the company application, which you need to install on your smartphone. Mint is able to detect very approximately only sulphur components and it is doubtful that this metrological scheme can be expanded for other chemical and biochemical groups of agents.

13.3.2 AI-Powered Breath Detector

Technion-Israel Institute of Technology has developed a smart breath sensor based on artificial intelligence (AI) abilities. The developed sensor [13] detects many different molecules and finds correlations of the tested biomarkers to 17 different diseases.

Before modern methods of medical laboratory research became available, doctors often diagnosed certain diseases, sniffing the patient's exhaled air. Scientists for many years worked on the creation of analytical tools that could simulate this ability to detect disease by smell. Recently, the researchers reported in the journal ACS Nano [13] that they identified the presence of a unique breathing spectrum of the 'breath print' = breath spectrum' for each disease. Using this information, they created a device analysing samples of exhaled air to classify and diagnose several different types of disease.

The researchers developed a *grating* made up of nanoscale sensors, which represents an array of specially prepared micro-sensors from gold nanoparticles, each of which contains a randomly organized network of carbon nanotubes. Based on the analysis of the results obtained using artificial intelligence methods, scientists were able to use this lattice to classify and diagnose various diseases. They used mass spectrometry to identify those respiratory components that are associated with diseases.

The exhaled air contains *nitrogen*, *carbon dioxide* and *oxygen*, as well as small amounts of more than 100 other chemicals. Researchers managed to find out that each disease has a unique volatile "breathprint", based on different concentrations of 13 components. The experiments also showed that the presence of one disease will not interfere with the detection of the other, which was the prerequisite for developing a practical device for detecting and diagnosing various diseases, and non-invasive, inexpensive and portable. The breath analyser diagnoses the disease by the exit. Already, the respiratory analyser is able to diagnose 17 diseases in just one exhalation.

In recent years, a number of specific tests have been developed to determine hypertension, tuberculosis, certain forms of cancer and other diseases by breathing the patient.

The device contains an array of specially prepared micro-sensors from gold nanoparticles, each of which contains a randomly organized network of carbon nanotubes. The device can be used to collect breath samples of numerous patients with various diseases, and to correlate the information obtained with certain diseases using software based on artificial intelligence, which revealed patterns and deviations in the concentration of various gases. As a result, researchers were able to accurately determine the disease in 90% of cases. As the test studies showed, the average accuracy of diagnosing this system was not less than 86% for 17 diseases, including several cancers, kidney failure, Crohn's disease, multiple sclerosis, two types of Parkinson's disease etc.

13.3.3 Laser Spectroscopic Techniques of Breath Analysis

The analysis of human breath biomarkers is a way for preliminary diagnostics of a set of diseases. Breath analysis, as a promising new field of medicine and medical instrumentation, potentially offers non-invasive, real-time, and point-of-care (POC) disease diagnostics and metabolic status monitoring. Recent advances in laser spectroscopic techniques and laser sources have driven breath analysis to new heights, moving from laboratory research to commercial reality. Unfortunately, laser spectroscopic techniques cannot be considered as a real-time technique [14].

There are 35 major breath biomarkers which were tested by the Laser Spectroscopic Techniques. However, only 14 biomarkers (*Spectral fingerprints: Acetaldehyde-1, Acetone-1, Ammonia-6, Carbon dioxide and C-isotope-4*) have been detected in actual human breath. The main problem that breath probes should be preliminary prepared for further spectroscopic measurements which are very far from direct bio-agent influences detection. The interpretation of optical spectra is not so easy and evident.

13.3.4 Detection of Sarin and Other Toxic Synthetic Organophosphorus Compounds

The most dangerous in an emergency and terrorist relationships are organophosphorus poisoning substances – sarin, soman, substance VX, related to nerve agents [15]. The extremely high toxicity of synthetic organophosphorus compounds, which exceeds many times the known poisonous substances, the rapid development of poisoning (within a few minutes) require the creation of devices capable of accurately determining the chemical nature of a hazardous substance within a short time interval (5–30 s). The solution of this chemical-analytical problem is possible with the use of a special high-speed equipment that can work stably at control sites (in the field).

When choosing the most suitable detection method, the following criteria are required:

(1) the ability of the method to provide a direct and specific measurement of the analytic signal of the compound being determined; sensitivity;
(2) the operating range of concentrations;
(3) the detection limit;
(4) information;
(5) the influence of interfering components and factors;
(6) the possibility of automation.

There are some basic groups of detection methods: chromatography, mass spectrometry IR spectroscopy and enzymatic tools.

Enzymatic methods. Enzymatic method is more preferable from the point of view of interpretation of direct correlations ‘agent-electric response’.

The use of enzymatic reactions is a particular case of kinetic analysis methods based on measuring the rate of the indicator catalytic reaction in the presence of various amounts of detectable substances. Only about 20 enzymes are of practical interest. They have high specificity and provide low detection limits. Among them, cholinesterases, related to hydrolase enzymes, are widely used [16].

Sarin detection. Let us consider some details of the detection of sarin as an example of a very agrressive and dangerous chemical agent.

Sarin ($C_4H_{10}FO_2P$) is an organophosphorous toxic agent, isopropyl ether of methylphosphonic acid fluoride, a liquid without colour and odour. At the room temperature, sarin is a colourless liquid with a faint smell of flowering apple trees. Miscible with water and organic solvents in all respects. The relatively high pressure of its vapour causes it to evaporate rapidly (about 36 times faster than, e.g., tabun). In the gaseous state, sarin is also colourless and odourless.

Sarin molecule is chiral because it contains four different chemical substituents combined by a phosphorus atom into a tetrahedron. Sarin is biologically active and has a great affinity for the enzyme acetylcholinesterase.

Note. *Acetylcholinesterase* (acetylcholine acetylhydrolase), enzyme class hydrolase, catalyzing hydrolyzacetylcholine, as well as other ethers of choline:

$$[CH_3COOCH_2CH_2N^+(CH_3)_3]OH^- + H_2O \rightarrow$$

$$CH_3COOH + [HOCH_2CH_2N^+(CH_3)_3]OH^-.$$

Optimal catalytic activity of acetylcholinesterase is observed at pH = 7.5–8.5. Sarin inhibits the enzyme acetylcholinesterase by forming a covalent compound with that part of the enzyme where acetylcholine undergoes hydrolysis. The general scheme of enzymatic sarin detection using the developed nanotrack sensor technique is presented in Fig. 13.7.

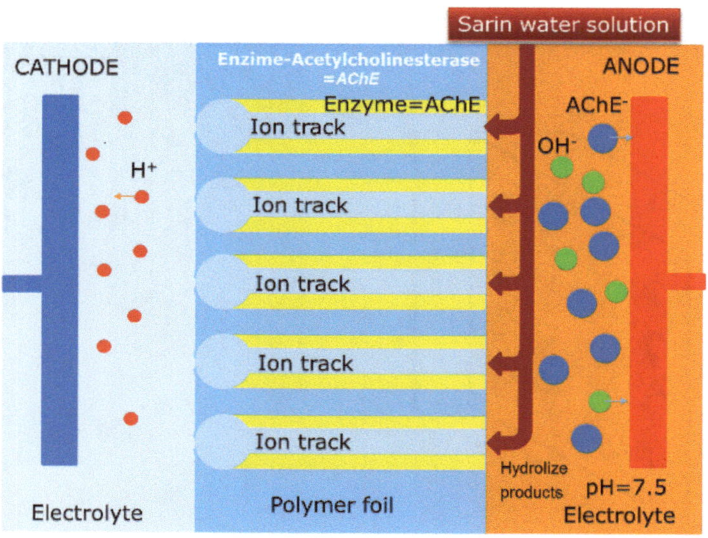

Fig. 13.7 General model of developed sarin detection process on the ion track-containing foils embedded in electrolyte, following the proposed measurements strategy (see Fig. 13.4)

13.4 Conclusion. Multiagent Nanotrack Based Nanosensor Model

The creation of novel biosensors and their further improvements require a careful study of the mechanisms of electrolytes passing through the tracks [17, 18]. Experimental and theoretical calibration dependences demonstrate similar trends. The proposed device can serve to detect physiologically relevant glucose concentrations. The catalytic sensor can be made re-usable due to the formation of diffusible products from the oxidative biomolecular recognition event. Moreover, we can develop a multi-agent packet nanosensor, which can be used as a human breathing analyzer in relation to cancer detection, hepatitis, and many other diseases (Fig. 13.8) [19].

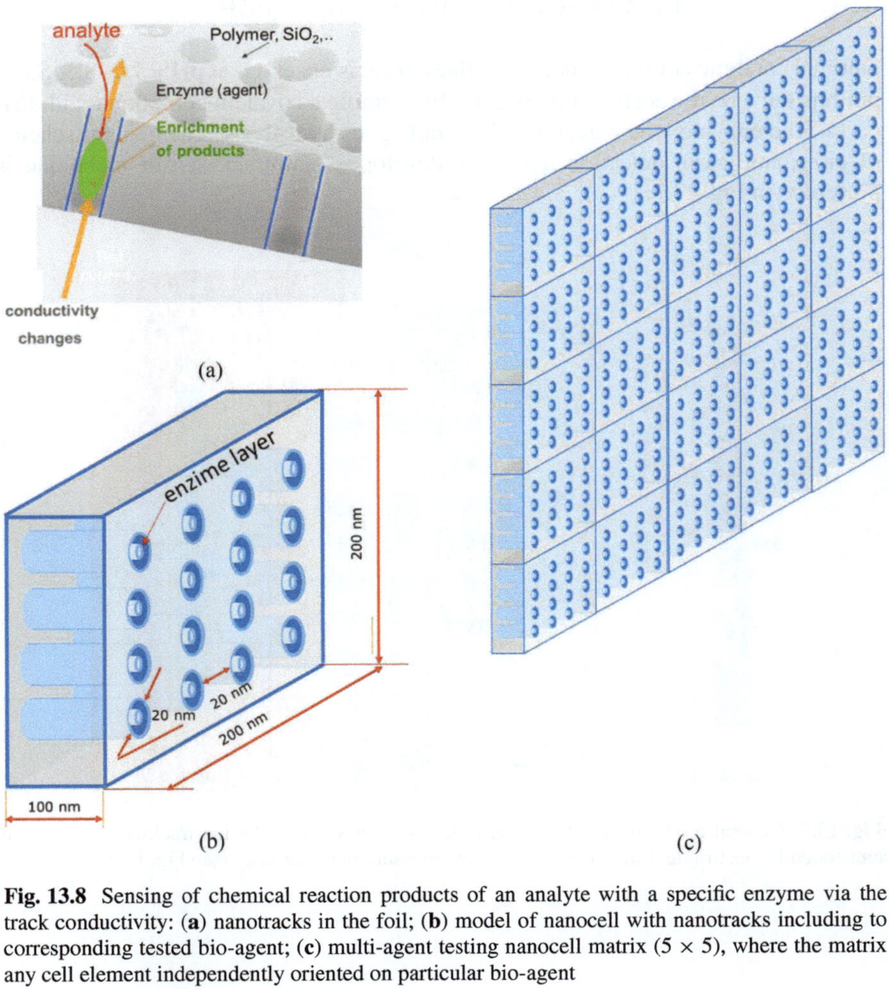

Fig. 13.8 Sensing of chemical reaction products of an analyte with a specific enzyme via the track conductivity: (**a**) nanotracks in the foil; (**b**) model of nanocell with nanotracks including to corresponding tested bio-agent; (**c**) multi-agent testing nanocell matrix (5 × 5), where the matrix any cell element independently oriented on particular bio-agent

Every nanocell (Figs. 13.2 and 13.4) has an independent electrical metrology scheme. Moreover, sensitive cells for particular agents can be distributed along the sensitive surface of multi-agents flux.

All electrical responses corresponding to a particular agent accumulated digitally as calibrated medical data on the special integrated screen. In this way, we can instantly obtain the information about various agent concentrations in real-time. This multi-agent detector can be used as a device for the analysis of the human breath supplying a spectrum of medical information in a non-invasive way. This is very important for the evaluation of post effects of aggressive agents (e.g. CBRN).

Acknowledgements This research has been partially supported by grant 'Nanostructures for bacteria detection and study' (NANOBAC) (01.10.2015-31.12.2017) the Ministry of Education and Science of the Republic of Kazakhstan (2015–2017), the Research Division of Ventspils University College (Latvia), Ben-Gurion University (Israel).

References

1. Kumar V, Goel R, Chawla R, Silambarasan M, Sharma RK et al (2010) Chemical, biological, radiological, and nuclear decontamination: recent trends and future perspective. J Pharm Bioallied Sci 2(3):220
2. Sinclair DW (2003) Relative biological effectiveness (RBE), quality factor (Q) and radiation weighting factor (Wr). Ann ICRP 33(4):92
3. Fink D, Klinkovich I, Bukelman O, Marks RS, Kiv A, Fuks D, Fahrner WR, Alfonta L (2009) Glucose determination using a re-usable enzyme-modified ion track membrane sensor. Biosens Bioelectron 24:2702–2706
4. Wang J (2008) Electrochemical glucose biosensors. Chem Rev 108:814
5. Fink D, Kiv A, Shunin Y, Mykytenko N, Lobanova-Shunina T, Mansharipova A, Koycheva T, Muhamediyev R, Gopeyenko V, Burlutskaya N, Zhukovskii Y, Bellucci S (2015) The nature of oscillations of ion currents in the ion track electronics. Comput Model New Technol 19(6):7
6. Fink D, Muñoz HG, Alfonta L, Mandabi Y, Dias JF, de Souza CT, Bacakova LE, Vacík J, Hnatowicz V, Kiv AE, Fuks D, Papaleo RM (2012) Status and perspectives of ion track electronics for advanced biosensing. In: Shunin Yu, and Kiv A (eds) Nanodevices and nanomaterials for ecological security. Series: NATO science for peace series B – physics and biophysics. Springer, Heidelberg, p 269
7. Shunin Yu, Alfonta L, Fink D, Kiv A, Mansharipova A, Muhamediyev R, Zhukovskii Yu, Lobanova-Shunina T, Burlutskaya N, Gopeyenko V, Bellucci S (2016) Modelling and simulation of electric response of nanocarbon nanocomposites and nanoporous polymer based structures for nanosensor devices. In: Theses of the 14th international scientific conference information technologies and management, 14–15 Apr 2016. ISMA University Riga Latvia, p 11. http://isma.lv/FILES/SCIENCE/IT&M2016_THESES/NN/01_IT&M2016_Shunin.pdf
8. Shunin Yu, Fink D, Kiv A, Alfonta L, Mansharipova A, Muhamediyev R, Zhukovskii Yu, Lobanova-Shunina T, Burlutskaya N, Gopeyenko V, Bellucci S (2016) Theory and modelling of physical and bio-nanosensor systems. In: Proceedings of the 5th international workshop nanocarbon photonics and optoelectronics, 1–6 Aug 2016. Lappeeranta, p 101. http://www.npo.fi/documents/409792/994899/Yury_Shunin.pdf/08431866-6f2c-4624-a4c9-9614dea5f747
9. Ermakov V, Kruchinin S, Fujiwara A (2008) Electronic nanosensors based on nanotransistor with bistability behaviour. In: Bonca J and Kruchinin S (eds) Proceedings of NATO ARW "Electron transport in nanosystems". Springer, Dordrecht, pp 341–349

10. Ermakov V, Kruchinin S, Hori H, Fujiwara A (2007) Phenomena of strong electron correlastion in the resonant tunneling. Int J Mod Phys B 11:827–835
11. Fink D, Muñoz HG, Alfonta L, Mandabi Y, Dias JF, de Souza CT, Bacakova LE, Vacík J, Hnatowicz V, Kiv AE, Fuks D, Papaleo RM (2012) Status and perspectives of ion track electronics for advanced biosensing. In: Shunin Yu, Kiv A (eds) Nanodevices and nanomaterials for ecological security. Series: NATO science for peace series B – physics and biophysics. Springer, Heidelberg, p 269
12. Fink D, Vacík J, Hnatowicz V, Muñoz HG, Garcia-Arrellano H, Alfonta L, Kiv A (2017) Diffusion kinetics of the Glucose/Glucose Oxidase system in swift heavy ion track-based biosensors. Nucl Instr Meth B 398:21
13. Nakhleh MK, Amal H, Jeries R et al (2017) Diagnosis and classification of 17 diseases from 1404 subjects via pattern analysis of exhaled molecules. ACS Nano 11:112
14. Wang Ch, Sahay P (2009) Breath analysis using laser spectroscopic techniques: breath biomarkers, spectral fingerprints, and detection limits. Sensors 9:8230
15. Types of Chemical Weapons. FAS-Federation of American Scientists. http://fas.org/cw/documents/cwagents.pdf
16. Aziz A, Palleschi G (2004) Phoshate, nitrate, and sulfate biosensors. Anal Lett 37(1):1
17. Shunin Yu, Kiv A (eds) (2012) Nanodevices and nanomaterials for ecological security. Springer, Heidelberg
18. Fink D (eds) (2004) Fundamentals of ion-irradiated polymers, vol 63. Springer, Berlin
19. Shunin Yu, Bellucci S, Gruodis A, Lobanova-Shunina T (2017) Nonregular nanosystems. theory and applications. Springer, Heidelberg. in print

Chapter 14
Improving the Design of Ion Track-Based Biosensors

D. Fink, A. Kiv, L. Alfonta, H. García-Arrellano, H. G. Muñoz, J. Vacik, V. Hnatowicz, Yu. Shunin, Yu. Bondaruk, A. Mansharipova, and R. Mukhamediyev

Abstract In the last decade we had developed new types of biosensors, by cladding the inner walls of transparent etched swift heavy ion tracks in thin polymer foils with enzymes. The enzymatic reaction products of appropriate analytes penetrating

---

D. Fink (✉)
Departamento de Fisica, Universidad Autónoma Metropolitana-Iztapalapa, México, DF, México

Nuclear Physics Institute, Řež, Czech Republic

A. Kiv
Department of Materials Engineering, Ben-Gurion University of the Negev, Beer-Sheva, Israel

South-Ukrainian National Pedagogical University after K.D. Ushinskij, Odessa, Ukraine

L. Alfonta
Avram and Stella Goldstein-Goren Department of Biotechnology Engineering, Ben-Gurion University of the Negev, Beer-Sheva, Israel

H. García-Arrellano
Departamento de Ciencias Ambientales, Division de Ciencias Biológicas y de la Salud, Universidad Autónoma Metropolitana-Lerma, Lerma de Villada, Municipio de Lerma, Estado de México, México

H. G. Muñoz
Departamento de Fisica, Universidad Autónoma Metropolitana-Iztapalapa, México, DF, México

J. Vacik · V. Hnatowicz
Nuclear Physics Institute, Řež, Czech Republic

Yu. Shunin
Institute of Solid State Physics of the University of Latvia, Riga, Latvia

Ventspils University College, Ventspils, Latvia
e-mail: yu_shunin@inbox.lv; jurjs.sunins@venta.lv

Yu. Bondaruk
South-Ukrainian National Pedagogical University after K.D. Ushinskij, Odessa, Ukraine

A. Mansharipova · R. Mukhamediyev
Kazakh-Russian Medical University, Almaty, Kazakhstan

J. Bonča, S. Kruchinin (eds.), *Nanostructured Materials for the Detection of CBRN*, NATO Science for Peace and Security Series A: Chemistry and Biology,
https://doi.org/10.1007/978-94-024-1304-5_14

into narrow tracks are enriched in the track's confinement, and they change the electrical track properties if their charge states differ from those of the analytes. It was yet unknown how to design these sensors so that their best efficiency and highest possible sensitivity is achieved. This requires the accurate knowledge of the optimum track radius and the degree of product enrichment within the tracks. These questions were answered by appropriate experiments described here. The above studies restricted to transparent tracks only. After it became evident that also thin membranes can be formed within such tracks – thus separating transparent tracks into two adjacent semi-transparent segments each – we were curious in how far such structures might also become useful as sensors. In fact, two promising approaches emerged, one that uses these structures as polarization-induced capacitive biosensors, and another one that considers the semitransparent track segments as neighboring "electrostatic bottles" which can be discharged by pulse-wise product emission. Preliminary results indicate that the latter sensor type is superior in its performance to all other ones.

Keywords Biosensors · Nanopores · Heavy ion tracks

14.1 Introduction: Biosensors Based on Transparent Swift Heavy Ion Tracks

Long micro- or nanopores in insulating or semiconducting materials are useful construction elements for cheap and sensitive biosensors. The large inner surface of such pores can host a large amount of reacting agents, resulting in strong sensor signals. The tiny volumes of such pores – femto-liters per nanopore or so – signify very low bio-liquid consumption. As the nanometric pore dimensions inhibit turbulent flow, there is no risk of mixing different bioliquids in the pores.

Usually we arrange parallel micro- or nanopores perpendicularly to the surface of substrates such as thin ($\sim$12 μm or so) self-carrying polymer foils (usually polyethylene terephthalate, PET) [1, 2]. The pores are created by irradiation of the foils with swift heavy ions (atomic masses between $\sim$40 (Ar) and $\sim$235(U)) at energies of some 100 MeV to few GeV and subsequent etching (4 M NaOH). Various concepts have been developed to use such nanoporous foils for biosensing [2]. For this sake, some agent E (e.g. an enzyme) is often attached to the wall of etched tracks within the substrate (enzymes can be attached to the polymer's dangling bonds (e.g. -COO- groups in the case of PET) on the inner track surfaces by application of EDC and S-NHS, which transform the carboxyl bonds to amino groups required for the enzyme bonding [3]). The presence of some analyte A in the tracks is then detected via changes in the pore's electrical signal (conductivity, capacity, impedance) that emerges from different charge states of the products P upon the reaction of the analyte A with the reagent E: $A+E=P+E$. For monitoring these electrical changes, some test voltage in the order of $\sim$5 V was applied across the tracks.

14.1.1 *Enzymatic Reaction Product Enrichment Within Etched Tracks*

Hitherto etched track-based biosensing was verified for the sensing of glucose (via the enzyme glucose oxidase, GO_x [3]), urea (via the enzyme urease [4]) and phenols (via the enzyme lactase [5]); in this work we restrict to glucose sensing (i.e., A = glucose, E = GO_x, P = gluconic acid, GA) as the model system. Usually we made use of the "Product Enrichment" strategy [2] where the enzymatic reaction products gradually enrich in the strong confinement of very narrow tracks and with poor escape possibilities due to their long extension (their "aspect ratios" or pore length/radius ratios are typically around 100–1000, depending on the used polymer type [6]).

If the production rate R_p of P (after the reaction of A with E) exceeds their escape rate R_e through the track openings on (one or) both sides of the carrier foil (i.e., $R_p > R_e$), their concentration C_p within the tracks will increase up to some certain upper limit $C_{p,\max,}$ thus enhancing the sensor's sensitivity. This situation had been simulated [6, 7] by a self-written code "DIFF" based on the Finite Differences approach [8]. Though this code – originally written for other purpose e.g., [9, 10] – did not yet include electrostatic interactions as usually present in nanofluidic channels due to wall charges, it is nevertheless applicable here as the originally present wall charges (stemming from the polymeric substrate – here: PET with -COO- dangling bonds after track etching) had vanished within $\sim$1% after bonding of the enzymes to those charges [3]. The simulation showed that $C_{p,\max}$ scales linearly with the concentration of the penetrating analytes C_a and the enzymes C_e bond to the track walls, quadratically with the etched track length d, and roughly with $\sqrt{D_P/D_A}$, the square root of the ratio D_p/D_a of the corresponding diffusion coefficients of both the products and the analyte within the tracks, respectively [7].

In order to obtain realistic values for $C_{p,\max}$ from these results, both D_p and D_a had been determined by a specially designed experiment [11]. The results showed that the enzymatic reaction product (gluconic acid, GA) migrates considerably faster than glucose through the etched tracks in spite of similar molecular weight and size. This is understood by taking into account that, in contrast to glucose, the acid GA is partly dissociated to GA^- and H^+ and hence is subject to electro-diffusion additionally to its thermal mobility. The ratio of both diffusion coefficients amounts to $D_{\text{gluc}}/D_{\text{GA}} = 0.75$, which is the desired parameter to be inserted into the previous diffusion simulation of [7]. It follows that the product enrichment within the etched tracks amounts to about factor 20 within typically 16 s or so. According to our experience, this number appears to be realistic.

14.1.2 Dependence of the Diffusion Coefficient on the Track Radius

According to Renkin's [12] empirically derived formula the diffusion coefficient D_{pore} in narrow cylindrical nanopores decreases below ~30 nm radius or so, as compared with that one without confinement, D_{free}, due to spatial restrictions of the random mobility in very narrow tracks. In Renkin's formula enters the so-called Stokes radius (or Stokes–Einstein radius, or hydrodynamic radius) $r_S = k_B T/(6\pi \mu D)$ of nanoparticles in a carrier liquid with viscosity μ, with k_B being the Boltzmann coefficient, T the absolute temperature, and D being the diffusion coefficient of the liquid within which the corresponding nanoparticle moves.

To verify in how far that prediction also holds for narrow etched tracks with different radii in polymers, we determined the diffusion coefficients of GA across them as described in [11]. The measurements show that there exist indeed deviations from Renkin's equation, which indicate the additional existence of a GA diffusion component outside the etched tracks across the polymeric bulk: $D_{\text{etched track}} = D_{track} + D_{bulk}$, with D_{track} following Renkin's correlation and $D_{bulk} = 1 \times 10^{-9}\,\text{cm}^2\text{s}^{-1}$. With this correction, Renkin's prediction matches reasonably well with our measurements, under the condition that a hydrodynamic radius r_S of GA of (1.5 ± 0.3) nm is chosen as the best fit [11].

14.1.3 Molecular Dynamics Simulation in Tracks with Charged Walls

To understand the passage of ions through nanotracks in polymers better, we also undertook computer simulations of the passage of charged particles through nanopores of various diameters and lengths, using the classical methods of molecular dynamics [13–16]. A cylindrical volume is filled with a gas of charged particles of a certain density. Gas passes from one side of the tube through the latter at a certain rate in accordance with the laws of diffusion, while being accelerated by an electric field. Boundary conditions forbid the particles to penetrate across the walls of the allocated cylindrical volume (i.e., the track walls). In addition, the internal walls of the cylinder are charged (for example, due to existence of dangling bonds), so that charged particles are reflected from the walls via elastic Coulomb collisions. The interaction of the particles is described by a screened Coulomb potential. The screening constant characterizes the substance passing through the pore.

The program yields the dependence of the density of the particle flux through the pore on the pore diameter, the pore aspect ratio and the density and properties of the substance passing through the pore. The first simulation result indicates that the flux (and hence also the diffusion coefficient) *de*creases with decreasing pore radius (from 50 to 25 nm). This is due to a sharp increase in the contribution of multiple scattering of particles on the internal pore walls.

Another result relates to the sensitivity of the sensor: a number of "foreign" particles are included in the composition of the gas passing through the pore which differs from the main particle flux by the nature of the interaction with other particles (the screening parameter in the Coulomb potential). The program yields a change in the flux density of particles through the pore, depending on the percentage of "foreign" particles. In addition, it turns out how this dependence is affected by the magnitude of the tube radius and its aspect ratio. The simulation shows that, despite the decrease in the particle flux with decreasing pore radius in the nanoscale region, the sensitivity of the sensor *in*creases, which can be explained by the growing predominance of the interaction of particles with the pore walls (i.e., the increase of the contribution of multiple scattering of foreign particles on the internal walls of the pore). Details will be described in a forthcoming paper.

At present, these examinations restrict to GA only. Taking into account the similar size and mass of glucose molecules as compared with GA, one should assume that similar spatial restrictions in narrow tracks will also hold for glucose. If that is the case, then the diffusion coefficient ratio $D_{\text{gluc}}/D_{\text{GA}}$ as derived in the Sect. 14.1.2 for small track radii ($0 < r < 30\,\text{nm}$) will hardly differ from the ratios for large track radii ($r > 30\,\text{nm}$). This would signify that the measuring sensitivity should be more or less constant, rather independent of the track radius.

14.1.4 Influence of Track Radius and Analyte Concentration on Biosensor Efficiency

To check the validity of this conclusion, a corresponding measurement on the correlation between the biosensor amplitude and the track radius was performed [17]. It revealed that the sensor signal is indeed practically radius independent in the ~10–50 nm range. However, for pore radii below 10 nm or so (roughly corresponding to the track's central zone, the so-called "core"), the sensor's efficiency first decreases very strongly with *de*creasing radius, but then surprisingly *in*creases again for radii below some 3 nm (Fig. 14.1).

For explanation it is tentatively suggested that enzymes in solution can penetrate into irradiated polymer foils which were not yet etched but already treated with EDC and S-NHS. They are expected to diffuse somewhat around there till they are trapped at the active dangling amino bonds (stemming from the EDC/S-NHS treatment [3]) within isolated nanopores – the latter may either stem from the intrinsic polymeric free volume or from the swift heavy ion irradiation, preferably in the track's cores. Analyte molecules diffusing into the bulk may then react with the bonded enzymes, to produce the reaction products which are detected as usual by an increase of the overall sensor foil's conductivity.

During the onset of track etching, the irradiation-induced nanopores will merge with each other, thus reducing their inner total surface area; hence the amount of

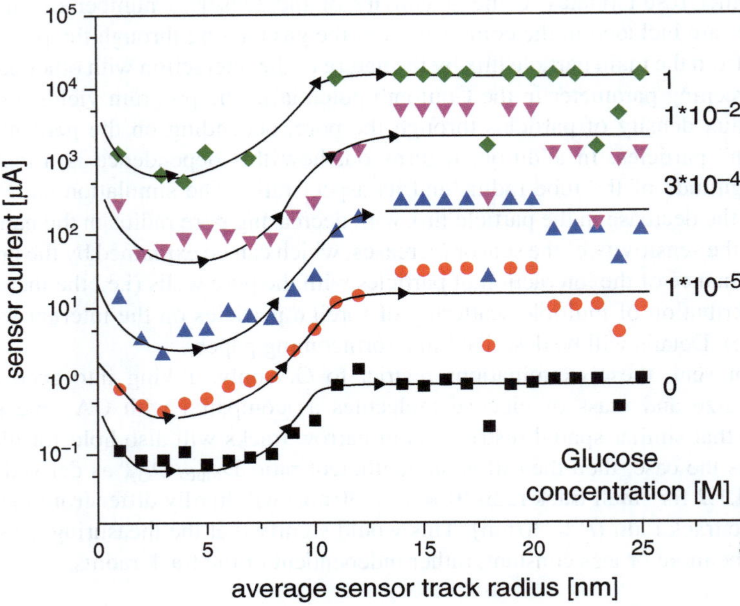

Fig. 14.1 The biosensor's signal amplitude as a function of the etched track radius (corrected by the thickness of the enzyme layer) [17]; for better overview, data for different concentrations are shifted vertically by a factor 10 each. Lines are to guide the eye

enzyme which can bond to the nanopore surfaces (and consequently the biosensing efficiency) is reduced. A minimum of that efficiency is reached when all the isolated nanopores merge to a continuous free pore, the so-called etched track. Our experiments indicate that this happens for a nominal average etched track radius of around 3 nm (as narrow etched tracks are conical or funnel-type structures rather than cylindrical ones, we deal here with average track radii rather than with real ones). With continuing etching, the etched track radius increases, thus also increasing the inner track surface, hence the enzyme coverage and the sensor efficiency.

However, the sensor efficiency does not grow infinitely with continuous track etching, as larger tracks enable the products to diffuse more and hence to reduce their enrichment. Therefore, above an etched track radius of ~10 nm the sensor current saturates for increasing radius [17], which means that the increase of enzymatic coverage of the etched track surfaces is compensated by the increase of enzymatic product diffusion out of the tracks.

Precondition for this study was to acquire an accurate knowledge of the enzyme-clad etched track radii. As after enzyme bonding to the etched tracks, 99% of the original etched track's dangling bonds were used up [3] and the tracks had lost most surface charges [7] (see above), they hardly exhibit nanofluidic behavior any longer, in spite of their nanometric dimensions. Then the conventional non-nanofluidic approach for large track radii can be applied here.

For very strong confinement or very low analyte concentrations, when the enzymes incorporated in our sensors show only little activity, the recorded sensor currents across the biosensors essentially stem from the carrier liquid only (here: ultrapure water). With increasing (neutral) analyte concentration, the water ions get increasingly diluted and thus the transmitted currents decrease [17]. Under this working condition, the sensor response is unspecific for the analyte as it works with any ionized liquid. The biosensors give rise to proper (i.e., analyte-specific) biosensing behavior only with less enzyme confinement and/or increasing analyte concentration [17]. The biosensor calibration curves derived here are rather similar to those ones measured earlier [3], their absolute values differing from each other at maximum by a factor 2 or so.

14.2 Biosensors Based on Swift Heavy Ion Tracks with Membranes

Sequential coupling of two (or more) biosensors based on transparent swift heavy ion tracks with each other towards sensor cascades improves the sensor's properties [18]. Having this experience in mind, we tried to combine two sensors each within one etched track, by separating enzyme-clad transparent tracks into two segments, by inserting thin membranes in their middle. This approach is indeed feasible by applying the "Coupled chemical-topological reactions (CCR)" strategy [19, 20]. Here, two components A and B that react with each other by forming a precipitate C are added to the etchants on both sides of a latent track. In the moment when both etch cones meet with each other (i.e. when a transparent etched track has just been formed), the emerging new component C separates both segments from each other, i.e. acts as a membrane.

After having examined the electronic properties of CCR structures [21, 22] with various types of membranes (such as Ag_2O, LiF, $CaCO_3$, BaF, etc. [22–24]), we started first tests with enzyme-clad CCR structures. It has turned out that especially two strategies emerged, the polarization-induced capacitive approach [24] and the "electrostatic bottle" concept [25]. They differ indeed markedly from the simple case of biosensors with transparent etched tracks – see above – where the concentration of ionized enzymatic reaction products enriches with increasing analyte concentration, so that passing electrical test currents *in*crease. But anyhow, all the findings worked out above for transparent tracks also apply to the semitransparent track systems discussed here – with the only two exceptions that leakage of P out of the tracks is reduced by a factor two and that the tracks interact with each other across the membranes.

14.2.1 Track Biosensors with Membranes: The Polarization-Induced Capacitive Approach

In this case both the charged dissociated enzymatic reaction products (e.g., GA^- and H^+ in the case of the glucose/GO_x system) enrich on opposite sides of the membranes due to the electrostatic attraction of an externally applied electric field. This leads to intrinsic polarization across the membranes [24]. The latter is the stronger the higher the analyte concentration is. For a sufficiently high frequency of an externally applied alternating voltage the enzymatic reaction products can no longer escape from the tracks in due time upon phase reversal, so that the intrinsic polarization across the membrane counteracts the externally applied electrical field, and thus makes the extracted capacitive current signal *de*crease more and more with increasing analyte concentration. The integration of all accumulated enzymatic reaction products during each half phase of the applied external voltage enhances this effect even more, thus improving the efficiency of this sensor type greatly, as compared with that of sensors without internal membranes [24]. At very high analyte concentrations, the extracted current signal of these segmented biosensors becomes independent of the concentration, as the density of the charged enzymatic reaction products comes to saturation due to their Coulomb repulsion within the tracks.

14.2.2 Track Biosensors with Membranes: The "Electrostatic Bottle" Approach

This approach makes use of the electrical analogue of an analyte-filled semipermeable track to a condenser: the integrated charge deposited on an electric condenser by charging it up for a long time by small currents, can be discharged thereafter in a very short and big current pulse. Comparing ever-present electronic noise with the small charge-up currents on the one hand, and the large pulse discharge current on the other hand, it is obvious that the signal/noise ratio of the short discharge pulses is by far superior to that of the slow charge-up currents.

Analogously, analyte-filled semipermeable tracks of biosensors with central membranes can be slowly charged up with enzymatic reaction products during a long time, and rapidly discharged in a pulse at a later moment, thus integrating all accumulated biomolecules and improving both the sensor sensitivity and accuracy. Tests have revealed that the sensor response depends somewhat on the position of the membranes within the tracks (such as surface or center); in general sensors with central membranes show a slightly better correlation with analyte concentration [25]. In any case, both the pulse width and area were found to *de*crease strongly with increasing analyte concentration, the pulse height to hardly *in*crease, and the pulse height/width ratio to *in*crease strongly, Fig. 14.2 [25].

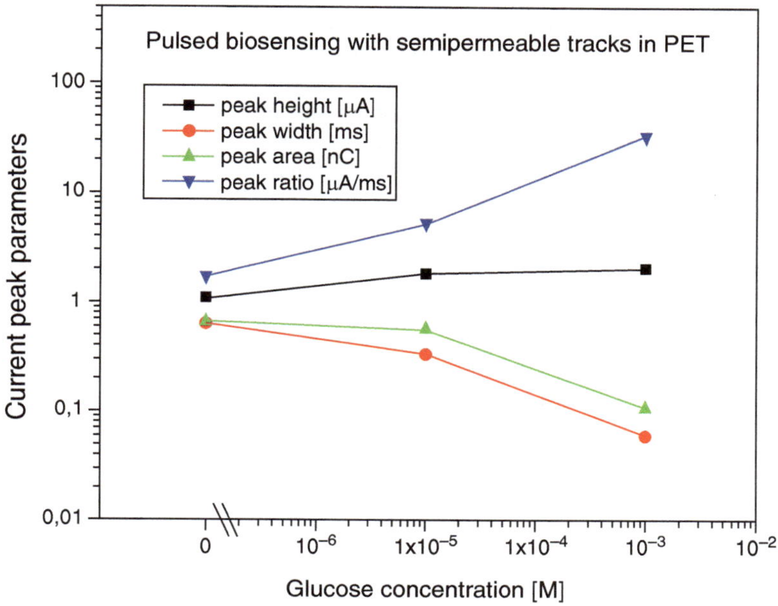

Fig. 14.2 Calibration curves for glucose sensing with track-based biosensor with membranes operating in the "electrostatic bottle" mode; shown are pulse height, width, area and height/width ratio as a function of the analyte concentration for tracks with surface-near membranes. Lines are to guide the eye

14.3 Comparison of the Different Track-Based Biosensor Concepts with Each Other

When comparing with each other the three principle sensing techniques – the current measurement across transparent tracks, the "polarization-induced capacitive" approach for membrane-containing tracks, and the "electrostatic bottle" approach for membrane-containing tracks, one finds that the slopes of all calibration curves just fall into three different categories, as shown in Fig. 14.3 (here we have normalized the biosensor currents at zero analyte concentration to each other and put all graphs into a log/log graph frame). Whereas sensors made of transparent ion tracks (first and second cases in Fig. 14.3) exhibit small positive slopes, all sensors made of etched tracks with embedded membranes (all other cases of Fig. 14.3) exhibit comparatively large positive or negative slopes. Note that the calibration curve referring to the polarization mode (third case of Fig. 14.3) is limited to concentrations below 10^{-2} M as signal saturation sets in above.

Both the biosensors based on transparent tracks with sufficiently large track diameters (Case 2 of Fig. 14.3 with $r \sim 30$ nm) and those ones with membranes based on the electrostatic bottle-approach (Case 4 of Fig. 14.3; pulse height evaluated) exhibit the largest measured (non-normalized) currents, hence are most

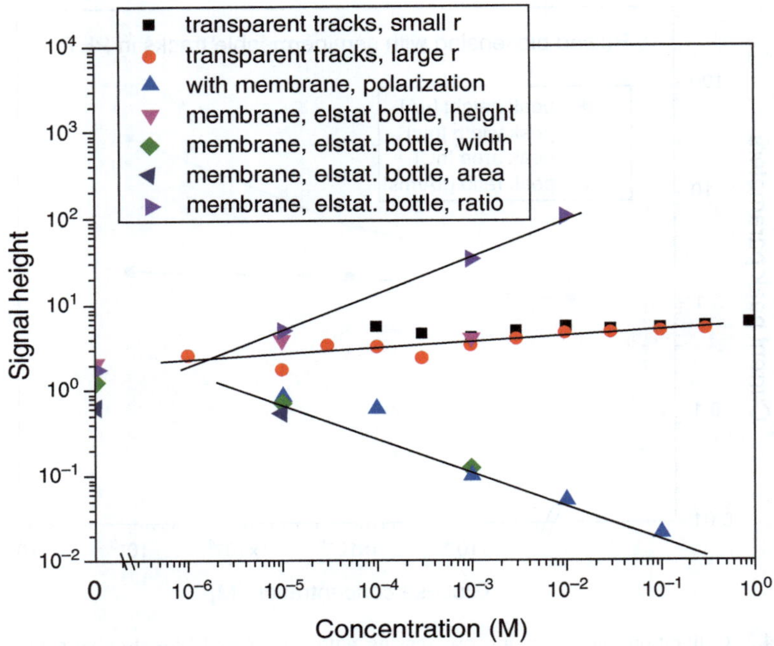

Fig. 14.3 Comparison of biosensor signal height as a function of the applied analyte concentration, for the different biosensor modes discussed here. Straight lines are to guide the eye

convenient for easy measurement. However, both cases have poor slopes of their calibration curves which mean that they do not enable precise distinguishing of neighbored analyte concentrations.

For contrast, the polarization-type strategy of biosensors with membrane-containing tracks (Case 3 of Fig. 14.3) exhibits the smallest measured currents in the order of pA only, however has steep calibration curves (with negative slopes), hence enables precise concentration measurements. The same also holds for using the pulse width (calibration curve with negative slopes), area (calibration curve with negative slopes), or height/width ratio (calibration curve with positive slopes), of the electrostatic bottle concept.

14.4 Results and Conclusions

Recent new details of ion track-based biosensors were reported, with the aim of optimization of future sensor production. The results can be summarized thus:

1. The simulation of diffusion of enzymatic reaction products in transparent etched ion tracks revealed that the maximum product enrichment correlates

with the ratio of the diffusion coefficients of the analytes to the products. The measurements of these diffusion coefficients revealed that this ratio amounts in large tracks ($r > 10$ nm) to 0.75. From this the maximum product enrichment is estimated to be in the order of a factor 20.

2. Renkin's empirical estimation showing that the diffusion coefficients decrease at transparent very narrow cylindrical pores could be reconfirmed for etched ion tracks in polymers below ~30 nm track radius. However, an addition had to be made, to also account for the contribution of the polymeric bulk to the overall diffusion. Supposing that this radial dependence accounts for both the analyte and the products in the same way, then the ratio of both would reduce to unity so that overall biosensor yield would be independent from the track radius.
3. The experimental check of this question reconfirmed that this expectation is indeed right for radii larger than ~10 nm. However, for smaller radii the sensor yield decreases down to a minimum at ~3 nm, below which it increases again. This is tentatively understood by the additional bulk contribution on the one hand, and by the increase of the (intrinsic and track core) free volume's surface with larger radii on the other hand.
4. Whereas for biosensors with large transparent tracks ($r > 30$ nm) the sensor signal always increases with increasing analyte concentration, the radial dependence of the yield differs for sensors with narrow tracks ($r < 10$ nm); here it consists of two branches, one which *in*creases as above with concentration at high analyte concentration, and another one at low concentrations which *de*creases and is not analyte-specific.
5. It is possible to design track-based biosensors not only with transparent etched ion tracks, but also with etched tracks into which membranes are embedded, thus separating the whole track into two adjacent semi-transparent compartments. Two construction principles emerge, one which we denoted as the "polarization-induced capacitive" approach, and another one which was denoted as the "electrostatic bottle" approach.
6. In the first case, a capacitive sensor current emerges where signals *de*crease with increasing analyte concentration. In the other case, where sensor pulses emerge from the discharge of the "electrostatic bottles", two pulse properties – the pulse width and the pulse area – *de*crease with increasing analyte concentration, one signal – the pulse height – is rather concentration independent, and another one – the pulse height/width ratio – *in*creases strongly.
7. The three biosensor types discussed here differ both in the height and accuracy of their sensor response signals, which is summarized in Table 14.1 (worst of all is to use sensors with transparent tracks only, with very low track diameters ($r < 10$ nm), due to their low current, limited region of applicability ($\sim 3 \times 10^{-4}$ to 1 mole only) and poor slopes of the calibration curves – therefore this case was not included in Table 14.1). It is obvious that the electrostatic bottle concept is the best one due to both high signals and steep calibration curves; best is to evaluate here the emitted pulse's height/width ratios. Though the production of these membrane-containing biosensors is more complicated than that one of simple transparent sensors, it pays out due to the measuring advantages (note

Table 14.1 Qualitative comparison of the different biosensor types discussed here

Track-based biosensor type	Signal height of sensor response	Slope of calibration curves
Transparent tracks with large track radii ($r > 30\,\text{nm}$)	High	Poor
Etched tracks with membranes; polarization-type approach	Small	Strong
Etched tracks with membranes; discharge pulses from electrostatic bottles	High	Strong

that the poor and steep slopes of calibration curves shown in the logarithmic comparison of Fig. 14.3 differ by about two orders of magnitude!). In all cases, the useful concentration interval is mostly similar: between $\{10^{-5}–10^{-6}\}$ and $\{0.1–1\}$ moles. It depends on the selected enzymatic system rather than on measurement details.

Acknowledgements D.F. is grateful to the Universidad Autónoma Metropolitana-Iztapalapa, Mexico City, for the guest professorship in the frame of the Cathedra "Alonso Fernandez". We are especially obliged to Prof. S.A. Cruz and Mr. N. Camarillo from UAM-Iztapalapa for their continuous help and discussions and for providing us with adequate working facilities. The project was supported by the Grant Agency of the Czech Republic (P108-12G-108) and the Nuclear Physics Institute, Řež near Prague. We are further obliged to Dr. P. Apel from JNRI Dubna, Russia for providing us with the ion-irradiated foils. L.A. acknowledges the support of the Edmond J. Safra Center for the Design and Engineering of Functional Biopolymers at Ben-Gurion University.

References

1. Fink D (ed) (2004) Fundamentals of ion-irradiated polymers, vol 63. Springer, Berlin
2. Fink D, Muñoz HG, García-Arellano H, Vacík J, Havranek V, Hnatowicz V, Kiv A, Alfonta L (2016) Nuclear track-based biosensing: an overview. Rad Eff Def Solids 171:173
3. Fink D, Klinkovich I, Bukelman O, Marks RS, Kiv A, Fuks D, Fahrner WR, Alfonta L (2009) Glucose determination using a re-usable enzyme-modified ion track membrane. Biosens Bioelectron 24:2702–2706
4. Fink D, Muñoz HG, Alfonta L (2011) Ion track-based urea sensing. Sens Actuators B 156:467
5. García-Arellano H, Fink D, Muñoz HG, Vacík J, Hnatowicz V, Alfonta L (2014) A nuclear track-based biosensor using the enzyme Laccase. Appl Surf Sci 310:66. In: Öztarhan A, Baglin J (eds) Selected manuscripts arising from the 18th international conference on surface modification of materials by ion beams (SMMIB-2013), 15–20 Sept 2013, Kusadasi
6. Fink D (2004) Transport processes in ion-irradiated polymers, vol 65. Springer, Berlin
7. Fink D, Vacík J, Alfonta L, Kiv A, Mandabi Y, Muñoz HG (2012) Optimization of transport processes in etched track-based biosensors. Rad Eff Def Solids 167:548
8. Smith GD (1978) Numerical solution of partial differential equations: finite difference methods. Clarendon Press, Oxford
9. Fink D, Wang L, Martan J (1991) Mobility of helium and nitrogen implanted at high fluences into solids, as derived from their concentration profiles. In: Donelly SE, Evans JH (eds) Fundamental aspects of inert gases in solids. Plenum Press, New York, pp 67–79

10. Fink D, Szimkoviak P, Hu X, Hnatowicz V, Vacík J, Chadderton LT (2000) Redistribution of boron in silicon after Ne^+ post-irradiation and thermal annealing. Rad Eff Def Solids 152:67
11. Fink D, Vacík J, Hnatowicz V, Muñoz HG, García-Arrellano H, Alfonta L, Kiv A (2017) Diffusion kinetics of the Glucose/Glucose Oxidase system in swift heavy ion track-based biosensors. Nucl Instr Methods B 398:21
12. Renkin EM (1954) Filtration, diffusion, and molecular sieving through porous cellulose membranes. J Gen Physiol 38:225
13. Mykytenko N, Fink D, Kiv A (2015) Computer modeling of ion current pulsations in track-containing foils. J Comput Sci 6:34
14. Kiv A, Mykytenko N, Fuks D, Dahan I, Meshi L (2016) Molecular dynamics probing of the energy spectrum of particles in radiation stimulated processes. Int J Adv Comput Technol 4(6):81
15. Ermakov V, Kruchinin S, Fujiwara A (2008) Electronic nanosensors based on nanotransistor with bistability behaviour. In: Bonca J, Kruchinin S (eds) Proceedings of NATO ARW "Electron transport in nanosystems". Springer, pp 341–349
16. Ermakov V, Kruchinin S, Hori H, Fujiwara A (2007) Phenomena in resonant tunneling through degenerated energy states with electron correlation. Int J Mod Phys B 21(11):1827
17. García-Arellano H, Muñoz HG, Fink D, Cruz SA, Vacík J, Hnatowicz V, Alfonta L, Kiv A (2018) Dependence of yield of nuclear track-biosensors on analyte concentration and track radius. Nucl Instr Methods B 420:69–75
18. Fink D, Muñoz HG, Alfonta L (2012) Highly sensitive ion track-based urea sensing with ion-irradiated polymer foils. Nucl Instr Methods B 273:164
19. Fink D, Vacík J, García-Arellano H, Muñoz HG, Alfonta L, Fahrner WR, Hoppe K, Kiv A (2014) Biosensors with nuclear tracks and embedded membranes. Key Eng Mater 605:83; Hristoforou E, Vlachos DS (eds) Materials and applications for sensors and transducers III
20. Muñoz HG, Cruz SA, Quintero R, Fink D, Alfonta L, Mandabi Y, Kiv A, Vacík J (2013) Coupled chemical reactions in dynamic nanometric confinement: I. Ag_2O formation during ion track etching. Rad Eff Def Solids 168:396
21. Fink D, Muñoz HG, García-Arellano H, Fahrner WR, Hoppe K, Kiv A, Vacík J, Alfonta L, Mandabi Y (2017) Coupled chemical reactions in dynamic nanometric confinement: II. Preparation conditions for Ag_2O membranes within etched tracks and integrated micro sensors. In: Iniewski K, Yallup K (eds) Trends in real world sensing and integrated micro sensors, Chapter 24 In: Technologies for smart sensors and sensor fusion. CRC Press, Boca Raton, p 451
22. Fink D, Muñoz HG, García-Arellano H, Fahrner WR, Hoppe K, Kiv A, Vacík J, Alfonta L, Mandabi Y (2017) Coupled chemical reactions in dynamic nanometric confinement: III. Electronic characterization of Ag_2O membranes within etched tracks and of their precursor structures. In: Iniewski K, Yallup K (eds) Technologies for smart sensors and sensor fusion; Chemical and environmental sensors. CRC Press, Boca Raton, p 161
23. Fink D, Muñoz HG, Ruiz N, Vacík J, Hnatowicz V, Alfonta L, Kiv A (2014) Coupled chemical reactions in dynamic nanometric confinement: V. The influence of Li^+ and F^- on etching of nuclear tracks in polymers. Rad Eff Def Solids 169:1
24. Fink D, Muñoz HG, García-Arellano H, Alfonta L, Vacík J, Kiv A, Hnatowicz V (2017) Coupled chemical reactions in dynamic nanometric confinement: VII. Biosensors based on swift heavy ion tracks with membranes. Rad Eff Def Solids 172:159 (Editor's Special)
25. Fink D, Vacík J, Hnatowicz V, Muñoz HG, Cruz S, García-Arellano H, Kiv A, Alfonta L, Shunin YuN, Mansharipova AT, Mukhamedyev RI (2016) Coupled chemical reactions in dynamic nanometric confinement: VIII. Capacitive discharges in nuclear track-based biosensing 3:62–75

Chapter 15
Development of a Piezoelectric Flexural Plate-Wave (FPW) Biomems-Sensor for Rapid Point-of-Care Diagnostics

A. Jupe, P. Livshits, S. Kahnert, M. Figge, S. Mross, M. Goertz, H. Kappert, H. Vogt, and A. Goehlich

Abstract A high rate of post-neonatal deaths is due to a Respiratory Syncytial Virus (RSV). An early diagnosis with the accurate patient stratification and forecast of disease outcome is vital for an effective treatment. Common diagnostics for the identification of unknown pathogens require large sample volumes and are laborious. In this work, a novel piezoelectric FPW Bio-MEMS sensor based on the frequency shift of a resonating membrane due to binding of an additional mass was developed. The experimental results show that the sensor, a critical part of point-of-care devices, can detect with a high degree of accuracy RSV chemokines in complex media and is much faster than standard methods.

Keywords Flexural plate wave (FPW) Bio-MEMS sensor · Ultra-sensitive point-of-care diagnostics · Detection of causative agent of the infection · Host response in blood and nasopharyngeal aspirate

15.1 Introduction

Standard diagnostics for the detection of unknown pathogens are based on pathogen cultivation, polymerase chain reaction and enzyme immunoassays, which are all laborious, time-consuming and expensive methods. They require large sample volumes, special apparatuses, well trained personnel in specialized laboratories and substantial effort in sample collection, preparation and treatment. The point-of-care (PoC) diagnostics is a promising candidate to address these challenges. This potentially low-cost, portable and patient-friendly diagnostics method can enable a

---

A. Jupe · P. Livshits (✉) · S. Kahnert · M. Figge · S. Mross · M. Goertz · H. Kappert · H. Vogt
A. Goehlich
Fraunhofer Institute for Microelectronic Circuits and Systems IMS, Duisburg, Germany
e-mail: pavel.livshits@ims.fraunhofer.de

J. Bonča, S. Kruchinin (eds.), *Nanostructured Materials for the Detection of CBRN*, NATO Science for Peace and Security Series A: Chemistry and Biology,
https://doi.org/10.1007/978-94-024-1304-5_15

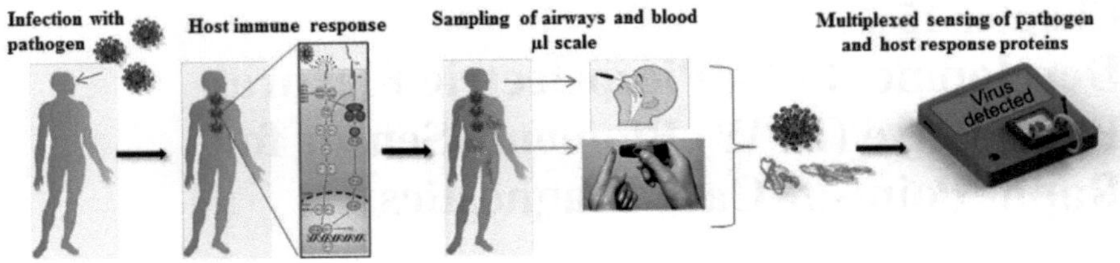

Fig. 15.1 A novel approach with a multiplexed measurement of pathogen and host response

rapid availability of data, immediate clinical decisions and effective care of patients [1–7]. Therefore, it is currently gaining a lot of interest and under intensive research.

In this work, a novel PoC device for detection of respiratory infectious viral diseases at newborns is being developed. Respiratory diseases are highly prevalent in young children. Almost 100% of all children are infected with a RSV virus before the age of two. On average, 12.9% of post-neonatal deaths in European countries are due to respiratory diseases and in lower income regions in Europe more than 5 of 1000 live births die as a result of it [8]. The PoC device is based on an innovative approach of a multiplexed measurement of clinical relevant combinations of host and pathogen (causative agent of the infection) derived biomarkers in complex media, such as blood and nasopharyngeal aspirate. The principle of the applied approach is schematically summarized in Fig. 15.1.

A biosensor is an analytical component and one of the key parts of a PoC device. The sensor actually defines the accuracy of the device performance. Biosensors can be of a potentiometric, amperometric, conductometric, optical, electrochemical, piezoelectric, calorimetric and several other types [9, 10]. For the proposed PoC device an optical biosensor and electrochemical biosensor cannot suit, since it would require a large area for multiplexed measurements and long testing-time in conjunction with a large quantity of expensive chemicals, respectively. On the other hand, a molecular mass sensitive piezoelectric biosensor can best fulfill the necessary requirements.

Recently, it has been reported about many acoustic sensors for molecular mass detection. However, the operating frequencies of sensors, such as shear horizontal surface acoustic wave (SH-SAW), surface transverse wave (STW), shear horizontal acoustic plate mode (SH-APM) and several other [11–15] are, usually, larger than 100 MHz, which substantially complicates the readout electronics and, consequently, increases the overall device costs. Only flexural plate wave (FPW) sensors have operating frequencies at the lower MHz region. Moreover, they achieve high mass sensitivity, one of the crucial sensor's parameters, and their phase velocity is less than the sound velocity in liquid, thus resulting only in minor energy dissipation into a testing liquid [15–17]. Thereby, a novel piezoelectric FPW-sensor has been developed for the PoC device in this work. The design specifications, fabrication process and characterization of the sensor as well as readout and biochemical experimental setups are detailed described in the following sections.

15.2 Design and Fabrication

15.2.1 Design

A FPW-sensor consists of a particular electrode configuration termed as an interdigital transducer (IDT). Specifically, IDTs comprise two identical comb-like structures, where respective fingers are placed on the surface in an interleaved alternating manner. The input IDT and output IDT are also denoted in the literature as exciting fingers and receiving fingers, respectively. These fingers, in combination with the underlying piezoelectric layer, are aimed at exciting and detecting the propagating acoustic waves.

The acoustic waves theory behind the design of a FPW-sensor as well as the way to select its major geometrical parameters, such as drift length (distance, by which two IDTs are separated), pitch (distance between centers of two consecutive fingers of the same comb) and fingers' length (acoustic aperture) have been widely studied and reported. Therefore, the description of selection procedure of these parameters, which are playing a crucial role in the sensor's performance and its frequency characteristics, is left out of scope of this paper. Based on the COMSOL Multiphysics simulations input and output IDTs of the prosed FPW BioMEMS sensor are constructed by 14 pairs of AlSi fingers. The fingers width and length are 13.5 and 320 μm, respectively. The pitch is 50 μm and drift length is 400 μm. The sensor is designed to output the center frequency of $\sim$25 MHz (in a dry state). The layout of the sensor has been realized by a commercially available layout tool.

15.2.2 Fabrication

15.2.2.1 Fabrication Process

The fabrication process of the FPW sensor applies 8 in. (100) SOI silicon wafers with a grown epitaxial layer (EPI-silicon). The silicon handle layer (doped with phosphorus) is thinned down from ca. 725 μm to ca. 425 μm. The buried oxide (BOX) is 0.4 μm thick, the device silicon layer (doped with phosphorus) is of a 0.2 μm thickness, and the epitaxial layer is ca. 1.8 μm thick.

The entire fabrication process of the FPW-sensor consists of five lithographic steps based on a standard photomask-set. At the first step, on top of the epitaxial Si-layer, a silicon dioxide (SiO_2) layer is thermally grown, that is serving as a diffusion barrier for the proper functioning of the bottom electrode, which in its turn consists of a thin titanium layer (Ti) as adhesion promoter and platinum layer (Pt) as a conducting layer. Subsequently, the LNO ($LaNiO_3$) is deposited, which is actually a seed layer for a further piezoelectric layer. In this work, the latter is PZT (Lead Zirconate Titanate). Subsequently, PZT layer and platinum layer are patterned in succession. This step is detailed discussed in the following section. The patterning is followed by the deposition of an electrical isolation film (in this study is

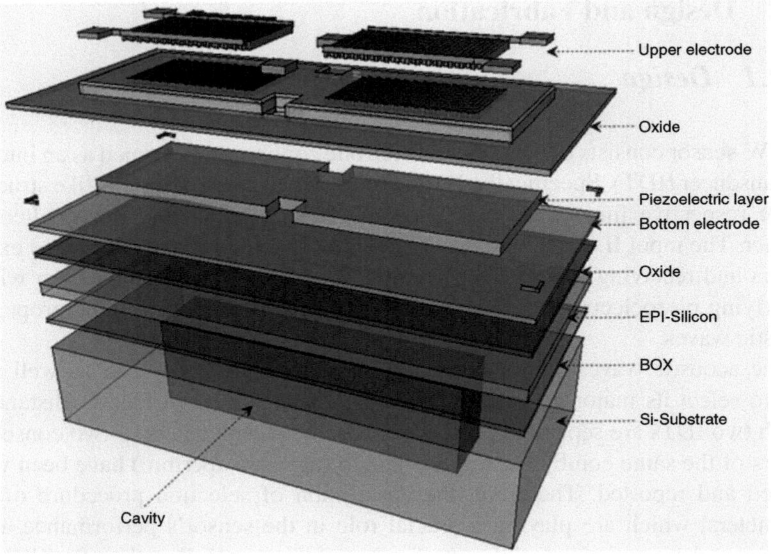

Fig. 15.2 An exploded schematic of the layers constituting the BioMEMS FPW-sensor

silicon dioxide (SiO_2) layer) utilizing the so called standard TEOS method (reaction where tetraethyl orthosilicate is converted into SiO_2). Afterwards the SiO_2-layer is patterned by a chemical plasma process and subsequently, the metallization in the FPW-sensor (i.e., exciting and receiving as well as contacts' pads) is realized through the sputter-deposition of the Al-Si layer. The metallic layer is patterned by plasma etching. At the last manufacturing step, the cavity is etched on the backside of a wafer by deep reactive ion etching (i.e., a 2 μm thick membrane is created). With the removal of the BOX (to enable an accurate bio-functionalization) the fabrication process ends.

Figure 15.2 shows an exploded view of the sensor. The schematically illustrated three-dimensional cross-sectional view of the sensor's membrane is shown in Fig. 15.3. The length, width, and height of the membrane are approximately 1925, 340 and 425 μm, respectively. Upon the completion of the fabrication process the wafers are diced using a standard mechanical sawing procedure. To protect very thin and fragile membranes in the course of the dicing process a commercially available dicing tape for securing the wafers are employed.

15.2.2.2 Piezoelectric Layer: Main Challenges

The excitation and the detection of acoustic waves are accomplished by input and output IDTs placed above a piezoelectric layer. FPW sensors with piezoelectric layers, such as zinc oxide (ZnO), aluminum nitride (AlN) or some crystalline materials have been intensively studied and their fabrication processes are well

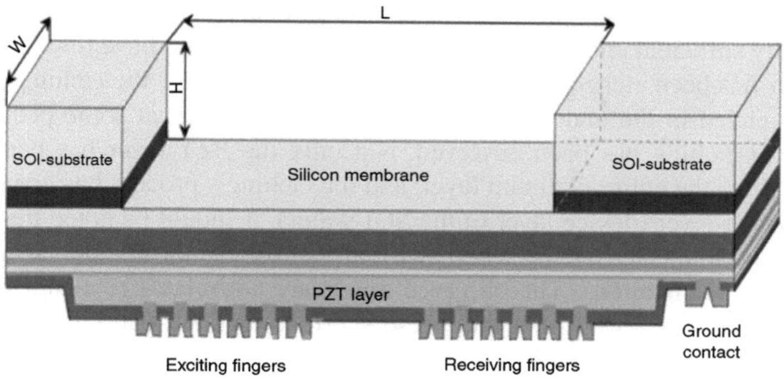

Fig. 15.3 A schematically illustrated configuration of the developed in this work BioMEMS FPW-sensor with an ultra-thin silicon membrane

established. The electromechanical coupling effect ($K2$) of the PZT is three to nine times and the dielectric constant 100 times over AlN and ZnO. This makes PZT as a much more attractive material for acoustic sensors. The proper deposition of the PZT-layer is playing a critical role in piezoelectric features of the sensor, and consequently in its future performances. For the accurate functioning of the proposed sensor, a PZT-layer with a columnar structure, that in its turn allows the sensor's membrane displacement on a z-axis as a result of the applied voltage, is required. However, the deposition processes of PZT are rather complex and entail some serious challenges for further technological steps. These challenges have been also arisen in our study and are elaborately discussed in the following paragraphs.

The stress measurements on thinned down to 425 μm SOI wafers, upon the deposition of the PZT-layer by a pulsed laser deposition (PLD) method, have revealed bow values within the range of ca. 200 μm (tensile stress). This rather substantial stress may cause technical problems on clean-room's machines as well as difficulties during the wafers' processing and even lead to an irreparable damage of the wafers. One of the well-known stress-reduction techniques is a stress-compensation by a deposition of another film. However, the development of an adequate compensation's layer is in the given case a rather complicated and not trivial procedure, since the stress is of a tensile type and not of a compressive type, which is a typical and well-studied one. Therefore, the stress-compensation procedure had to be developed and included the determination of several critical parameters, such as temperature, pressure, gas flow, RF-Power, and ICP-Power. These parameters have been successfully estimated taking into account that the removal of the BOX layer of SOI-wafers (this step is a part of the sensor's manufacturing) should lead to some further stress-reduction.

The patterning of the PZT layer is also a rather challenging task. Two approaches for the patterning can be employed, namely plasma assisted dry-etching and wet-etching processes. At the beginning, a dry-etching process has been applied, since it requires considerably less logistical efforts than its wet counterpart. The important criterion in each etching process is the resultant etch-profile that actually indicates the quality of the process. The cross-sectional SEM-images, obtained from statistically sufficient enough number of positions on a wafer, have revealed that the PZT-layer has been etched properly at some positions, where the etching's process has been ended in the underlying platinum layer. However, at some positions the strong over-etching has been observed. Not only the PZT layer has been etched away, but also the entire platinum layer, and the etching's process has been stopped only in the Si-layer (device layer of the SOI-wafer). It should be noted that in case, where the bottom electrode is etched away, the sensor cannot function according to the desired specifications. The observed trends are not related to center-periphery distribution. It has been found that these shortcomings are caused as a result of rather substantial variations in the PZT layer's thickness throughout a wafer. The thicknesses' difference (attributed to the PLD method) across the wafer can even reach 0.15 μm (the PZT-thickness is ~1 μm), which poses a serious challenge to be resolved.

To further optimise the plasma etching recipe, the optical endpoint detection methodology has been introduced. Specifically, the lead content in the plasma, which indicates the presence of the PZT layer, has been monitored photometrically by measuring plasmas' light emission at the wavelength of 406 nm. The etching process has been stopped as soon as no significant lead's concentration in the plasma has been detected. The undertaken adjustment of the etching recipe has led, to some extent, to better, but not satisfactory results. For that reason, the wet-etching procedure, which on the one hand requires much more laborious logistical efforts, but on the other hand is as a promising alternative to the dry plasma etching of the PZT-layers, since it offers much better selectivity with respect to the underlying layer, has been used. The realization of this etching method has involved the adjustment of available in the literature recipes in such a way that undercuts does not appear.

Figure 15.4 compares cross-sectional SEM-images obtained upon dry-etching and wet-etching processes of the PZT layer. It is evident that the wet-etching method has an excellent selectivity with respect to the underlying platinum layer. This trend was also confirmed by front-view optical microscope images (see Fig. 15.5). The achieved etching quality definitely cannot be accomplished by a dry-etching method

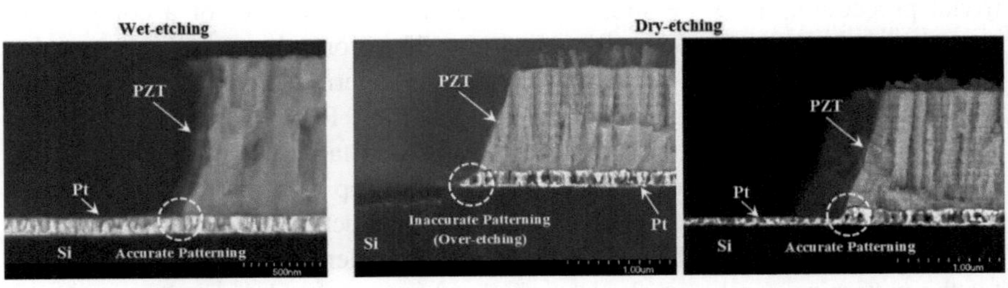

Fig. 15.4 Comparison of cross-sectional SEM-images received upon the completion of the PZT patterning performed by wet-etching and dry-etching methods. For the dry-etching process two cases are displayed. Specifically, one can see an accurate etching with a proper stop on the underlying layer and an inaccurate etching with a substantial over-etching

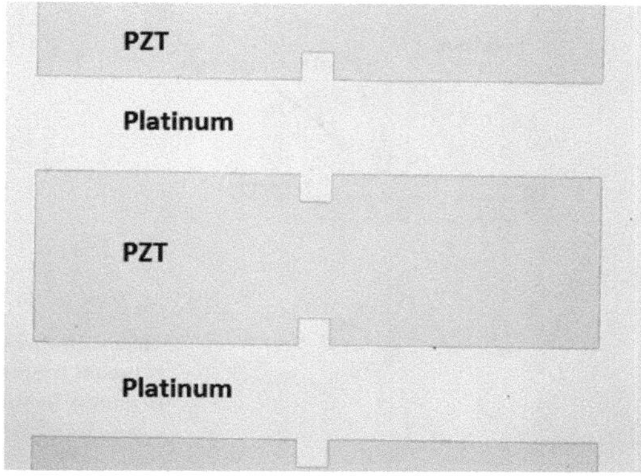

Fig. 15.5 An optical microscopic front-view image of the FPW BioMEMS sensor upon the completion of the patterning process of the PZT layer

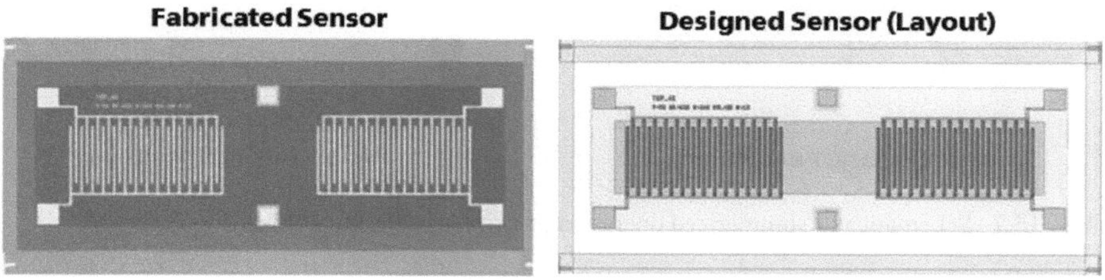

Fig. 15.6 The comparison of snapshot of layout of the FPW-sensor (right) with an optical microscope image (top view) of the fabricated FPW-sensor (left)

even in case, where the etching process is stopped in the platinum layer, since some part of the latter is also etched away. The comparison between designed FPW-sensor (snapshot of the sensor's layout) and a fabricated one (a front view obtained by an optical microscopy) is displayed in Fig. 15.6. A very close match can be clearly seen, thus indicating a successful fabrication of the sensor.

15.3 Experimental

15.3.1 Operating Principles of the Sensor

Upon the successful fabrication, the BioMEMS sensor has to be functionalized, to be capable to detect the biomolecules, when they bind to the surface of the sensor membrane. To detect and measure the amount of bound biomolecules using the piezoelectric effect, self-resonant oscillations are induced in the sensor membrane in the range of 20–30 MHz. With the binding of biomolecules to the membrane's surface, the oscillating mass naturally increases and the resonant frequency accordingly decreases. This frequency downshift is related to the amount of bound biomolecules and can be, therefore, converted to the concentration of biomolecules in the sample solution.

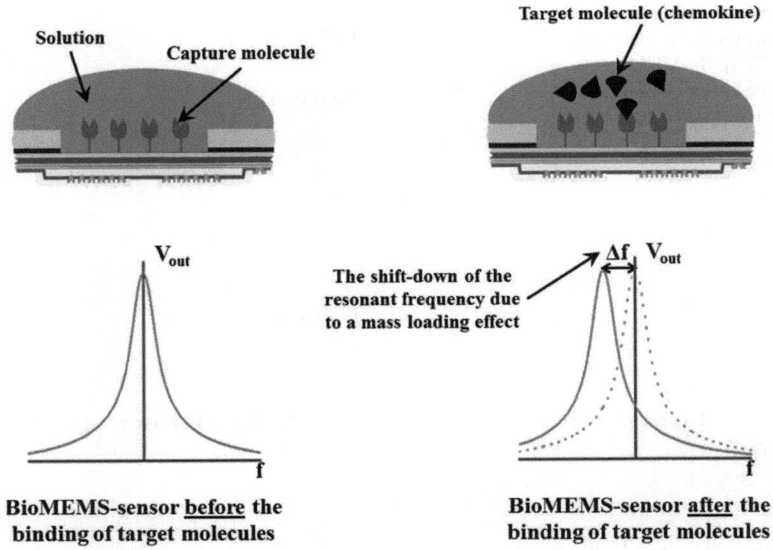

Fig. 15.7 Schematically summarized operation principles of the BioMEMS FPW-sensor

The sensor membrane itself is not able to specifically bind the biomolecules of interest. On the contrary, it adsorbs a broad range of molecules from the sample (e.g. molecules related to other diseases or any protein and sugar in blood of the patient), thus leading to an unspecific frequency shift. To make the sensor specific to a certain biomolecule, capture molecules are attached to the sensor membrane. The capture molecules are designed to bind only one biomolecule and repel other molecules. Through this mechanism the sensor is made specific to the chemokine that is to be detected. The mechanism used to attach the capture molecules to the sensor membrane is described in details in the Sect. 15.3.3. The operating principle is schematically shown in Fig. 15.7.

15.3.2 Readout Electronics

The readout of the BioMEMS sensor is based on a network analyzer (NWA) exciting the piezoelectric layer of the sensor and recording its transmission spectrum. The NWA is designed for measurements in the lower frequency domain beneath 100 MHz, which addresses well the requirements of this study. The NWA consists of two main parts: a signal-generator that produces an alternating harmonic voltage with a variable frequency for excitation of a 'device under test' and a very sensitive receiver that amplifies and detects the output signal of the examined device. To enhance the sensitivity to a feasible degree, a lock-in technique is employed. This is realized by a demodulation of the sensor signal that in its turn is phase synchronous to the generator signal.

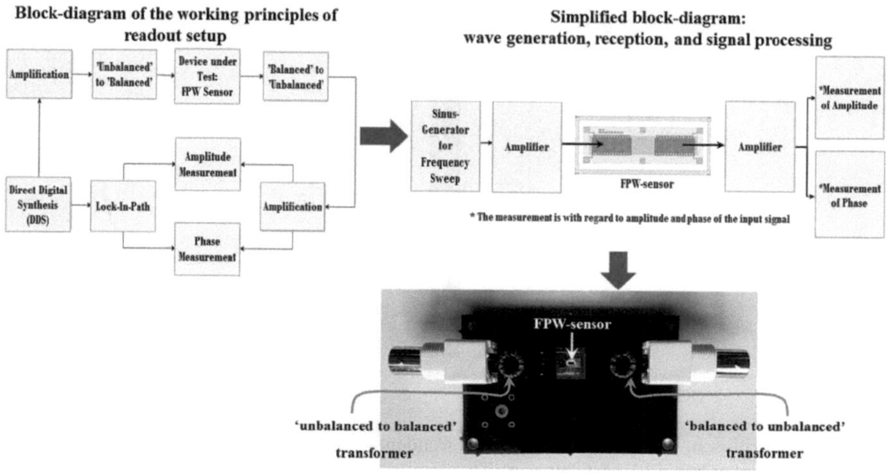

Fig. 15.8 Schematically shown function block of the frequency-downshift readout setup

The NWA feeds a single sided sine signal, the frequency of which is varied in a sweep, to the input port of the BioMEMS. The FPW-sensor, however, needs a symmetrical excitation. Therefore, before the signal is applied to the piezoelectric layer, it is converted from a strictly positive ground-related signal to one that is symmetrical around the ground level. This necessitates the adequate adaption. For this purpose, an 'unbalanced to balanced transformer' is used. For the reverse adaption of the output signal, a second transformer 'balanced to unbalanced' is required. These transformers are not a part of the NWA itself, and have to be incorporated into a separate printed circuit board (PCB), on which the FPW-sensor had been assembled. The image of the PCB with above described transformers is shown in Fig. 15.8 (beneath). The entire measurement principle is schematically summarized in Fig. 15.8 (two block diagrams at the upper part). At the output port, the transmitted signal is converted back from symmetrical to ground related and its amplitude and phase are measured and compared to a reference signal, which is directly fed from the second NWA output port to its reference port. Repeating these measurement steps for every frequency in the sweep gives the transmission spectrum of the BioMEMS. In this spectrum the most prominent resonance peak is identified and its frequency monitored over time to detect shifts induced by the binding of biomolecules to the sensor membrane.

The signal-integrity is a very important pillar of a successful and accurate functioning of the sensor. Many factors are influencing the signal quality. To guarantee a very accurate calibration of the experimental setup, a so called 'dummy-PCB' is used. There are no FPW-sensors assembled on this dummy board. Instead of it, short circuits between the input and output pads are created by two short wires. A special firmware of the NWA offers an automatic calibration routine. In the course of following measurements on a "real" PCB containing the FPW-sensor, the amplitude- and phase-responses are correspondently corrected. This way, the

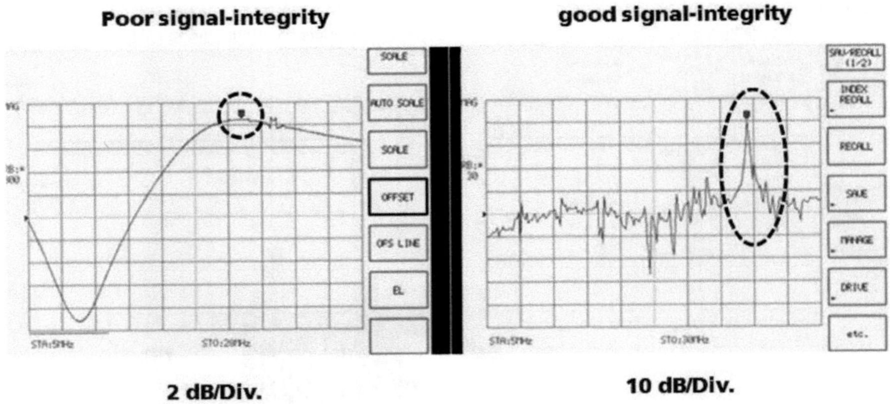

Fig. 15.9 Oscilloscope snapshots of measured frequency response of the BioMEMS FPW-sensor in case of a poor and good signal-integrity. The demonstrated measurements have been conducted in a dry state

influence of all used cables, the transformers, and the layout of the printed circuit board is properly compensated, and only the pure amplitude- and phase-response of the sensor chip is measured. Another factor that was found to play a critical role is the depolarization of the piezoelectric layer, which in its turn results in a subsequent degradation of the piezoelectric properties. This technological drawback is caused by stimulation with a pure AC voltage. The addition of a suitable DC voltage prevents this unwanted deleterious side effect. Another critical factor is a wrong signal outcoupling at the sensor's output. Figure 15.9 demonstrates the extent of impact of the poor signal integrity. It should be noted, that the scaling of the vertical axis in the left diagram is only 2dB/Div., whereas in the right one is 10 dB/Div. Furthermore, in the right diagram, the desired peak of resonance is the indubitable paramount in the entire spectrum, whereas in the left one it does not dominate the spectrum (i.e., the spectrum can be inaccurately interpreted).

15.3.3 Biochemical Functionalization

To allow the specific detection of target molecules, such as chemokines, capture molecules are immobilized on the backside of the BioMEMS sensor's membrane, which has a very thin natural SiO_2 surface. To this end, the membrane surface is first cleaned using the alkaline cleaning agent Hellmanex (Hellma Analytics, Germany). Later on, the surface is incubated for 30 min with an aqueous solution of 238 mM 3-azidopropyltriethoxysilane (abcr Chemicals, Germany), the silane group of which forms bonds with the hydroxy groups of the SiO_2 surface. This leads to a self-assembled alkyl monolayer (SAM) with azide (N_3) head groups. After silanization, the BioMEMS sensor is heat-treated for 6 h at 110 °C to remove water molecules from the surface and promote the cross-linking of the silane layer.

The capture molecules are coupled to the SAM on the membrane by Click Chemistry. This involves a bonding reaction between a dibenzocyclooctyne (DBCO) group at the capture molecule and the azide group on the SAM. This reaction takes place in aqueous conditions in 30 min, thus avoiding harsh chemicals and reaction conditions, which could damage both the BioMEMS membrane and the capture molecules.

15.4 Results and Discussion

For the detection of chemokines by the BioMEMS sensors a setup was used that included a network analyzer Anritsu MS4630B (Anritsu, Japan) for measurement of transmission spectra, microfluidics, a 12-port valve and tubing for application of a constant fluid stream with and without biomolecules. The FPW BioMEMS sensor was mounted on a PCB with fluid access port to the sensing membrane. The picture of the setup is demonstrated in Fig. 15.10.

To examine the proper functioning two similar sensors have been functionalized in a different way. The sensing membrane of the first BioMEMS sensor was functionalized with a capture molecule for chemokine detection. The second sensor was functionalized with a non-specific capture molecule and it was used as a so called reference sensor, whose signal is subtracted from the signal of the first sensor. This approach minimizes the adverse influences of unspecific adsorption, temperature or pressure changes. The NWA applied a sinus signal in an adjustable frequency range to the piezoelectric layer of the sensor and recorded the transmission spectrum.

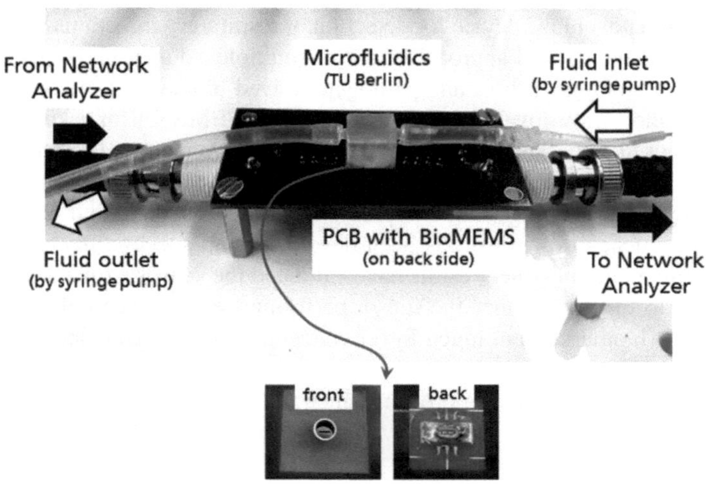

Fig. 15.10 An experimental setup, used in this work, for detection of biomarkers by the BioMEMS FPW-sensor with application of buffer flow and sample volumes

A typical measurement began with identifying the resonance peak with the smallest half-value width in the transmission spectrum of the sensor by recording the spectrum in a wide range from 1 MHz to approximately 30 MHz. The frequency span of the measurement was then reduced to a section of 10 kHz or 20 kHz around this peak to increase the measurement resolution. In the following, the peak frequency and its changes on the application of biomolecules were recorded over time.

As a second step, a stable baseline was established by applying a flow of Tris buffer (Sigma Aldrich, USA) to the sensor at a flow rate of 20 μl/min and letting the resonance frequency settle for ca. 45 min. Then, sample volumes of 50 μl Tris with 1 μM chemokine were added to the buffer stream via the 12-port valve. After each application of the chemokine, the capture molecules on the sensor surface were regenerated by letting 50 μl of 4 M NaCl flow over the surface.

The measurement graphs in Fig. 15.11 show the frequency behavior of two BioMEMS sensors functionalized with (a) a functional capture molecule and (b) a non-functional capture molecule on application of a chemokine and NaCl. It can be seen that the sensor with the functional capture molecule showed a frequency shift of up to 1.7 KHz on the application of the chemokine in concentrations of 1 μM. In comparison, the sensor with the non-functional capture molecule showed only a weak shift of up to 0.35 KHz. These results substantiate the functionality of the BioMEMS sensor and the direct functionalization procedure on silicon dioxide. It should be noted that a good repeatability of results has been observed.

15.5 Summary

In this work, a novel PoC device for detection of respiratory infectious viral diseases at newborns using a novel approach based on multiplexed measurements of clinical relevant combinations of host and pathogen derived biomarkers in complex media is being developed. A novel piezoelectric FPW BioMEMS sensor, a key component of the proposed PoC device, has been designed, fabricated and tested. The main challenges in the fabrication route, such as stress management and patterning of the PZT layer have been effectively resolved. The setup for readout of the sensor's resonant frequency downshift has been developed and successfully implemented. The experimentally obtained results have validated the sensor's functionality.

In order to further enhance the sensor performance (e.g. enable the detection of chemokines (biomarkers) at much lower concentrations) several steps are going to be undertaken in future works. First, the sensor readout could be modified to an oscillator/frequency counter approach, which allows a more precise measurement

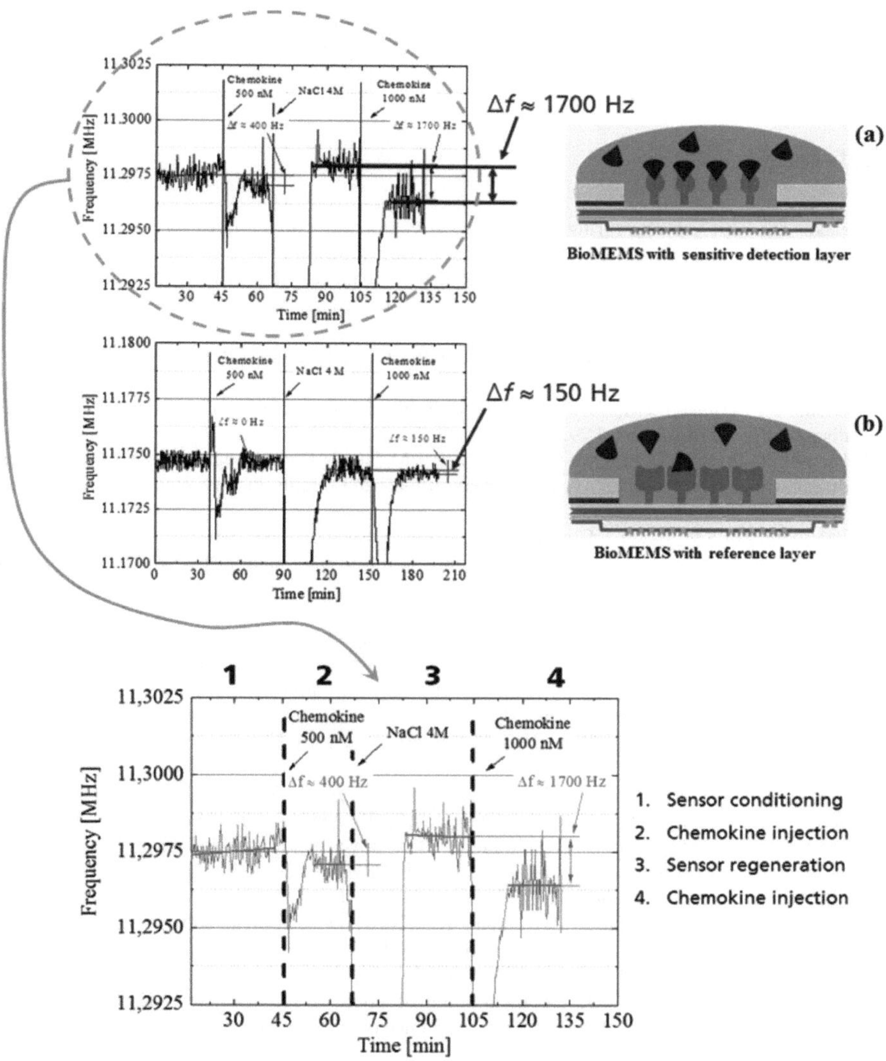

Fig. 15.11 The behavior of the resonance frequency of BioMEMS FPW-sensors functionalized with (**a**) a sensitive detection layer that binds chemokines and (**b**) reference layer that does not bind chemokines

of the resonant frequency. Second, the mass change induced by the binding of the chemokines to the capture molecules can be increased by implementing a so called competitive binding method. With these adjustments a significant improvement of the sensor performance is likely to be achieved.

Acknowledgements This project has received funding from the European Union's Horizon 2020 research and innovation programme under grant agreement no. 634415.

References

1. Baker BK (2013) Empowering patients and strengthening communities for real health workforce and funding targets. Bull World Health Organ 91:889
2. Chan C, Mak W, Cheung K, Sin K, Yu C, Rainer T, Renneberg R (2013) Evidence-based point-of-care diagnostics: current status and emerging technologies. Annu Rev Anal Chem (Palo Alto Calif) 6:191
3. Point of care testing: applications of microfluidic technologies. Report May 2014 (Yole-Developement, 2014)
4. Niemz A, Ferguson TM, Boyle DS (2011) Point-of-care nucleic acid testing for infectious diseases. Trends Biotechnol 29:240
5. Pai NP, Vadnais C, Denkinger C, Engel N, Pai M (2012) Point-of-care testing for infectious diseases: diversity, complexity, and barriers in low- and middle-income countries. PLoS Med 9(9):1
6. Palamountain KM, Baker J, Cowan EP, Essajee S, Mazzola LT, Metzler M (2012) Perspectives on introduction and implementation of new point-of-care diagnostic tests. J Infect Dis 205(2):181
7. Peeling RW, McNerney R (2014) Emerging technologies in point-of-care molecular diagnostics for resource-limited settings. Expert Rev Mol Diagn 14:525
8. WHO (2009) Infant mortality from respiratory diseases. Fact Sheet CODE: RPG3_Air_E2
9. Bjurstrom J, Yantchev V, Katardjiev I (2006) Thin film Lamb wave resonant structures – the first approach. Solid State Electron 50:322
10. Shruthi GS, Amitha CV, Mathew BB (2014) Biosensors: a modern day achievement. J Instrum Technol 2(1):26
11. Rocha-Gaso MI, March-Iborra C, Montoya-Baides A, Arnau-Vives A (2009) Surface generated acoustic wave biosensors for the detection of pathogens: a review. Sensors 9:5740
12. Huang IY, Lee MC (2006) Development of a novel flexural plate wave biosensor for immunoglobulin-E detection by using SAM and MEMS technologies. In: Proceedings of the 5th IEEE conference on sensors (IEEE sensors 2006), Daegu, 22–25 Oct 2006
13. Huang IY, Lee MC, Chang YW, Huang RS (2007) Development and characterization of FPW based allergy biosensor. In: Proceedings of the 2007 IEEE international symposium on industrial electronics (IEEE ISIE 2007), Vigo, 4–7 June
14. Huang IY, Lee MC (2008) Development of a FPW allergy biosensor using NEMS technology. In: Proceedings of the 2008 Asia-Pacific conference on transducers and micro-nano technology (APCOT 2008), Tainan, 22–25 June 2008
15. Huang IY, Lee MC (2008) Development of a FPW allergy biosensor for human IgE detection by MEMS and cystamine-based SAM technologies. Sens Actuators B 132:340–348
16. Ballantine DS Jr, White RM, Martin SJ, Ricco AJ, Frye GC, Zellers ET, Wohltjec H (1997) Acoustic wave sensors: theory, design, and physicochemical application. Academic, New York
17. Bjurstrom J, Yantchev V, Katardjiev I (2006) Thin film Lamb wave resonant structures – the first approach. Solid State Electron 50:322

Chapter 16
Optical Polarization Characteristics of Zeolite Deposited on Different Substrates for Perspective Modulation Biosensor Systems

S. P. Rudenko, M. O. Stetsenko, L. S. Maksimenko, S. B. Kryvyi, B. K. Serdega, A. S. Fiorillo, and S. A. Pullano

Abstract This work reports the investigation of Zeolite 3A layers obtained by spin coating process at low temperature for registration alcohol vapor. Phase composition of Zeolite 3A layers was studied with the help of X-ray diffraction (XRD). Optical polarization properties and sensorics application of Zeolite 3A layers deposited onto different substrates were investigated by modulation polarimetry. Optical polarization properties were observed by measuring of angular characteristics of V-component of Stokes Vector. The influence of substrate materials for Zeolite 3A on amplitude parameter of angular characteristics of V-component of Stokes Vector was studied. The dielectric and conductor materials as Glass slide, Indium Tin Oxide nanofilm and Gold nanofilm were used. The highest amplitude value of V-component of Stokes Vector corresponds to the Zeolite layer on gold nanofilm. The response to ethanol vapor was registered for the Zeolite layer on gold nanofilm.

Keywords Zeolite · Modulation polarimetry · V-component of Stokes vector · Sensor application

16.1 Introduction

Zeolite layer is of a great interest as a promising material for applications in sensorics Zeolites possess regular porous structures, high surface areas, shape-selectivity, and unique solid acidity, having extensive applications in catalysis, ionic

S. P. Rudenko (✉) · M. O. Stetsenko · L. S. Maksimenko · S. B. Kryvyi · B.K. Serdega
V. Lashkaryov Institute of Semiconductor Physics, National Academy of Sciences of Ukraine, Kyiv, Ukraine

A. S. Fiorillo · S. A. Pullano
Department of Health Sciences, University Magna Græcia, Catanzaro, Italy

J. Bonča, S. Kruchinin (eds.), *Nanostructured Materials for the Detection of CBRN*, NATO Science for Peace and Security Series A: Chemistry and Biology, https://doi.org/10.1007/978-94-024-1304-5_16

exchange, and adsorption [1, 2]. Besides the different pore size and shape, the hydrophilic/hydrophobic nature of zeolites render them as useful selective sorbents and hosts for guest molecules (organic or inorganic) that are stable in gas and liquid phase [3].

Several methods for the preparation of zeolite films on surfaces have been described [4–7]. Most of these methods involve the direct nucleation and crystallization of the desired zeolite on the substrate surface [8, 9].

The structures of the zeolites consist of three-dimensional frameworks of SiO^4 and AlO^4 tetrahedra with pores. The aluminum ion is small enough to occupy the position in the center of the tetrahedron of four oxygen atoms, and the isomorphous replacement of Si^{4+} by Al^{3+} produces a negative charge in the lattice [10]. The net negative charge is balanced by the exchangeable cation such as sodium, potassium, calcium or proton, a counterion that is present in the pores [11]. The cationic exchange property is a function of the radio of Si to Al. This capacity is expressed as the number of cations per mass or volume unit available for exchange. The zeolite A is usually synthesized with sodium as the changeable cation [12] and the chemical composition of that zeolite is represented by $Na_{12}Al_{12}O_{48} \times 27H_2O$ [13].

Due to the position of the cations in the zeolite structure, the effective pore diameter may vary according to the type of compensation cation. In the case of a potassium (K^+) cation, the effective diameter of the pore is approximately 3 Å and the zeolite is known as 3A.

Zeolites have attracted increasing attention in the fields of biology [14] and chemical [15] sensor applications, and also optical sensors structures based on surface plasmon resonance [16, 17].

The future of our planet will indeed rely on the ability to detect dangerous agents (chemical, biological and radionuclear = CBRN) as well as to provide a clean environmental, health, and energy and information safety. The surface to volume ratio of a material increases drastically if it is nanoscaled. This has a tremendous effect of the working (e.g. of sensors and detectors utilizing nanoscaled materials). The large surface to volume ratio is decisive for many detectors. Since zeolite has a good mechanical stability and porosity it is a good candidate layer to increase the strength of the sensor element and increase its sensor sensitivity to detect of CBRN.

The zeolite-based materials used in gas sensors have received great attention in review [18]. The high and selective absorption properties of zeolites towards a specific gas greatly enhance the sensing selectivity for the gas. The compound or cluster sensing towards a gas assembled into the cages or channels of zeolites results in its high stability, and maximally elevates the sensing property of the materials.

Unique properties of zeolite in combination with sensitivity *o*f the modulation polarimetry and prospects of plasmonics are perspective for optical biosensor systems. That is why one can expect sensorial properties in porous zeolite layers. The knowledge about optical properties for zeolite layers in UV-Vis-NIR spectral region is limited. The gas applications of zeolite materials have been reviewed previously [19], but there have been no reports covering the uses of zeolite-based materials combined with modulation polarimetry for gas sensors.

The aim of our work was to study the optical polarization properties of zeolite 3A on different substrates using modulation polarimetry for future applications in low-cost devices for CBRN gas sensor. In this case, alcohol vapor detection is an area of intense interest. Detection of ethanol vapor is also an area of particular interest because of the need for responsive and reversible sensors to detect alcohol in human breath.

The modulation polarimetry technique is effective for diagnostics and characterization of the SPR features in nanosized films of noble metals and metal-dielectric nanocomposites [20–26]. In our work [27], the optical polarization properties was detected for zeolite 3A deposited onto (111) p-type silicon using this technique.

For investigation of the structural and optical polarization properties we used a combination of different analytical techniques such as: X-ray diffraction (XRD) and modulation polarimetry (MP).

16.2 Samples

Zeolite layers have been fabricated by spin coating process at low temperature for avoiding the process of synthesis of zeolite and its calcination process at very high temperature. This method is proposed by Fiorillo et al. [28, 29]. Zeolite type 3A powder with a grain dimensions ranging from 1 to 5 μm has been mixed with vegetable oil in order to create a thin layer composed by zeolite and a supporting matrix on a rigid substrate. The mixture was spun onto different substrate materials (glass, Gold and Indium Tin Oxide (ITO) nanofilms onto glass slides) and then cured at a temperature of 120 °C. This process partially activates the zeolite while the oil creates a supporting matrix for the nanoporous crystals in contact with the substrate. After washing with isopropyl alcohol, the sample undergoes another annealing process. This process is characterized by a relatively low curing temperature, compatible with standard integrated circuit technology, while the oil choice were made according to oil Iodine Value.

The obtained zeolite layers are characterized by thickness of 10 μm good degree of adhesion. Typical morphology structure of such zeolite films possess a grainy structure with a grain dimensions ranging from 1–5 μm and pores dimensions about 3 Å which was demonstrated earlier [27].

Thin gold (Au 99.999%) and Indium Tin Oxide films were deposited on a glass substrate maintained at room temperature by thermal evaporation from a molybdenum heater in vacuum (10^{-5} Pa). Deposition rate and film thickness were kept in the range of 1.0–1.5 nm/s, and $d = 45$ nm for Au nanofilm [23] and $d = 100$ nm for ITO nanofilm, respectively. The film thickness was controlled by deposition time. Glass slides (10×15 mm^2) with one-millimeter thickness were cut from commercially available cover slips (SCHOTT).

16.3 Experimental Details

16.3.1 X-Ray Diffraction

The study of the structural properties and phase composition of the films was carried out with the help of PANalytical X'Pert Pro MRD XL (X'Pert, PANalytical B.V., Almelo, The Netherlands) equipped with a $Cu_{K\alpha}$ source of radiation ($\lambda = 0.15406$ nm). W/Si parabolic mirror was used to creates high intensity parallel beam. The diffracted beam was collimated by parallel plate collimator with the acceptance angle 0.27°, used in combination with a 0.04 rad Soller slit.

16.3.2 Modulation Polarimetry

The optical scheme of the setup for the registration of an internal reflection of V – component of the Stokes vector in the Kretschmann geometry is shown in Fig. 16.1. This portable modulation polarimetry device was developed in Modulation Polarization Spectroscopy laboratory in cooperation with Yu.V. Ushenin science group at the V. Lashkaryov Institute of Semiconductor Physics NAS of Ukraine [30].

The semiconductor GaAs laser with a wavelength of $\lambda = 650$ nm has been used as a radiation source. The azimuth of linearly polarized radiation E is set up at the angle of 45° relative to the modulator axes. Consequently, the values of orthogonal components of the linearly polarized wave (E_s and E_p) become equal. The value of 45° between the azimuth of polarization and the optical axis of the modulator is the optimal one due to the effective transformation of polarization state. In this case, the modulator represents a dynamic phase plate [31], in which the compression–tension optical anisotropy is produced by a connected quartz (crystalline) resonator.

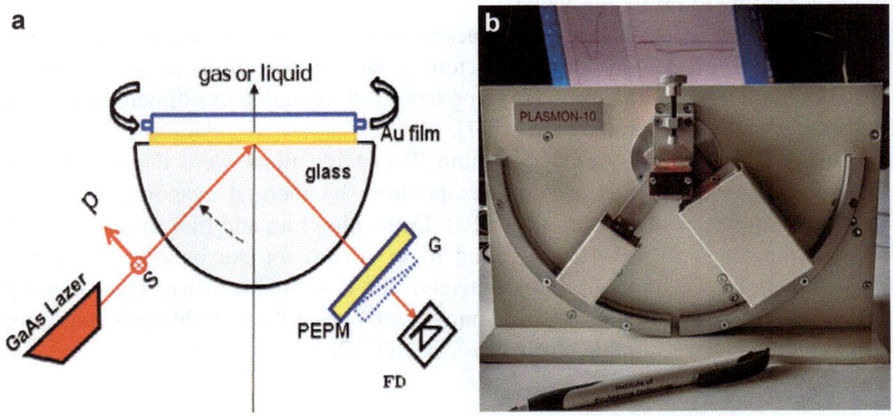

Fig. 16.1 Optical scheme and portable setup for measuring the phase characteristics of internal reflection

As it is shown in Fig. 16.1a, the photoelastic polarization modulator (PEPM), along with the linear polarizer G, serves as the polarization analyzer. These optical elements are located behind the investigated sample.

The investigated samples were deposited on a glass half cylinder of internal reflection with a refractive index $n_{\text{glass}} = 1.45$ on which linearly polarized laser radiation fells with the azimuth of the wave field 45° relative to the incidence plane. When this wave is reflected from the working prism surface due to its various reflection coefficients of s- and p-polarization both in amplitude and phase, it turns into an elliptically polarized one. The state of an elliptically polarized reflected wave is analyzed by the polarization modulator in conjunction with the linear polarizer. The resulted signal is recorded by the photodetector (FD). This signal is registered by a selective amplifier equipped with the Lock-in detector tuned to the modulation frequency ($f = 200\,\text{kHz}$).

In this case, the circular polarization is characterized by the quantity of recorded signal $\text{V}(\theta) = E_s \cdot E_p \cdot \sin \Delta$. This signal is determined by the phase difference Δ between the orthogonal components E_s and E_p and their product of electric field intensity (E_s and E_p, respectively).

The use of the Fresnel formulas for a multilayer structure [32] and the matrix formalism based on the Stokes vector and the corresponding 4×4 Muller matrix for each optical element [33, 34] of the Kretschmann geometry allows to examine the changes in the polarization state of the radiation, passed through all optical elements.

16.4 Experimental Results and Discussion

The X-ray diffractogram of the 3A zeolite can be observed in Fig. 16.2, which presents high crystallinity. The plot shows all the characteristic peaks matching with the reported in the diffraction equipment library, with corresponding formula to Zeolite A ($Na_{12}(AlSiO_4)_{12}Al_{12}Na_{12}O_{48}Si_{12}$, Ref. Code 010-71-0370) [35].

To study the dependencies of optical polarization properties of zeolite 3A on different substrates using modulation polarimetry, it was used the next substrate materials with dielectric and conductor properties as Glass slide, Indium Tin Oxide nanofilm and Gold nanofilm.

The dependencies of the V-component of the Stokes vector on the incidence angle of light are shown in Fig. 16.3 for zeolite 3A layer with thickness $10\,\mu\text{m}$ deposited on glass, gold and ITO substrates. Among obtained characteristics of $\text{V}(\theta)$ the highest values corresponds to the zeolite layer on gold nanofilm against presented characteristics both for glass slide and ITO nanofilm. This can be explained by the presence of gold nanosized film and it resonant interaction with incident radiation due to the surface plasmons excitations. It is known, if the thickness of Au film in the direction of wave propagation is far less its length and less skin-layer thickness, the surface plasmons can be resonant excited with formation of surface electromagnetic waves, which decay exponentially away from

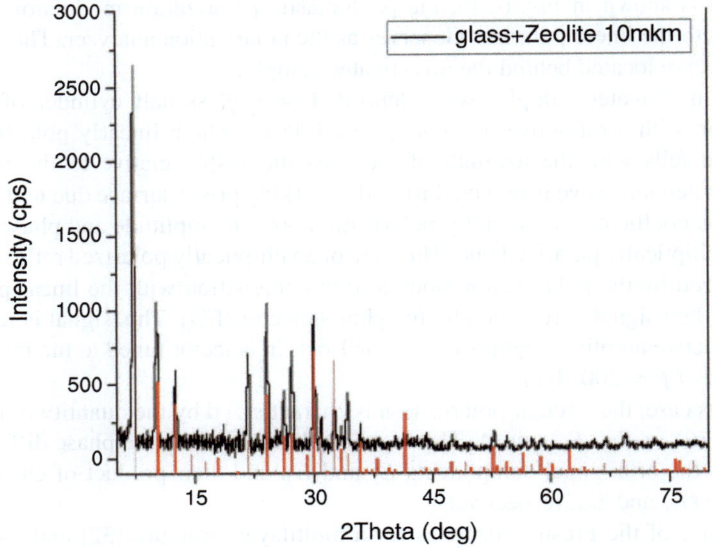

Fig. 16.2 X−ray diffraction of Zeolite 3A layer deposited on Glass slide

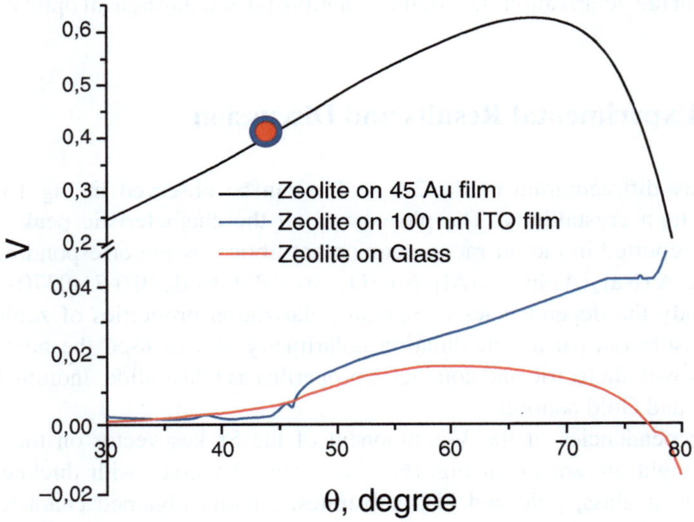

Fig. 16.3 Angular dependences of V-component of Stokes vector for different substrate

the surface [36]. The ability to manifest itself the surface plasmon resonance in gold nanofilm with thickness of 45 nm at a wavelength of 650 nm leads to increase the intensity of V-component in the appropriate angular characteristic of V(θ) when covering of zeolite layer.

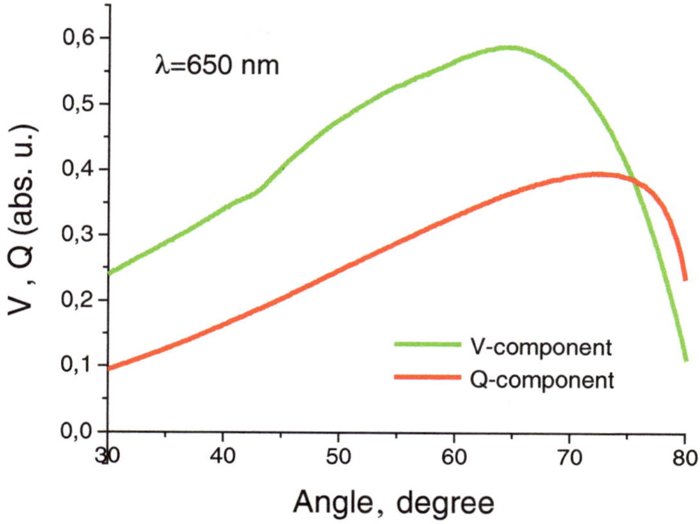

Fig. 16.4 Angular dependences of Q and V -components of Stokes vector for structure Zeolite on Gold nanofilm

The surface plasmon resonance is a polarization-dependent effect and demonstrates both the properties of the amplitude anisotropy and the phase anisotropy [37]. Modulation polarimetry technique has a considerable experience in the study of interaction of light and matter by operating with the analog of Q- and V- the Stokes components [38].

Figure 16.4 shows the comparison of the angular characteristics of the polarization difference $\rho(\theta)$ and the V(θ) – component of the Stokes vector for zeolite layer on gold nanofilm at a wavelength of 630 nm. It is seen that in all measurement range of angles the intensity of V-component is greater than the appropriate for parameter of polarization difference ρ, which is an analog of Q-component.

Moreover, for sensor applications it is preferable to perform the registration of changing the intensity of measurement signal at the greatest slope of the corresponding curve. This feature is more pronounced for the angular characteristic of V(θ), which will be used for further study the kinetic process for gas evaporation. When the incidence angle of light is fixed, the phase difference Δ is as a function of surrounding medium. Thus any change in the dielectric properties of the surrounding detected medium led to change the intensity of the V-component of the Stokes vector.

With the help of combination of unique properties of the zeolite as a sensor chip with the advantages of the modulation polarimetry technique, the detection of gas evaporation was performed. The ethanol vapors of 96% CH_3CH_2OH (SIGMA-ALDRICH) with refractive index of ethyl alcohol $n = 1.000878$ were used as the detectable substance. Figure 16.5 shows the difference signal in the V-components

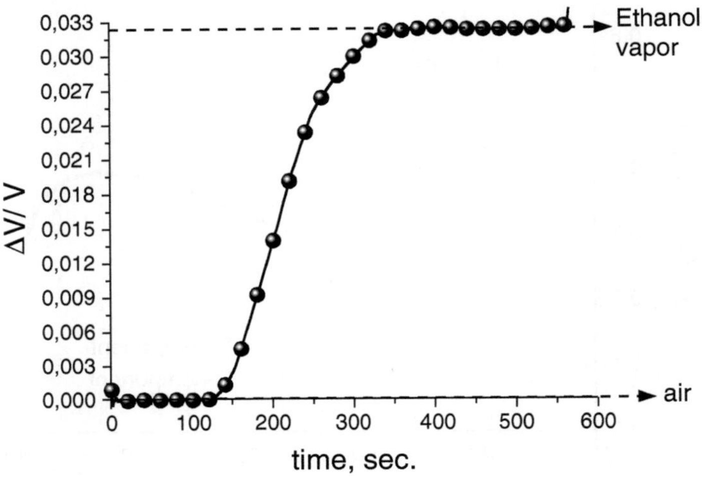

Fig. 16.5 Kinetic characteristic of V-component of Stokes vector response on ethanol vapor

in the beginning and end-point study, e.g. absorption of ethyl alcohol in the zeolite layer led to ΔV shift to 0.033 under ambient conditions. The angle of incidence is fixed at $\theta = 47°$ and is pointed in the Fig. 16.5.

Recovery of the zeolite chip after sensing measurements required purging with heat.

We are planning to investigate structure of zeolite layer on ITO as plasmonic chip for detection liquid or gas mediums.

16.5 Conclusions

In this work the sensors properties of zeolite layers with 3A type were tested using their unique surface properties combined with high sensitivity of modulation polarimetry in sensors applications.

Characterization of zeolite 3A layers obtained by spin coating deposition have been performed by the modulation polarimetry technique and X-ray diffraction.

The influence of different substrate materials as glass, ITO and gold nanofilms on the optical polarization properties of zeolite 3A layers were obtained using of V-component of the Stokes vector.

It was found that response to ethanol vapors makes the zeolite films perspective as the chips for sensors devices based on modulation polarimetry. For this task the structures of zeolite 3A on gold nanofilm are better suited.

Zeolites are often referred to as molecular sieves, since the crystalline nature allows for strict size discrimination. Change of optical polarization properties due to presence of any guest species upon interaction with medium can be used for sensing. The results obtained in this work are a framework for future determination of the characteristic features of sensoric applications for modified porous surface of zeolite thin layer in gas sensors.

References

1. Cejka J, Corma A, Zones S (2010) Zeolites and catalysis: synthesis, reactions and applications. Wiley-VCH, New York
2. Zaarour M, Dong B, Naydenova I, Retoux R, Mintova S (2014) Progress in zeolite synthesis promotes advanced applications. Microporous Mesoporous Mater 189:11–21
3. Ismail AA, Mohamed RM, Ibrahim IA, Kini G, Koopman B (2010) Synthesis, optimization and characterization of zeolite A and its ion-exchange properties. Colloids Surf A 366:80–87
4. Valtchev V, Mintova S, Konstantinov L (1995) Influence of metal substrate properties on the kinetics of zeolite film formation. Zeolites 15:679–683
5. Geus E, Mulder A, Vischjager D, Schoonman J, van Bekkum H (1991) Synthesis of a ceramic zeolite membrane by means of a dip-coating technique. Key Eng Mater 61–62:461–464
6. Geus E, van Bekkum H, Bakker W, Moulijn J (1993) High-temperature stainless steel supported zeolite (MFI) membranes: preparation, module construction, and permeation experiments. Microporous Mater 1(2):131–147
7. Mintova S, Edlund J, Schoeman B, Valtchev V, Sterte J (1997) Continuous films of zeolite ZSM-5 on modified gold surfaces. Chem Commun (1):15–16. https://doi.org/10.1039/A606906A
8. Balkus KJ Jr, Scott AS (1999) Zeolite coatings on three-dimensional objects via laser ablation. Chem Mater 11(2):189–191
9. Fiorillo AS, Pullano SA, Tiriolo R, Vinko JD (2016) Iono-electronic interface based on innovative low temperature zeolite coated NMOS (circuits) for bio-nanosensor manufacture, nanomaterials for security. NATO Sci Peace Secur Ser A Chem Biol 201–214. https://link.springer.com/chapter/10.1007/978-94-017-7593-9_16
10. Erdem E, Karapinar N, Dona R (2004) The removal of heavy metal cations by natural zeolites. J Colloid Interface Sci 280(2):309–314
11. Rondón W, Freire D, de Benzo Z, Sifontes AB, González Y, Valero M, Brito JL (2013) Application of 3A zeolite prepared from venezuelan kaolin for removal of Pb (II) from wastewater and its determination by flame atomic absorption spectrometry. Am J Anal Chem 4:584–593
12. Melo CR, Riella HG, Kuhnen NC, Angioletto E, Melo AR, Bernardin ALM, da Rocha MR, da Silva L (2012) Synthesis of 4A zeolites from kaolin for obtaining 5A zeolites through ionic exchange for adsorption of arsenic. Mater Sci Eng B 177(4):345–349
13. Loiola AR, Andrade JCRA, Sasak JM, da Silva LRD (2012) Structural analysis of zeolite NaA synthesized by a cost-effective hydrothermal method using kaolin and its use as water softener. J Colloid Interface Sci 367(1):34–39
14. Kucherenko I, Soldatkin O, Kasap BO, Kirdeciler SK, Kurc BA, Jaffrezic-Renault N, Soldatkin A, Lagarde F, Dzyadevych S (2015) Nanosized zeolites as a perspective material for conductometric biosensors creation. Nanoscale Res Lett 10(209):1–9
15. Vilaseca M, Coronas J, Cirera A, Cornet A, Morante JR, Santamarıa J (2003) Use of zeolite films to improve the selectivity of reactive gas sensors. Catal Today 82:179–185
16. Lopez-Bastidas C, Smolentseva E, Petranovskii V, Machorro R (2013) Plasmon features of coinage metal nanoparticles supported on zeolites. Plasmonics 8(4):1551–1558

17. Bottoni U, Tiriolo R, Pullano SA, Dastoli S, Amoruso GF, Nisticò SP, Fiorillo AS (2016) Infrared saliva analysis of psoriatic and diabetic patients: similarities in protein components. IEEE Trans Biomed Eng 63(2):379–384
18. Xu X, Wang J, Long Y (2006) Review, Zeolite-based materials for gas sensors. Sensors 6:1751–1764
19. Zheng Y, Li X, Dutta PK (2012) Review, exploitation of unique properties of zeolites in the development of gas sensors. Sensors 12(4):5170–5194
20. Stetsenko MO, Maksimenko LS, Rudenko SP et al (2016) Surface plasmon's dispersion properties of porous gold films. Nanoscale Res Lett 11(1):116
21. Stetsenko MO, Rudenko SP, Maksimenko LS, Serdega BK, Pluñhery O, Snegir SV (2017) Solid-state synthesis and optical properties of gold nanoparticle assemblies on a glass surface. Nanoscale Res Lett 12:348
22. Stetsenko MÎ, Voznyi AA, Kosyak VV, Rudenko SP, Maksimenko LS, Serdega BK, Opanasuk AS (2016) Plasmonic effects in tin disulfide nanostructured thin films obtained by the close-spaced vacuum sublimation. Plasmonics 12:1213–1220
23. Rudenko SP, Maksimenko LS, Matyash IE, Mischuk OM (2016) Diagnostic of surface plasmons resonances in nanosized gold films by modulation polarization spectroscopy. Plasmonics 11(2):557–563
24. Ermakov V, Kruchinin S, Fujiwara A (2008) Electronic nanosensors based on nanotransistor with bistability behaviour. In: Bonca J, Kruchinin S (eds) Proceeding of NATO ARW "Electron transport in nanosystems". Springer, pp 341–349
25. Rodionov VE, Shnidko IN, Zolotovsky A, Kruchinin SP (2013) Electroluminescence of Y_2O_3:Eu and Y_2O_3:Sm films. Mater Sci 31:232–239
26. Ermakov V, Kruchinin S, Hori H, Fujiwara A (2007) Phenomena of strong electron correlastion in the resonant tunneling. Int J Mod Phys B 11:827–835
27. Fiorillo AS, Rudenko SP, Stetsenko MO et al (2016) Optical polarization properties of zeolite thin films: aspects for medical applications. In: IEEE international symposium on medical measurements and applications, Benevento, pp 332–335. https://doi.org/10.1109/MeMeA.2016.7533757
28. Fiorillo AS, Tiriolo R, Pullano SA (2015) Absorption of urea into zeolite layer integrated with microelectronic circuits. IEEE Trans Nanotechnol 14(2):214–217
29. Fiorillo AS (2012) Deposition of zeolite thin layers onto silicon wafers for biomedical use. IEEE Trans Nanotechnol 11(4):654–656
30. Stetsenko MO, Matyash IE, Rudenko SP, Minailova IA, Maksimenko LS, Serdega BK (2017) New type of plasmonic biosensors based on modulation polarimetry technique. In: IEEE international young scientists forum on applied physics and engineering, Lviv, pp 100–103
31. Jasperson SN, Schnatterly SE (1969) An improved method for high reflectivity ellipsometry based on a new polarization modulation technique. Rev Sci Instrum 40(6):761
32. Born M, Wolf E (1999) Principles of optics. Cambridge University Press, Cambridge
33. Gerrard A, Burch JM (1975) Introduction to matrix methods in optics. Wiley, London
34. Azzam RMA, Bashara NMI (1989) Ellipsometry and polarized light. North Holland, New York
35. PDF-ICDD. Power Diffraction File (Set-1-S1) International Centre for Diffraction Data
36. Maier SA (2007) Plasmonics: fundamental and application. Springer, New York
37. Serdega BK, Rudenko SP, Maksimenko LS, Matyash IE (2011) Plasmonic optical properties and the polarization modulation technique. In: Mishchenko MI, Yatskiv YaS, Rosenbush VK, Videen G (eds) Polarimetric detection, characterization and remote sensing. Springer, Dordrecht, pp 473–500
38. Maksimenko LS, Matyash IE, Minailova IA, Mishchuk ON, Rudenko SP, Serdega BK (2010) Polarization stokes polarimetry of the amplitude and phase characteristics of surface plasmon polariton resonance. Opt Spectrosc 109(5):808–813

Chapter 17
Design, Analysis and Application of Dynamic Visual Cryptography for Visual Inspection of Biomedical Systems

A. Palevicius, G. Janusas, M. Ragulskis, P. Palevicius, and A. Sodah

Abstract Health care industry has a growing need for advanced tools that enables the discovery the existence or concentration of biological analytes. In this article a concept of novel cantilever-type microsystem platform in the field of biomechanics using optical technique of image encryption and communication scheme based on computer generated holography is proposed. An image hiding technique based on computer generated holography and dynamic visual cryptography is employed. Dynamic visual cryptography is a visual cryptography scheme based on time-averaging geometric moiré. It is used together with Gerchberg–Saxton algorithm and 3D microstructure manufacturing techniques to design the optical scheme. Stochastic moiré grating is used to embed the secret image into a cover image. The image can visually decoded by a naked eye – the secret is revealed if the amplitude of harmonic oscillations in the Fourier plane corresponds to an accurately preselected value. The phase information of computer generated hologram is then formed on a piezoelectric cantilever-type microsystem platform using electron beam lithography. It serves as an optical element for visual inspections of dynamical changes in investigated biological environment.

Keywords Computer generated holography · Dynamic visual cryptography · Biomedical systems

---

A. Palevicius (✉) · G. Janusas · A. Sodah
Faculty of Mechanical Engineering and Design, Kaunas University of Technology, Kaunas, Lithuania
e-mail: arvydas.palevicius@ktu.lt; giedrius.janusas@ktu.lt; amer.sodah@ktu.lt

M. Ragulskis · P. Palevicius
Faculty of Mathematics and Natural Sciences, Kaunas University of Technology, Kaunas, Lithuania
e-mail: minvydas.ragulskis@ktu.lt; paulius.palevicius@ktu.lt

J. Bonča, S. Kruchinin (eds.), *Nanostructured Materials for the Detection of CBRN*, NATO Science for Peace and Security Series A: Chemistry and Biology,
https://doi.org/10.1007/978-94-024-1304-5_17

17.1 Introduction

It is difficult to imagine health care industry without advanced tools for inspection and biological analysis of the biological structures without using functional elements containing smart materials where created sensitive element provides functionality to analyze, manipulate, or assembly of chemical and/or biological entities in micro scale such as, pacemakers [1], vascular stents [2–4], hearing aids [5], cochlea implants [6], microelectrodes [7], microendoscopes [8], microcatheter systems [9], biosensors [10, 11], viscometric sensors [12–15], etc.

The mentioned techniques are based on the interaction of a sensing element with biological environment. We can evaluate and characterize the processes taking place in investigated biological environment because of the changes in behaviour of the mechanical characteristics of used smarts materials in microsystems or optical properties of surfaces. Smart materials based on the piezoelectricity provides a lot of opportunities in enabling inspection and analysis of biological objects in real time. Well known piezoelectric films Zirconium Titanate (PZT), Polyvinylidene fluoride (PVDF) and Barium Titanate (BaTiO3) are used and are applied as a sensing and actuating element in majority microelectromechanic systems. Therefore, they are useful in designing novel tools for the inspection in the environmental, industrial and defense areas. Furthermore, they may be used for biological environment investigation. The main benefit of application of such materials are compact size, simplicity in use and control, high sensitivity, absence of radioactive, and biocompatibility. Therefore, it makes them very attractive in design of biomedical devices.

Visual cryptography is a cryptographic method for the encryption of visual information in such a way that a decryption can be performed by human visual system without the aid of computers. Visual cryptography was first introduced by Naor and Shamir [16]. They introduced a visual secret sharing scheme, where an image was broken up into n shares in such a way that only a person with all n shares could decrypt the secrect image. Dynamic visual cryptography is another image hiding technique. It combines two different methods – visual cryptography and time-averaging geometric moiré by exploiting time-averaging moiré for synthesis of a predefined pattern of time-averaged fringes [17].

Computer generated holography is the technique which enables digital generation of holographic interference patterns. The main difference to classical holography is that diffractive optical element is optimized mathematically by using a computer rather than experimentally [18]. Computer generated holograms are being used for construction of high spatial-frequency gratings [19], direct laser beam writing [20], gray-tone lithography [21]. Different computational techniques can be used to compute a computer generated hologram of a virtual object. Fourier transform method and point source holograms are two different techniques that most of the computation algorithms relay on. Adaptive–Additive algorithm [22] and Gerchberg–Saxton algorithm [23] are two common methods based on Fourier transform algorithms and are used to calculate a distribution of phase hologram. Computer generated holography has been employed in image hiding and encryption schemes. A method of optical image encryption using pixel-scrambling technology and binary Fourier transform computer generated holography was proposed in [24], where orders of the pixel scrambling as well as the encrypted image are used as the keys to decrypt the original image. Computer generated holography has been exploited in cryptography by using its theory with the 4f double random phase encryption system in order to implement optical image encryption [25]. Computer generated holograms has been also used for three-dimensional information hierarchical encryption [26]. One of the most recent studies proposes symmetric and asymmetric hybrid cryptosystem based on compressive sensing and computer generated holography [27].

A concept of novel cantilever-type microsystem platform in the field of biomechanics using optical technique of image encryption and communication based on computer generated holography is proposed in this article. The article is structured as follows. The theory and experimental results on the image encryption scheme based on time-averaging moiré and computer generated holography are given in Sect. 17.2. Concept and design of a cantilever-type microsystem platform are given in Sect. 17.3. Finally, concluding remarks are given in Sect. 17.4.

17.2 The Optical Image Hiding Scheme

The optical image hiding scheme based on computer based holography and time-averaging moiré and its experimental implementation has been introduced in [28–32]. In this article it is being applied for visual inspection of dynamical changes in biological environments. During the encryption process of this optical scheme, an image hiding technique based on optical time-averaging moiré is used to encrypt the secret image. Furthermore, Gerchberg–Saxton algorithm is used to obtain the phase data from the previously encrypted secret image. In the end, different 3D microstructure manufacturing techniques are used to fabricate the computer based hologram that corresponds to the retrieved phase data. The decryption is performed by illuminating the computer generated hologram by a coherent laser beam. Light distribution of the encrypted data is obtained in the Fourier plane. Oscillation of obtained image by a predefined amplitude decrypts the encrypted image and the secret image is retrieved.

The methods used for encrypting the image by using dynamic visual cryptography and retrieving the phase distribution of the computer generated hologram are given in the following sections.

17.2.1 Image Hiding Based on Optical Time-Averaging Moiré

In this section the image hiding technique based on optical time-averaging moiré that is used to encrypt the secret image is given [17]. Let us define a one-dimensional harmonic moiré grating:

$$F(x) = \frac{1}{2} + \frac{1}{2}\cos\left(\frac{2\pi}{\lambda}x\right), \tag{17.1}$$

where λ is the pitch of the grating; $F(x)$ is a grayscale level ranging from 0 to 1. Here 0 represents black and 1 represents white color. This one-dimensional harmonic moiré grating is performing non-deformable oscillations around the state of equilibrium. A deflection from the state of equilibrium does not depend on variable x:

$$u(x,t) = u(t) = a\sin(\omega t + \varphi), \tag{17.2}$$

where ω is the angular frequency; φ is the phase and a is the amplitude of the oscillation. Thus, the time-averaged image is defined as follows [33]:

$$\bar{F}(x) = \lim_{T\to\infty}\frac{1}{T}\int_0^T F(x - a\sin(\omega t + \varphi))\,dt = \frac{1}{2} + \frac{1}{2}\cos\left(\frac{2\pi}{\lambda}x\right)J_0\left(\frac{2\pi}{\lambda}a\right), \tag{17.3}$$

where T is the exposure time; J_0 is the zero-th order Bessel function of the first kind. The original moiré grating is mapped into a time-averaged fringe ($\bar{F}(x) = 0.5$) when J_0 becomes equal to zero. Thus, the relationship between the pitch of the moiré grating λ, the amplitude of harmonic oscillations a and the consecutive number of the time-averaged moiré fringe k reads:

$$\frac{2\pi}{\lambda}a_k = r_k;\ k = 1, 2, \ldots, \tag{17.4}$$

where r_k is the k-th root of J_0; a_k is the discrete value of the amplitude which results in the k-th time-averaged fringe in the time-averaged image. The phase matching algorithm is used to eliminate discontinuities at the boundaries between the background and the secret image and the stochastic initial phase deflection algorithm is used to encrypt the secret.

17.2.2 Formation of Computer Generated Hologram

The Gerchberg–Saxton algorithm is used to estimate the phase data of computer generated hologram from the encrypted image (discussed in the previous section)

[23, 34]. Initially, a random number generator can be used to generate a initial phase distribution from a stochastic variable distributed in the interval $[-\pi; \pi]$. Alternatively, initial phase can be set to uniform distribution such as:

$$\varphi_0^H = 0. \tag{17.5}$$

The following equations describe one iteration of the algorithm, during which an improved phase distribution in the hologram plane is calculated:

$$u_n^H = A(I_H)\exp\left[i\varphi_{n-1}^H\right], \tag{17.6}$$

$$\varphi_n^T = P\left(FFT\left(u_n^H\right)\right), \tag{17.7}$$

$$u_n^T = A(I_T)\exp\left[i\varphi_n^T\right], \tag{17.8}$$

$$\varphi_n^H = P\left(FFT^{-1}\left(u_n^T\right)\right), \tag{17.9}$$

where u_n^H and u_n^T are light fields represented by 2D array of complex numbers in the hologram and target planes receptively; I_H is the intensity distribution of the unshaped light beam in the hologram plane; φ_n^T is the phase distribution in the target plane; I_T is the target intensity distribution in the observation plane; φ_n^H is the phase distribution in the hologram plane. The iterative algorithm is repeated until the decision to terminate is made – the difference between the approximated intensity and the expected intensity in target plane is below the specified threshold with every new iteration. Here φ_n^H is a retrieved phase data of a complex 3D microstructure which is used to fabricate a computer generated hologram.

17.2.3 Computational Simulation of Optical Image Hiding Scheme

The results of the computational simulation of the optical image hiding scheme based on computer based holography and time-averaging moiré are shown in Fig. 17.1. The secret image containing a black solid circle on white background is given in part (a). The original encrypted image using image hiding technique based on optical time-averaging moiré is given in part (b). The computational simulations were performed using 10×10 composite pixel size. This results in reduction of noise influence that is related to diffraction and interference theoretical limits of computer generated holography. The phase data of computer generated hologram is given in part (c). The reconstructed image of computer generated hologram in observation plane is given in part (d). The decrypted image without

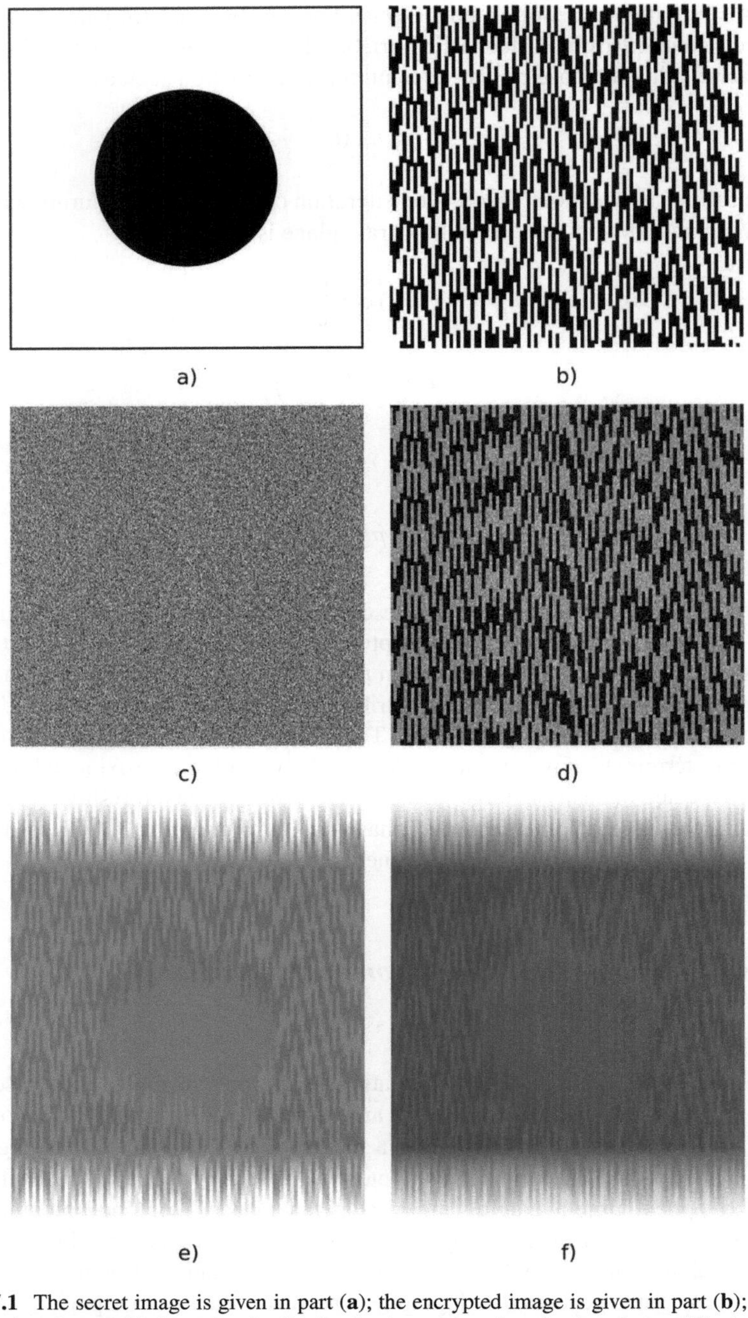

Fig. 17.1 The secret image is given in part (**a**); the encrypted image is given in part (**b**); a phase data is given in part (**c**); intensity distribution in observation plane is given in part (**d**); decryption results with and without Gerchberg–Saxton algorithm are given in parts (**e**) and (**f**) respectively [29]

Gerchberg–Saxton algorithm is given in part (e). The decrypted image with the applied Gerchberg–Saxton algorithm is shown in part (f). One should note, that various contrast enhancement techniques (such as the one given in [17]) can be used to increase image quality and readability.

17.3 The Cantilever-Type Microsystem Platform

The designed biomechanical system is a piezoelectric based cantilever. Schematic view of the designed system is given in Fig. 17.2. The cantilever is composed of 50 μm thick copper film (Young's modulus $E = 117$ GPa and density $\rho = 8940\,\text{kg/m}^2$), 20 nm thick aluminum electrode (Young's modulus $E = 70$ GPa and density $\rho = 2700\,\text{kg/m}^2$), and 50 μm thick piezoelectric composite (Young's modulus $E = 3.9$ GPa and density $\rho = 6298\,\text{kg/m}^2$) – polyvinyl butyral 20% mixed together with PZT powder. Cantilever is then coated with thin film (500 nm) of polyethylene (Young's modulus $E = 0.8$ GPa and density $\rho = 940\,\text{kg/m}^2$) working as an insulator from environmental conditions. Finally, a computer generated hologram is attached to a cantilever. The procedure and materials for the production of the computer generated hologram are given in detail in [29]. Computer generated hologram is designed for exposition on polymethyl methacrylate (Molecular mass of polymethyl methacrylate is 35 K and refractive index is 1.49) and reconstruction by a HeNe laser (wavelength $\lambda = 632.8$ nm). Pixels of designed computer generated hologram have eight grayscale levels ranging from black to white. Black color represents the biggest thickness of polymethyl methacrylate (it is not exposed) and white color represents the smallest thickness of polymethyl methacrylate under maximum exposition. Raith e-LiNEplus high resolution system was used for electron beam lithography process. Image of computer generated hologram exposed on polymethyl methacrylate is shown in Fig. 17.3. The proposed technique can be employed for monitoring of biomedical systems when the reconstructed image performs oscillations around the state of equilibrium in the observation plane – the changes in amplitude of oscillations due to changes in environmental conditions would result in changes of the target image in the observation plane. The

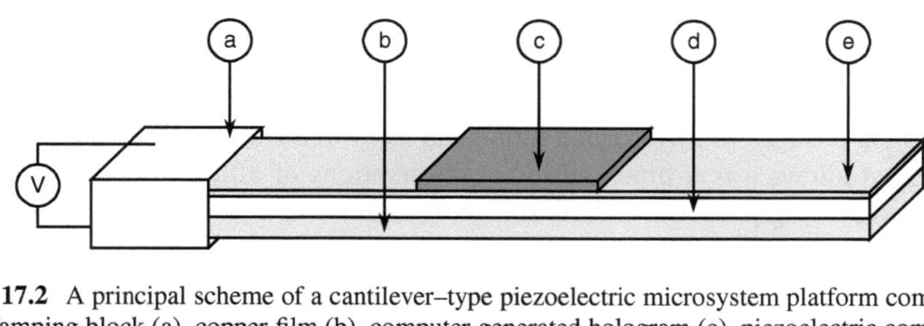

Fig. 17.2 A principal scheme of a cantilever–type piezoelectric microsystem platform composed of clamping block (a), copper film (b), computer generated hologram (c), piezoelectric composite (d), and electrode (e)

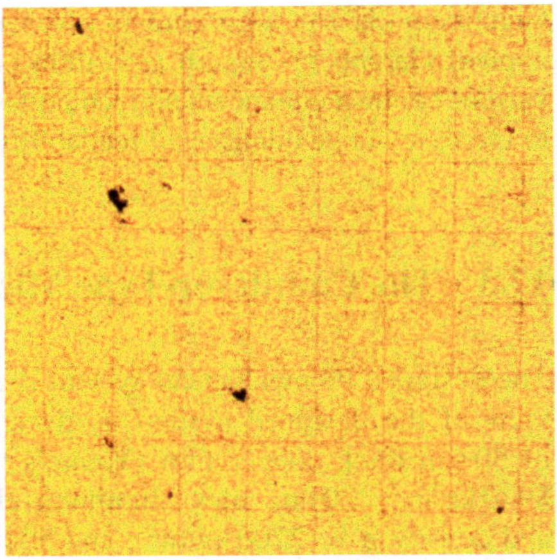

Fig. 17.3 Computer generated hologram exposed on polymethyl methacrylate

proposed technique could be employed for monitoring of glucose levels in diabetes patients. Cantilever could be immersed in the sensing fluid, which is made of protein concanavalin-A and polymer dextran. Furthermore, this sensing fluid would be exchanging glucose with the fluid outside this setup by a cellulose acetate semi-permeable membrane. Changes in the viscosity of the sensing fluid (the result of the interaction of glucose with concanavalin-A) introduce changes in the amplitude of the oscillations of the cantilever. As mentioned before, changes in amplitude of oscillations result in changes of the target image in the observation plane. This approach can be used to detect changes in any other biological environment.

17.4 Concluding Remarks

A concept of novel cantilever-type microsystem platform in the field of biomechanics using optical technique of image encryption and communication based on computer generated holography is proposed in this article. An image hiding technique based on computer generated holography and dynamic visual cryptography is employed to retriever the phase information of computer generated hologram. This computer generated hologram is then formed on a piezoelectric cantilever-type microsystem platform. Electron beam lithography is exploited for physical 3D patterning. It serves as an optical element for visual inspections of dynamical changes in investigated biological environment. Combination of unique properties allows a real time and direct observations of affinity interactions. The proposed sensing platform with piezoelectric effect employs the active method for measurements in medical or pharmaceutical fields.

Acknowledgements This research was funded by a grants (No. MIP-081/2015 and S-MIP-17-102) from the Research Council of Lithuania.

References

1. Yeknami AF (2016) A 300-mV modulator using a gain-enhanced, inverter-based amplifier for medical implant devices. J Low Power Electron Appl 6(1):1
2. Koenig O, Zengerle D, Perle N, Hossfeld S, Neumann B, Behring A, Avci-Adali M, Walker T, Schlensak C, Wendel HP, Nolte A (2017) RNA-eluting surfaces for the modulation of gene expression as a novel stent concept. Pharmaceuticals 10(1):23
3. Su J, Xu H, Sun J, Gong X, Zhao H (2013) Dual delivery of BMP-2 and bFGF from a new nano-composite scaffold, loaded with vascular stents for large-size mandibular defect regeneration. Int J Mol Sci 14(6):12714
4. Xu H, Su J, Sun J, Ren T (2012) Preparation and characterization of new nano-composite scaffolds loaded with vascular stents. Int J Mol Sci 13(3):3366
5. An P, Yuan W, Ren S (2009) MEMS biomimetic acoustic pressure gradient sensitive structure for sound source localization. Sensors 9(7):5637
6. Ilik B, Koyuncuoglu A, Ulusan H, Chamanian S, Isik D, Sardan-Sukas O, Kulah H (2017) Thin film PZT acoustic sensor for fully implantable cochlear implants. Proceedings 1(4):366
7. Bonanno A, Sanginario A, Marasso SL, Miccoli B, Bejtka K, Benetto S, Demarchi D (2016) A multipurpose CMOS platform for nanosensing. Sensors 16(12):2034
8. Risi MD, Makhlouf H, Rouse AR, Tanbakuchi AA, Gmitro AF (2014) Design and performance of a multi-point scan confocal microendoscope. Photonics 1(4):421
9. Lu Q, Yadid-Pecht O, Sadowski DC, Mintchev MP (2014) A catheter-based acoustic interrogation device for monitoring motility dynamics of the lower esophageal sphincter. Sensors 14(8):14700
10. Pasinszki T, Krebsz M, Tung TT, Losic D (2017) Carbon nanomaterial based biosensors for non-invasive detection of cancer and disease biomarkers for clinical diagnosis. Sensors 17(8):1–32
11. Yamanaka K, Vestergaard MC, Tamiya E (2016) Printable electrochemical biosensors: a focus on screen-printed electrodes and their application. Sensors 16(10):1761
12. Prabhakaran PN, Renugai M (2012) Design and analysis of capacitive MEMS viscometric sensor for CGM. Int J Emerg Technol 2(11):135
13. Ermakov V, Kruchinin S, Fujiwara A (2008) Electronic nanosensors based on nanotransistor with bistability behaviour. In: Bonca J, Kruchinin S (eds) Proceeding of NATO ARW "Electron transport in nanosystems". Springer, pp 341–349
14. Ermakov V, Kruchinin S, Hori H, Fujiwara A (2007) Phenomena of strong electron correlastion in the resonant tunneling. Int J Mod Phys B 11:827–835
15. Yang Z, Wang M, Bai Y, Chen X (2014) A differential capacitive viscometric sensor for continuous glucose monitoring. Springer International Publishing, Cham
16. Naor M, Shamir A (1995) Visual cryptography. Springer, Berlin, pp 1–12
17. Ragulskis M, Aleksa A (2009) Image hiding based on time-averaging moire. Opt Commun 282(14):2752
18. Turunen J (1997) Diffrctive optics for industrial and commercial applications. Akademie Verlag, Berlin
19. Suleski TJ, Baggett B, Delaney WF, Koehler C (1999) Johnson EG, Fabrication of high-spatial-frequency gratings through computer-generated near-field holography. Opt Lett 24(9):602
20. Schilling A, Herzig HP, Stauffer L, Vokinger U, Rossi M (2001) Efficient beam shaping of linear, high-power diode lasers by use of micro-optics. Appl Opt 40(32):5852
21. Kley EB, Wittig LC, Cumme M, Zeitner UD, Dannberg P (1999) Fabrication and properties of refractive micro-optical beam-shaping elements. SPIE 3879:20–31
22. Dufresne ER, Spalding GC, Dearing MT, Sheets SA, Grier DG (2001) Computer-generated holographic optical tweezer arrays. Rev Scient Instrum 72(3):1810
23. Gerchberg RW, Saxton WO, Owen W (1972) A practical algorithm for the determination of the phase from image and diffraction plane pictures. Optik 35:237

24. Wang YY, Wang YR, Wang Y, Li HJ, Sun WJ (2007) Optical image encryption based on binary Fourier transform computer-generated hologram and pixel scrambling technology. Opt Lasers Eng 45(7):761
25. Xi S, Sun X, Liu B, Tian W (2013) Spread spectrum characteristics of {CGH} for double random phase encrypted image. Optik 124(12):1260
26. Kong D, Shen X, Cao L, Zhang H, Zong S, Jin G (2016) Three-dimensional information hierarchical encryption based on computer-generated holograms. Opt Commun 380:387
27. Ma L, Jin W (2018) Symmetric and asymmetric hybrid cryptosystem based on compressive sensing and computer generated holography. Opt Commun 407(Suppl C):51
28. P. Palevicius, M. Ragulskis (2015) Image communication scheme based on dynamic visual cryptography and computer generated holography. Opt Commun 335:161
29. Palevicius P, Ragulskis M, Janusas G, Palevicius A (2017) Image encryption scheme based on computer generated holography and time-averaged moire. Proc SPIE 10453:10453
30. Rodionov VE, Shnidko IN, Zolotovsky A, Kruchinin SP (2013) Electroluminescence of Y_2O_3:Eu and Y_2O_3:Sm films. Mater Sci 31:232–239
31. Sugahara M, Kruchinin SP (2001) Controlled not gate based on a two-layer system of the fractional quantum Hall effect. Mod Phys Lett B 15:473–477
32. Kruchinin S, Klepikov V, Novikov VE (2005) Nonlinear current oscillations in a fractal Josephson junction. Mater Sci 23(4):1009–1013
33. Ragulskis M, Navickas Z (2009) Time average geometric moire-back to the basics. Exp Mech 49(4):439
34. Whyte G, Courtial J (2005) Experimental demonstration of holographic three-dimensional light shaping using a Gerchberg–Saxton algorithm. New J Phys 7(1):117

Chapter 18
Information Processing in Chemical Sensing: Unified Evolution Coding by Stretched Exponential

B. A. Snopok and O. B. Snopok

Abstract Multicomponent biochemical environments (MBE) of various composition and aggregate state are an integral part of the world around us (living organisms, ponds, atmosphere, products of the food and pharmaceutical industry, etc.). Therefore, the possibility of identifying (i) individual target components in MBE (including potentially dangerous ones); (ii) the occurrence of some specific reactions in them; (iii) the identification or monitoring of the MBE state as a whole is a problem of significant interest, and this interest is constantly growing as a result of the acceleration of informatization of modern industrial society. At the same time, the presence of low-informative MBE components containing a large number of different, often unknown, compounds determines uncertainty in the interpretation of the results of their analysis by (bio)chemical sensors due to non-specific sorption and, accordingly, limits their widespread use. Making decisions in such uncertain conditions requires additional information regarding the fact that the sensor response is the result of the "selected" specific recognition process. In this paper, this important scientific problem is solved by analyzing the dynamics of the sensory response and setting unique kinetic process markers at the interface of phases. This not only allows to further characterize the interaction of the analyte with the sensitive layer, which facilitates the identification of the analyte, but also provides an effective tool for the development and optimization of sensor elements and systems based on physical transducers of the surface type.

Keywords Multicomponent biochemical environments · (Bio)chemical sensors · Stretched exponential function

B. A. Snopok (✉) · O. B. Snopok
V.E. Lashkaryov Institute of Semiconductor Physics, National Academy of Sciences of Ukraine, Kyiv, Ukraine
e-mail: snopok@isp.kiev.ua

J. Bonča, S. Kruchinin (eds.), *Nanostructured Materials for the Detection of CBRN*, NATO Science for Peace and Security Series A: Chemistry and Biology,
https://doi.org/10.1007/978-94-024-1304-5_18

18.1 Introduction

The informatization of modern industrial society implies constant use of elements and systems for measuring the parameters of physical, chemical or biological processes. The purpose of informatization is to improve the quality of control and management in various areas of human activity, and at the same time reduce material costs, risks and occupation of qualified staff. Possible ways to achieve and the limits of application of informatization are determined by the development of not only more and more powerful means of information processing, but also by the level of scientific and technical knowledge in the field of analytical instrumentation, whose task is to develop methods for obtaining primary information about the world around us. The basis of the primary sensors of automation and control systems are functional elements and constructive components of the sensory equipment. The effectiveness of using such systems directly depends on the possibility of an adequate analysis of their response, formalized in a form acceptable for the use of information processing algorithms. Therefore, the more deeply developed and formalized the theory of the selection of the most informative component of the sensory response is, the objectively greater opportunities exist for the creation of certain devices, elements and systems necessary for the continuous and safe development of the industrial society.

18.2 Problem Analysis

The most common amongst the modern commercially available control systems are sensors of "physical" type, which determine intramolecular markers of the target analyte: specific absorption band, mass or mass/charge ratio, etc. An alternative approach is based on (bio-) chemical sensory systems, which allow obtaining quantitative or semi-quantitative analyte-specific information thanks to a recognition element of a (biological) chemical nature, which is in direct contact with a physical transducer (PT) sensitive to changes of the "amount of matter" on its surface. This approach significantly broadens the spectrum of possible analytes and conditions in which they can be determined, which is of particular importance for highly-molecular objects, which usually include the most dangerous substances [1, 2].

Taking into account the fact that the sensor element passes and encodes only a fraction of the information on the quantitative and qualitative composition of the sample, one can distinguish two main ways that can be used to determine, identify, and ultimately quantitatively analyze the analyte in the MBE. First of all, there are selective direct-acting sensors capable of establishing the presence of the analyte in the MBE at the "hardware" level by binding it to the functional components of the sensitive layer and uniquely connecting the analyte concentration in the sample to the magnitude of the output signal. In the case when the analyte is MBE itself, the identification is performed by forming a unique virtual "chemical image" (CI, some

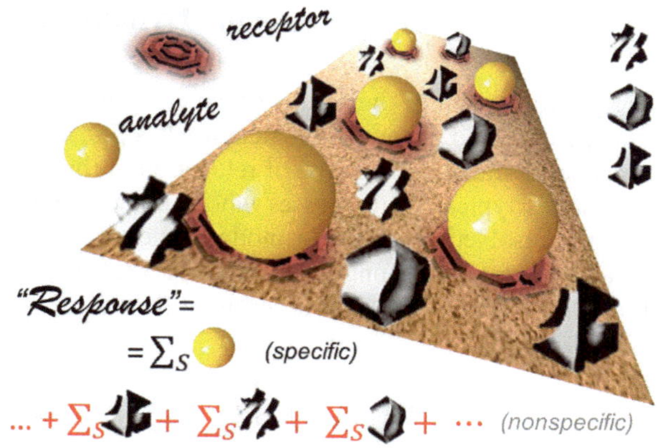

Fig. 18.1 Schematic diagram of core reactions illustrating the response generation in affine (bio)sensors (label free technology) inspired by the adsorption of both specific analyte and unspecific components

mathematical, geometric, etc. representation) of the MBE, using an array of cross-reactive, i.e. weakly selective sensor elements with selectivity profiles ("image" systems of the "electronic nose" type, EN) [3–6].

In both cases, typically, surface type PTs are used, which transform the changes in the amount of any substance on their surface into a convenient for further processing signal; that is, the appearance of the analyte or any other compound on the surface is interpreted as the fact of analyte presence in the mixture [7, 8]. For this reason, despite significant difference, the effectiveness of sensory systems of both types under MBE conditions is limited by the same reason: the presence of low-informative components of the mixture, which can significantly affect the analytical determination result due to non-specific sorption (Fig. 18.1). Indeed, the fact of the presence of a sensory response is only an "evidence" of the formation of some surface complex, which significantly complicates the analysis of MBE using surface-type sensors. Making decisions in such uncertain conditions requires additional information regarding the fact that the sensor response is the result of the "chosen" selective process on the surface of the sensing element.

The fundamental advantage of most (bio)chemical sensors, in particular those using optoelectronic and acoustic physical effects, is the ability to monitor the formation of a surface complex, i.e. the sensory response kinetics [9, 10]. The presence of a specific dynamics of the formation of the adsorbed layer provides additional information that can facilitate the identification of the analyte itself and which cannot be obtained in other ways. In accordance with the above, it makes sense to assume that if one takes into account the sensory response kinetics, which is a cumulative product of the processes at the interface of a complex multicomponent mixture and a heterogeneous surface, and unifies its manifestations by parametric formalization, then it is possible to further characterize the interaction of the analyte with the sensitive layer and obtain an effective tool for developing and optimizing sensor elements and systems based on surface type PTs.

18.3 Concept of SEF Specific Dynamic Analysis Approach

The generalization of the results of experimental studies [11–19], numerical modeling [20, 21] and phenomenological models [20, 22] suggests that non-exponential dynamics is the dominant type of relaxation in (bio)chemical sensor systems. Such behavior is typical for processes whose kinetics is directly related to "one of the oldest unresolved problems in modern science": non-exponential relaxation (NR, stretched exponential function (SEF), Fig. 18.2) in environments with nano-micro-dimensional topological non-ordering [23]:

$$u(t) = \exp\left(-\left(\frac{t}{\tau}\right)^{\beta}\right), \quad u(t) = \frac{(\Theta(\infty) - \Theta(0)) - \Theta(t)}{\Theta(\infty) - \Theta(0)},$$

$$\Theta(t) = (\Theta(\infty) - \Theta(0))\left(1 - \exp\left(-\left(\frac{t}{\tau}\right)^{\beta}\right)\right), \tag{18.1}$$

where $u(t)$ is relative deviations of the measured system parameter u(t) from equilibrium, $0 < \beta < 2$ and τ is the parameter that has the time dimension (Fig. 18.1). In general, one can say that at the phenomenological level, non-exponential relaxation of the form (18.1) is the general form of the evolution of surface type complex systems, and β is a kinetic marker of a certain surface process, which allows it to be distinguished from a series of similar ones [23].

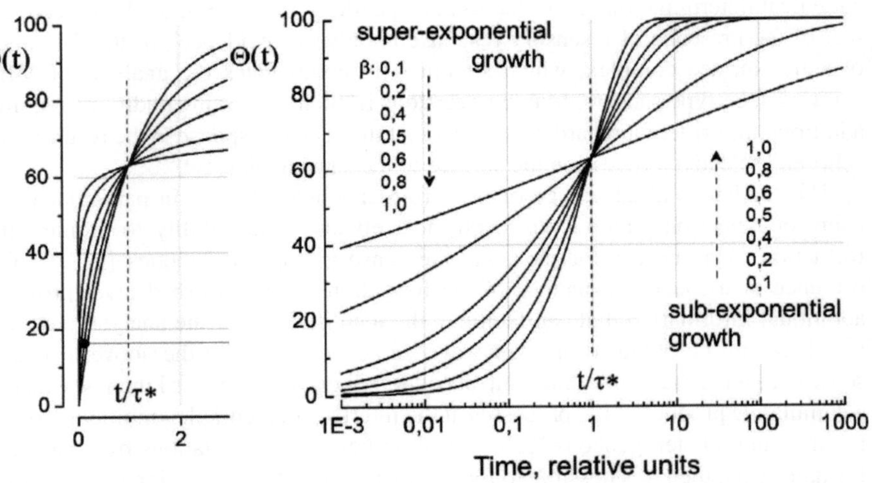

Fig. 18.2 Typical dependencies of non-exponential relaxation in the most frequently used coordinate representations ($\tau \equiv \tau^*$)

The obtained result is not surprising. Indeed, the complexity of the MBE and the heterogenity of the surface determine the diversity of the dynamic processes at their interface and do not allow using traditional deterministic methods to study the development of such complex distributed systems, whose acting forces are chaotic processes. The sensory element is an open nonlinear system that is in a state of stable non-equilibrium with the environment, which causes multivariance and unpredictability of the analyte motion processes at the microscopic level. At the same time, the response of the sensory element is probable (that is, logical, predictable) and is the result of the formation of spatially extended weakly organized structures as a result of the processes of surface self-organization, induced by single processes of selective binding. Consideration of such systems is possible only with the use of the synergistic approach methodology, when the chaotic motion is considered as a factor of evolution, which harmonizes the interconnection of individual components to maintain the system integrity under the influence of external factors [24, 25].

The microscopic model is formalized in the form of an appropriate statistical construct, if the formation of the adsorbed layer from the stream of reaction space particles is considered as a cumulative result of individual "successful" binding acts. Given the fractality of the reaction space, the Gaussian distribution cannot be used. In this case, the sum of individual "successful" binding acts converges not to the normal distribution, but to the Levi distribution (Paul Pierre Levy), which is a stable boundary distribution of the sum of random variables with inverse-power asymptotics (which are the ones that are usually observed in Nature [26–31]). Figure 18.3 shows the functions of the distribution of the probabilities density corresponding to the relaxation processes (18.1) in Fig. 18.1 for the same values of β.

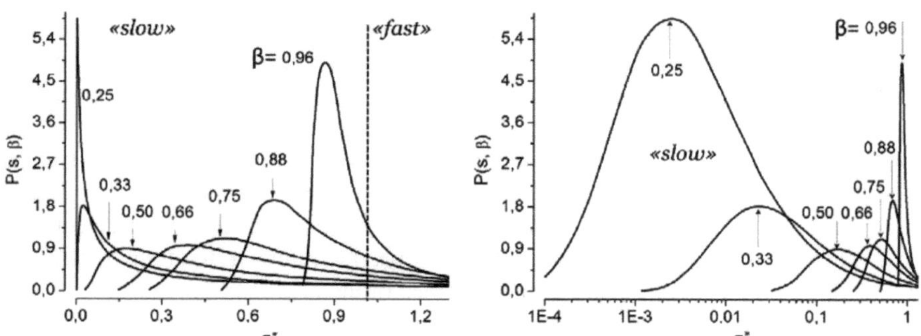

$$s \equiv \frac{\tau^*}{\tau}$$

$$\exp(-(\tfrac{t}{\tau^*})^\beta) \sim \sum_i \exp\left(-\left(\tfrac{t}{\tau_i}\right)\right) \rightleftharpoons \langle u(t)\rangle \rightleftharpoons \int_0^{+\infty} \exp(-s\cdot\tfrac{t}{\tau^*})\cdot P(s,\beta)\cdot ds$$

Fig. 18.3 Asymmetric (one-sided) Levy distribution in the most frequently used coordinate representations for different values of β

Thus, for complex systems, β defines that subset of the full set of possible states of the system, relaxation through which corresponds to the principle of least action. This set includes processes with different temporal scales, which are interconnected in the sense that they are coordinated within the system framework while exchanging energy, matter, etc. This means that there are dependencies between these processes, constant in a certain range of the spatial scale (spatially invariant), since spatial invariance is directly related to the temporal. In other words, the set of elementary relaxation processes in complex systems is not arbitrary and is defined by a Levy distribution with the corresponding value of β. Therefore, the change of the value of β is a redistribution of the whole set of elementary processes in the system, some of which become active, while others stop.

Sensory response generation is the result of a long and unlikely process of a "successful" analyte binding. In order to get to the receptor, the analyte must not only go a long way, but also face it with the correct side, approach it with an acceptable speed, etc. This is what explains the need for a wide range of elementary processes, which determine the way of surface self-organization. However, due to the need for their simultaneous action and effective spatio-temporal interconnection on different scales, the surface binding of the analyte and the response generation can only be realized with a well-defined set of them. The more difficult the conditions in the near-surface layer are, the greater the diversity of processes is needed to ensure the maximum probability of the realization of informative binding at the interface of the phases. The sensor is an open distributive nonlinear system, in which the superposition principle does not take place due to selective but coordinated at different organization levels redistribution of energy/substance between components from the allowed set of states. The connection between slow processes of the "upper level" (analyte concentration gradient change) and rapid chaotic motion (usually, indistinguishable in detail) is made through distributive nonlocal "control" factors. It is these factors that reflect long-term collective correlations dependent on the generalized structure features (symmetry, currents, etc.) and provide the connection between micro- and macrolevels, resulting in a mutually coordinated co-existence of different system components (Fig. 18.4). Ultimately, in non-equilibrium conditions initiated by the analyte appearance, the independence of system elements ("close order") is replaced by their "synchronous" behavior ("distant order"), aimed at achieving a new state of dynamic equilibrium. The nature of such reconstruction of a complex system depends on the analyte concentration, surface centers selectivity, the values of their adsorption and desorption barriers, the surface topography and many other factors taking place at the interface of the phases.

To increase the analysis efficiency in the MBE, that is, the probability of binding, it is necessary to increase the diversity of "small-scale movements" both in direction and intensity, in order to pass through the "labyrinth" and reach the receptor in the required state. Thus, the realization of unlikely slow binding processes is possible only in the presence of a wide variety of processes with small characteristic time. The more complex the system (the smaller the value of β), the greater the need for various "fast" processes in order to level out "ineffective" processes by stochastizing the analyte trajectory. At the same time, due to the greater "detail",

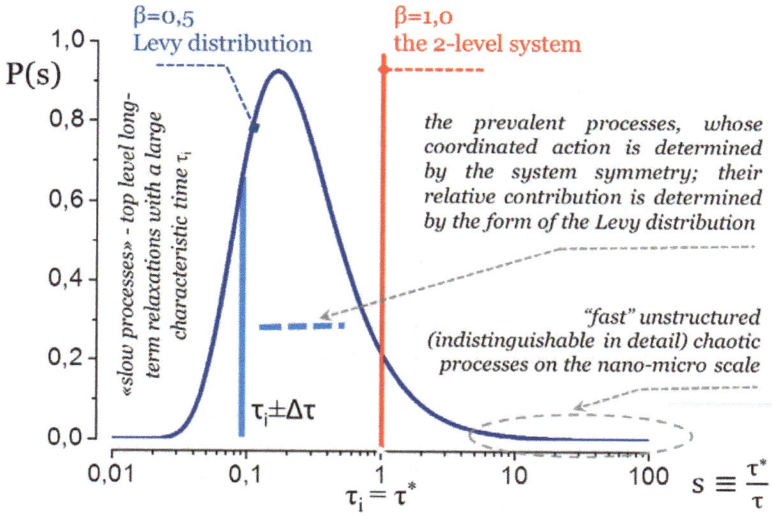

Fig. 18.4 The Levy distribution sets the allowable interconnection between a fixed set of processes on various spatial-temporal scales in line with the intrinsic symmetry of the system

the informative binding processes become more probable; the progress of these processes is mutually coordinated on different spatial scales. Thus, in a complex distribution system, there is a tight link between slow and fast processes, and only such sets can develop in the system effectively, determining its evolution path in accordance with the least action principle.

18.4 Concept of Kinetic Discrimination as Basis for Sensor Development and Optimization

The methodology of dynamic methods of MBE analysis is an effective tool for the development of biochemical sensory systems for MBE analysis, due to the following:

1. it establishes a universal relaxation form of the sensory response in the MBE with a fixed set of parameters;
2. it establishes a single procedure (SEF approximation) for the formalization of the informative primitive specific to a certain pair of sensitive coating and analyte; this allows for a formation of universal databases of corresponding standards;
3. it opens up the possibility of establishing the processes mechanism at the interface of the phases by comparing the value of β with the corresponding statistical distributions, model representations or computer simulations;

4. it allows to purposefully optimize sensitive coatings, using the given value of β as a measure (for example, non-specific sorption is characterized usually by $\beta = 1.0$);
5. it allows to identify factors that have the greatest impact on the analysis process and to optimize them depending on the task at hand; this is due to the fact that the inverse to β value gives an overall estimate of the number of degrees of freedom for the analyte;
6. in specific cases it allows to develop ways of optimization; for example, for cross-reactive sensor systems (e.g. "electronic nose") of the "image" type the use of sensors with $\beta \sim 0.68$ corresponds to the greatest variety of surface processes, enabling the coverage of the largest possible compounds spectrum;
7. it allows to use the value of β as a parameter directly related to the analyte concentration; this is possible for those analysis procedures where the analyte concentration changes the distribution of surface processes (see, for example, [12, 18]); such a procedure significantly improves the analysis accuracy, since β is calculated by approximating a large number of relaxation curve points;
8. it allows to increase the analysis efficiency or reduce its time by choosing the optimal time interval for the parametrization of the sensory response, due to the effects of kinetic discrimination [4];
9. it allows to form unique chemical images of the MBE by the parameterization of the entire response curve (β, τ), as well as by using the intersection of the response surface with a plane in the selected time interval;
10. and finally, in a number of cases, the analysis of NR allows to obtain information about the analyte that cannot be obtained by other methods (see, for example, [18]); this is due to the fact that the dynamic relaxation of the system implicitly contains information about all possible system states.

The methods of dynamic passportization allow highlighting the optimal ways of practical actions, while the topological model of adsorption [23] provides theoretical basis for solving the applied aspects of the development of biochemical sensory systems for MBE analysis. This approach opens up new opportunities and provides researchers or engineers with tools for building a reactionary space corresponding to the problem at hand, and further purposeful optimization of the desired instrumentation for achieving the analytical goal [32]. This methodology allows to equally consider the processes of both reversible and irreversible adsorption/sorption from the gas phase and solutions (analysis of the adsorption and/or desorption phase, etc. is possible) containing one analyte or ones that are a multicomponent mixture, etc. It should be reminded that the degrees of freedom include not only spatial coordinates, but also the reactivity of the surface, external fields or concentration gradients, etc. Thus, by varying the process parameters and determining the changes of β value (and of other variables in (18.1), if necessary), it is possible to optimize the process according to the given criterion. The basic principles of the concept of dynamic methods of MBE analysis and their relationship are presented in Fig. 18.5. At the heart of the concept are the principles of kinetic discrimination, which combine the main groups of principles of constructing the biochemical systems of MBE analysis. Following of the presented principles allows making optimal decisions

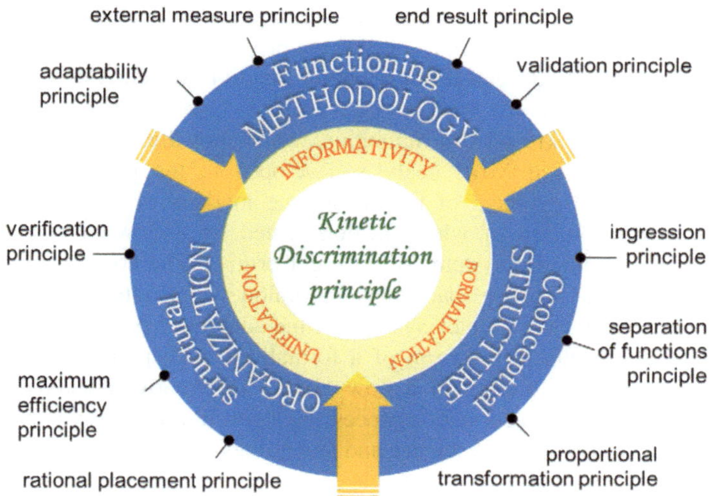

Fig. 18.5 The generalized structure of the concept of dynamic methods of MBE analysis, built on the principles of kinetic discrimination of the sensory response evolution

while developing biochemical sensory systems of MBE analysis in dynamic mode, using a set of methods and procedures focused on finding the best options. The most important practical result is the ability to estimate the number of independent factors that determine the evolution of a complex system, by analyzing the experimentally observed process kinetics. This allows to avoid a full sorting through possible options amongst endless alternatives, to minimize errors and wrong technical decisions and, accordingly, to accelerate the development process and improve its efficiency.

18.5 Concluding Remarks

By the generalization of obtained experimental and theoretical results, a formalized conceptual structure suitable for analyzing the evolution of complex adaptive systems whose dynamics is determined by isolated informative events in multi-parametric space-invariant systems of different origin is developed. It is established that the response of (bio)chemical sensory systems of the surface type is the result of the self-organization of a complex nonlinear open-type system at the interface of the phases heterogeneous surface – multicomponent chemical environment, which leads to the formation of the adsorbed layer under non-equilibrium conditions initiated by the analyte presence. The macroscopic kinetics of such systems in general occurs in accordance with the stretched exponential function, whose parameters (β and τ) are unique identifiers of the processes at the interface of the phases; the non-exponential nature of the relaxation is a reflection of the low probability of single successful analyte adsorption acts and of the multiscale coordination of the processes that satisfy the least action principle; a set of these processes can be obtained from the corresponding probability density distribution function, which is described by the Levy distribution for a given β value.

The principles of kinetic discrimination, which are the basis of the concept of dynamic methods of MBE analysis developed in the paper, are a part of the general physical problem of finding unique markers of nonlinear statistical ensembles, the evolution of which is determined by a flow of rare events in complex multi-parametric systems of different origins. The concept of the scale-invariant topological reaction space allows to uniquely characterize the features of the dynamic behavior of complex systems of the adaptive type. The generalization of the obtained results in the form of a formalized conceptual structure can be used to analyze the evolution of a flow of rare informative events in complex systems, to develop new information processing algorithms and to synthesize highly informative devices in various science and technology fields.

Acknowledgements This research was partly funded by the NATO Science for Peace and Security Programme under the Grant G5140 and Projects of the National Academy of Sciences of Ukraine.

References

1. Grimes CA, Dickey EC, Pishko MV (eds) (2005) Encyclopedia of sensors. American Scientific Publishers, Valencia
2. Snopok BA (2008) Rapid methods for multiply determining potent xenobiotics based on the optoelectronic imaging. In: Bonca J, Kruchinin S (eds) Proceeding of NATO ARW "Electron transport in nanosystems". Springer, pp 331–339
3. Snopok BA, Kruglenko IV (2002) Multisensor systems for chemical analysis: state-of-the-art in electronic nose technology and new trends in machine olfaction. Thin Solid Films 418(1):21–41
4. Burlachenko J, Kruglenko I, Snopok B, Persaud K (2016) Sample handling for electronic nose technology: state of the art and future trends. Trends Anal Chem 82:222–236
5. Ermakov V, Kruchinin S, Fujiwara A (2008) Electronic nanosensors based on nanotransistor with bistability behaviour. In: Bonca J, Kruchinin S (eds) Proceeding of NATO ARW "Electron transport in nanosystems". Springer, pp 341–349
6. Ermakov V, Kruchinin S, Hori H, Fujiwara A (2007) Phenomena of strong electron correlastion in the resonant tunneling. Int J Mod Phys B 11:827–835
7. Snopok BA, Kostyukevich KV, Lysenko SI, Lytvyn PM, Lytvyn OS, Mamykin SV, Zynyo SA, Shepelyavyj PE, Venger EF, Kostyukevich SA, Shirshov YuM (2001) Optical biosensors based on the surface plasmon resonance phenomenon: optimization of the metal layer parameters. Semicond Phys Quantum Electron Optoelectron 4(1):56–69
8. Boltovets PM, Snopok BA (2009) Measurement uncertainty in analytical studies based on surface plasmon resonance. Talanta 80:466–472
9. Gromashevskii V, Tatyanenko N, Snopok B (2015) Effect of the formation of silicon oxide on the sign, magnitude and formation of surface charge upon water adsorption on a silicon surface. Theor Exp Chem 51(3):170–176
10. Snopok BA (2012) Theory and practical use of surface plasmon resonance for analytical purposes (Review). Theor Exp Chem 48(5):265–284

11. Boltovets PM, Boyko VR, Snopok BA (2013) Surface capturing of virion-antibody complexes: kinetic study. Mater Sci Eng Tech 44:112–118
12. Boltovets PM, Polischuk OM, Kovalenko OG, Snopok BA (2013) A simple SPR-based method for the quantification of the effect of potential virus inhibitors. Analyst 138:480–486
13. Snitka V, Naumenko DO, Ramanauskaite L, Kravchenko SA, Snopok BA (2012) Generation of diversiform gold nanostructures inspired by honey's components: growth mechanism, characterization and shape separation by the centrifugation-assisted sedimentation. J Colloid Interface Sci 386:99–106
14. Snopok BA, Kruglenko IV (2005) Nonexponential relaxations in sensor arrays: forecasting strategy for electronic nose performance. Sensors Actuators B Chem 106(1):101–113
15. Snopok BA, Yurchenko M, Szekely L, Klein G, Kasuba E (2006) SPR based immuno-capture approach for in vitro analysis of protein complex formation: mapping of MRS18-2 binding site on retinoblastoma protein. Anal Bioanal Chem 386:2063–2073
16. Boltovets P, Shinkaruk S, Bennetau-Pelissero C, Bennetau B, Snopok B (2011) The effect of low pH on the glycitein-BSA conjugate interaction with specific antiserum: competitive inhibition study using surface plasmon resonance technique. Talanta 84(3):867–873
17. Naumenko D, Snopok BA, Serviene E, Bruzaite I, Snitka V (2013) Confocal Raman spectroscopy of biological objects in the face of photoinduced luminescence self-quenching. Theor Exp Chem 49:215–221
18. Snopok B, Kruglenko I (2015) Analyte induced water adsorbability in gas phase biosensors: the influence of ethinylestradiol on the water binding protein capacity. Analyst 140:3225–3232
19. Snopok BA, Darekar S, Kashuba EV (2012) Analysis of protein–protein interactions in a complex environment: capture of an analyte–receptor complex with standard additions of the receptor (CARSAR) approach. Analyst 137:3767
20. Manoilov EG, Kravchenko SA, Snopok BA (2017) Features of near-surface layer at monomolecular isotropic adsorption: nonequilibrium molecular dynamics simulation. Ukr J Phys 62(8):717–726
21. Manoilov EG, Kravchenko SA, Snopok BA (2017) Features of cooperative adsorption described by sticking probability that depends on the number of neighbors. Theor Exp Chem 53(1):17–24
22. Volkenshteyn FF (1987) Electronic processes on surface of semiconductor at chemisorptions. Nauka, Moskaw
23. Snopok BA (2014) Nonexponential kinetics of surface chemical reactions (Review). Theor Exp Chem 50(2):67–95
24. Haken H (1984) Synergetics – theory of nonequilibrium phase transitions and formation of spatio-temporal patterns. In: Grünewald H (ed) Chemistry for the future. Pergamon Press, Oxford
25. Haken H (1981) Erfolgsgeheimnisse der Natur: Synergetik, die Lehre vom Zusammenwirken. Deutsche Verlags-Anstalt
26. Clauset A, Shalizi CR, Newman MEJ (2009) Power-law distributions in empirical data. SIAM Rev 51:661–703
27. Eni-Olorunda I, Sadana A (2010) Kinetics of chemo/biosensors. In: Zourob M (ed) Recognition receptors in biosensors. Springer, New York
28. Sworakowski J, Matczyszyn K (2007) Non-exponential decays in first-order kinetic processes. The case of "Squeezed exponential". Acta Phys Pol A 112:S153–S159
29. Laherrere J, Sornette D (1998) Stretched exponential distributions in nature and economy: "Fat tails" with characteristic scales. Eur Phys J B 2:525–539
30. Brouers F, Sotolongo-Costa O (2006) Generalized fractal kinetics in complex systems (application to biophysics and biotechnology) Phys A 368:165–175
31. Schweiss R, Mirsky VM, Wolfbeis OS (1998) Capacity study of self-assembled alkylthiol monolayers: surface charge effect and kinetics of surfactants adsorption. Mater Sci Forum 287–288:427–430
32. Boltovets P, Shinkaruk S, Vellutini L, Snopok B (2017) Self-tuning interfacial architecture for Estradiol detection by surface plasmon resonance biosensor. Biosens Bioelectron 90:91–95

Chapter 19
Toxicology of Heterocarbon and Application of Nanoheterocarbon Materials for CBRN Defense

G. Kharlamova, O. Kharlamov, M. Bondarenko, P. Silenko, O. Khyzhun, and N. Gubareni

Abstract Nanoheterocarbon materials are widely used in the field of national security (in particular, chemical, biological, radiological, nuclear, explosive (CBRNE) defense), defense, environmental, food and health protection. The growing examples of international terrorist threats against citizens and major infrastructures push to strengthen security measures and the complication of sensing technologies. It requires the development of newer and newer nanomaterials based on carbon molecules and nanostructures modified (doped) by atoms of other elements. However, the toxicity of the most promising nanomaterials for nanosensors, catalysts and other nanodevices based on heterocarbon has not yet been studied enough. Therefore, along with the need to develop new methods for modifying carbon nanostructures with the aim of expanding the scope of their application, the most important task is to study their biocompatibility and potential toxicological effects on the human body.

Keywords Heterocarbon · Nanotoxicology · Nanoheterocarbon materials · CBRN defense · Nanosensors

19.1 Introduction

The discovery of carbon molecules and nanostructures, like any epoch-making discovery, is a "double-edged sword" for a future civilization through which mankind can, on the one hand, make a technological breakthrough, on the other hand, open a Pandora's box with unpredictable consequences – creating of new

---

G. Kharlamova (✉)
Taras Shevchenko National University of Kyiv, Kyiv, Ukraine
e-mail: akharlamova@ukr.net

O. Kharlamov · M. Bondarenko · P. Silenko · O. Khyzhun · N. Gubareni
Frantsevich Institute for Problems of Materials Science of NASU, Kiev, Ukraine

J. Bonča, S. Kruchinin (eds.), *Nanostructured Materials for the Detection of CBRN*, NATO Science for Peace and Security Series A: Chemistry and Biology,
https://doi.org/10.1007/978-94-024-1304-5_19

weapons of mass destruction. The most striking example of such a historical event was the discovery of the uranium nucleus splitting and the subsequent development of the nuclear chain reaction theory. As a result of this discovery, the mankind could partially solve energy problems by creating atomic energy. Almost simultaneously, the most deadly weapon in the history of mankind was created – an atomic bomb. The sad consequences of the Chernobyl accident or the tragedy of Hiroshima and Nagasaki will always remind scientists of the whole world that a criminal plan or negligence, an irresponsible approach to the results of scientific discovery in the future can turn into a tragedy that threatens the existence of all mankind and the planet as a whole.

It is assumed that the further development of nanotechnology in the 21st century will produce an equally large revolution in the manipulation of matter. The same scale as in the 20th century computers produced in the "manipulation of information". However, on the path to the development of nanotechnology, humanity faces a number of potential dangers: from uncontrolled toxicological, mutagenic effects on the human population to the creation on the basis of nanodevices of weapons of mass destruction of a new generation. Nevertheless, despite existing warnings, nanotechnologies and nanodevices are a logical step towards improving technical systems.

Large-scale use of nanoobjects, in particular carbon molecules and nanostructures, provides modern development of electronics (especially, by reducing the size of devices) and other high-end industries. So, for example, nanosensors based on nanocarbon materials are widely used in areas of national security (in particular, in the fields of chemical, biological, radiological, nuclear, explosive (CBRNE) defense), defense, environmental protection, food and healthcare. Biological nanosensors make possible the selective identification of very low concentrations of toxic compounds in industrial products, chemical materials, air, water, soil and biological systems. Based on growing examples of international terrorist threats against citizens and critical infrastructures, it is necessary to create increasingly sophisticated sensing technologies that provide early warning of the presence of toxins in the atmosphere, water, beverages and food.

The tendency to complicate sensing technologies, in turn, dictates the need to develop new, modern nanomaterials to create arrays of self-organizing nanosensors that are programmable to detect the presence of thousands of chemicals simultaneously. To expand the scope of application of nanocarbon materials in sensor systems, it becomes necessary to develop new methods for modifying carbon molecules and nanostructures by molecules or atoms of other elements [1–5]. Heteroatomic substitution significantly modifies the structure at the atomic level, its surface energy, chemical activity and mechanical properties of carbon nanostructures. That allows the significant expanding in the field of potential usage of materials based on them (heterocarbon materials). Heterocarbons are heteroatomic carbon molecules and nanostructures, in which one or more carbon atoms in the graphene layer are replaced by atoms of other elements, such as boron, nitrogen and silicon [4–6]. The nitrogen and boron atoms as electron donors and acceptors are the most typical substitute elements (dopants) in carbon materials.

This chapter describes methods for obtaining various heterocarbon nanoforms that can be promising targets for the development of selective and highly effective nanosensors for CBRN defense and other technological devices. We discuss the first results of toxicological studies and the biocompatibility of heterocarbon molecules and nanostructures.

19.2 Application of Nanoheterocarbon Materials for CBRN Defense and Others Technological Devices

In the nano-epoch, there is no longer possible to guarantee the security of society and a human being without using nanoscale systems for information, economic, military technologies and human health monitoring [3, 5]. The development based on nanomaterials (particularly carbon and heterocarbon) and nanosensor nanosystems is certainly a revolution in the security industry, especially for CBRN defense (Fig. 19.1). All types of carbon and heterocarbon nanomaterials are used for nanosensor manufacturing: single-walled and multi-walled nanotubes, fullerenes, nanocarbon (in particular, onion-like carbon) and graphene.

Thus, graphene and graphene isomers of a carbon single-walled nanotube (armchair or zig-zag), capable to contact by all its atoms with the molecules of the environment, are the most effective in CBRN defense for in the detecting of minimal amounts of explosive (TNT) and harmful substances such as nitric oxide and toxic gases of various types. The principle of function of such sensors is based on a change in the natural fluorescence of a carbon nanostructure when virtually all of its carbon atoms interact with various not only reagents but also adsorbates [3, 5].

The first results on the use of certain nanoforms of heterocarbon as nanosensors proved to be extremely significant. It was noted in [7] that, in contrast to monoatomic CNTs, N-substituted CNTs are more effective nanosensors for detecting gases (NH_3, NO_2, O_2, N_2, CO_2, CH_4, H_2O, H_2) and vapors (ethanol, acetone, chloroform, gasoline, pyridine and benzene) [8]. Sensors based on N-doped CNTs have advantages over undoped nanotubes, since they have higher sensitivity and response time (0.1–1 s) [7]. Based on N-doped nanotubes, a supersensitive nanosensor was created to detect the extremely toxic microcystin LR (most toxic cyanotoxins, microcystin-LR) in water samples [9]. A single-layer N-graphene sensor can detect rhodamine B (RhB), to which pristine graphene is insensitive [10]. RhB – rhodamine B ($C_{28}H_{31}ClN_2O_3$) is an indicator and dye. It is often used as a biomarker in oral rabies vaccines [4]. The usage N-doped graphene in a biosensor to detect glucose has increased the sensitivity of the sensor to 0.01 mM of the substance concentration [11].

Carbon nitride is the promising material for the creation of nanosensors. Graphite-like carbon nitride g-C_3N_4, in fact, can be considered as saturated with nitrogen heterocarbon, in which the degree of substitution of carbon atoms by nitrogen atoms is maximum and C/N ratio corresponds to 0.75. Carbon nitride

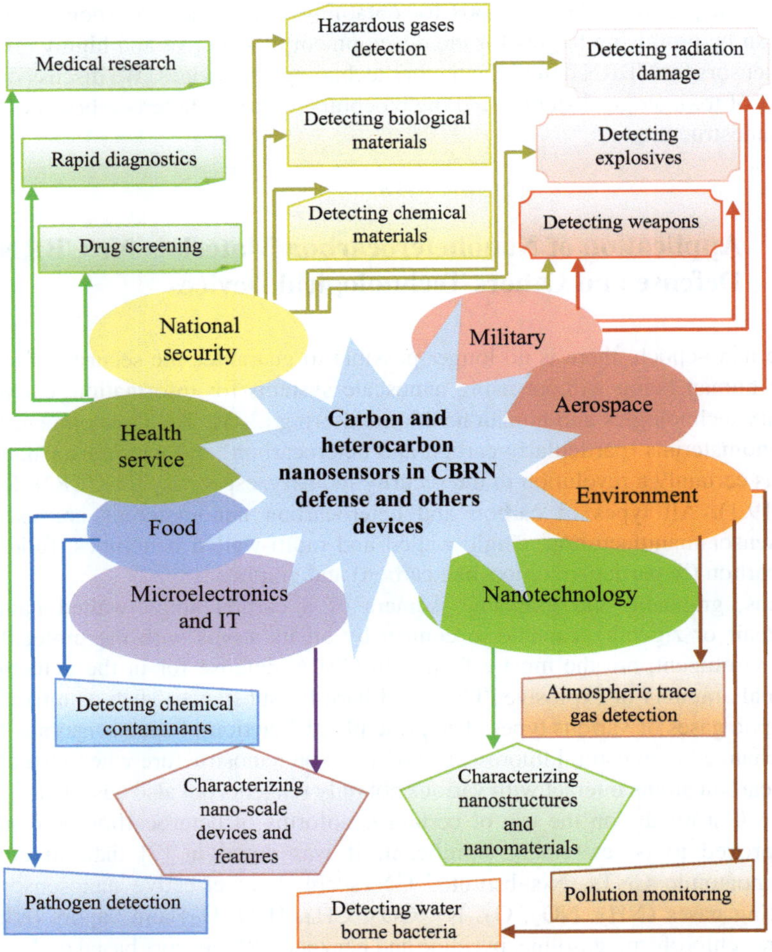

Fig. 19.1 Application of nanosensors based on carbon and heterocarbon in CBRN defense and others devices

has a wide range of homogeneity, within which the graphite-like structure is preserved, and the C/N ratio can vary from 0.75 (C_3N_4) to 3.0 (C_3N) [3, 6, 12–16]. Dispersing g-C_3N_4 to several (6-12) monolayers and modifying it with various elements, in particular with oxygen (O-doped carbon nitride O-g-C_3N_4 and its derivatives: carbon nitride oxide (g-C_3N_4)O and reduced carbon nitride [3, 14–16]), can significantly increase the sensitivity of nanosensors based on carbon nitride material.

Some synthesized n-forms of N-heterocarbon have already been successfully used as catalysts and carriers of catalysts [6, 17], sorbents for gas [18] and liquids [19] purification from pollutants, and are also used in electrochemical processes to create supercapacitors and fuel cells [20]. Thus, based on N-heterocarbon, cathode catalysts [6], alternative to expensive traditional platinum catalysts, have been synthesized. The catalytic activity of N-heteronanosheets (synthesized by collagen carbonization) in the oxygen reduction reaction is much higher than the activity of Vulcan XC-72 industrial catalyst containing 20% platinum [21]. N-heterocarbon is also used in fuel cells with a proton-exchange membrane [22]. Carbon N-nanostructures as catalysts reduce from 1.3 to 0.9 eV [23] the energy barrier of dissociative hydrogen adsorption and are therefore very promising for hydrogen storage.

The substitution of carbon atoms by nitrogen atoms in a graphene network of carbon nanostructures and molecules is a promising direction in creating a new generation of photonic and electronic devices, as well as functionally active adsorbents and catalytic systems [8]. The filling of single-walled carbon nanotubes (CNTs) with azafullerenes (($C_{59}N$) 2) [24, 25], as wellas their hydroxylated and hydrogenated derivatives [4-hydroxy-3,5-dimethyl-phenyl-hydro-azafullerene] [26] opens the probability of its usage as ambipolar field-effect transistors.

N-heterographene, its obtaining methods were developed quite recently, already has wide application in fuel cells, lithium ion batteries, biosensors, field effect transistors, ultracapacitors and photocatalysts [27]. N-graphene also exhibits high electrocatalytic activity in the hydrogen peroxide reduction reaction [11].

The large-scale usage of onion-like carbon structures (or onions) as effective lubricants [28], adsorbents and catalysts is considered very promising [24, 29]. According to the results of the available studies, the onion structure has a combination of such mechanical properties as high microhardness and superelastic reduction [30]. Moreover, it has been shown that nitrogen atoms contribute to the curvature of the layers of the carbon graphene-like layered structure, forming an onion or onion-like structure, since they facilitate the formation of pentagons, and also create active sites for the formation of interlayer bonds between the shells of the onion [31]. The compact material of N-doped onion-carbon is characterized by high microhardness and superelasticity. This allows the usage of N-onion-carbon, in particular, for the manufacture of damping elements, shock absorbers, friction pairs and wear-resistant parts of micromechanisms [32, 33].

Shu et al. [34] reported the usage of a mesoporous, nitrogen-doped, onion-like carbon (N-OLC) with a large surface area as a new electrocatalyst for a Li-O_2 battery. The catalyst contains multicomponent quasispherical concentric graphite shells sp^2. This remarkable feature of mesoporous N-OLC not only facilitates the diffusion of lithium and oxygen ions, but also provides many electrocatalytic sections and ultimately improves the characteristics of the Li-O_2 battery, including high turn-on efficiency, high specific capacity, good speed characteristics and excellent cycle stability at almost 200 cycles [34]. Carbon nanocomposites based on N-doped onion carbon (as well as carbon-doped nanotubes and graphene) demonstrated significantly improved (in comparison with expensive Pt or Ir) catalytic activity and stability in alkaline and acid electrolytes, particularly in the oxygen reduction reaction ORR [35–39].

Highly active and reliable base metal catalysts are in great demand for solving the issues of cost and durability of fuel cells, metal air batteries and water electrolyzers. Catalysts based on N-doped carbon nanostructures most reliably replace these precious metals because of its low cost, extremely high surface area, excellent mechanical and electrical properties, sufficient stability under severe conditions and high functionality [35]. The higher polarity of the surface of nitrogen-doped carbon nanostructures was experimentally established as compared to the undoped ones [40]. This result indicates that graphene layers doped with nitrogen show a faster charge transfer rate than undoped graphene layers [41, 42], and thus improve the transfer of protons and electrons in the oxygen reduction reaction (ORR). N-doped onion-like carbon is also used in binary catalysts from base metals (CoFe) for oxygen reduction [39].

Compared to conventional Fe-catalyst, the N-onion-based binary catalyst significantly increases the life of the fuel cells and shows a good combination of activity and longevity [39]. The composite material based on nitrogen-doped carbon onions encapsulated with metal alloys (ONCMAs) such as $FeNi_3$ ($ONCFeNi_3$) or FeCo (ONCFeCo) has excellently proven itself as an effective and stable catalyst for dye-sensitized solar cells (DSSCs) [43]. Such composite catalysts show excellent catalytic activity in the triiodide reduction reaction and exhibit low resistance to charge transfer between the electrode surfaces and electrolytes. As a result, DSSCs based on ONC@FeCo and $ONCFeNi_3$ achieve outstanding power conversion efficiency (PCE) of 8.26 and 8.87% respectively, which can compete with 8.28% DSSC based on Pt. In addition, the electrochemical impedance and cyclic voltammetry (CV) spectra (electrochemical impendence spectra and cyclic voltammetry (CV)) demonstrated excellent electrochemical stability for both catalysts. The TEM investigation showed that N-doped graphitic carbon onions have high structural stability in the iodine-containing medium, even in hundreds of CV-scans [43]. These results make N-doped onion-like carbon a promising candidate for replacing costly Pt as an effective and stable catalyst.

The widespread use of heterocarbon (or heteroatomic molecules (fullerenes) and nanostructures (graphene, nanotubes and onions) of carbon) is constrained by the lack of effective methods for its production. Therefore, after the discovery of spheroidal carbon molecules, planar and cylindrical nanostructures and the development of methods for the synthesis of these objects in preparative quantities, the main focus of researchers in the field of nanochemistry of carbon is now concentrated on the development of effective methods for modifying its graphene network with various elements.

19.3 Basic (Main) Methods of Synthesis of Heterocarbon Nanostructures

The discovery of a unique state of carbon in the form of spheroidal molecules and nanostructures and a huge interest of researchers to study and predict the properties of these objects stimulated new attempts to obtain various nanoforms of heterocarbon in this state.

Since the characteristics of carbon molecules and nanostructures, including CNTs and graphene, are closely related to the delocalized electronic subsystem, it is obvious that any chemical modification (a modification like this is commonly called *doping* [44]) will have a significant effect while using atoms of other elements of the periodic table as dopants [44]. Therefore, by the appropriate choice of the dopant type, it is possible to control the electronic properties. It is important to note that completely different types (and methods) of modification of the graphene network of carbon molecules and nanostructures are generally called *doping* (Fig. 19.2).

Thus, the *exohedral doping* should be more correctly termed if there is the fictionalization of the surface of a graphene network due to an addition reaction to adjacent carbon atoms (due to the opening of a double C=C bond), for example, atoms of hydrogen, fluorine, oxygen (oxygen-containing functional groups) or organic molecules. The introduction of atoms or molecules into a closed graphene structure (fullerene or nanotube) is an *endohedral doping*. The introduction of atoms or molecules between graphene layers (for example, in a bundle of nanotubes or a fullerene crystal) can be called *intercalation doping*. The subject (object) of our review is *substitution doping*, when in a graphene network the carbon atom is replaced by the atom of another element.

The first attempts to obtain N-doped carbon were carried out 50 years ago [45, 46], when carbon was obtained containing up to 0.5 at.% N. In more recent works [32] it was shown that the degree of substitution of carbon atoms in graphene sheets by nitrogen atoms essentially depends on the method of nitriding and the structure of the carbon nanoform. With the use of the catalyst, the degree of nitriding of the carbon structure is usually increased. The product formed, however, is substantially contaminated with the catalyst.

A detailed review of the methods for the synthesis of heterocarbon molecules (heterofullerenes) was presented by us earlier in [4, 6]. In this article, we will dwell in details on the methods of obtaining heterocarbon nanostructures – heteronanotubes, heterographene and N-doped onion-like structures.

19.3.1 Heteronanotubes

It is believed that the unique discovery of carbon nanotubes (CNTs) was held in 1991 by Iijima [47] because of the usage of a non-standard method for observing the already well-studied cathode carbon condensate by means teh newly created

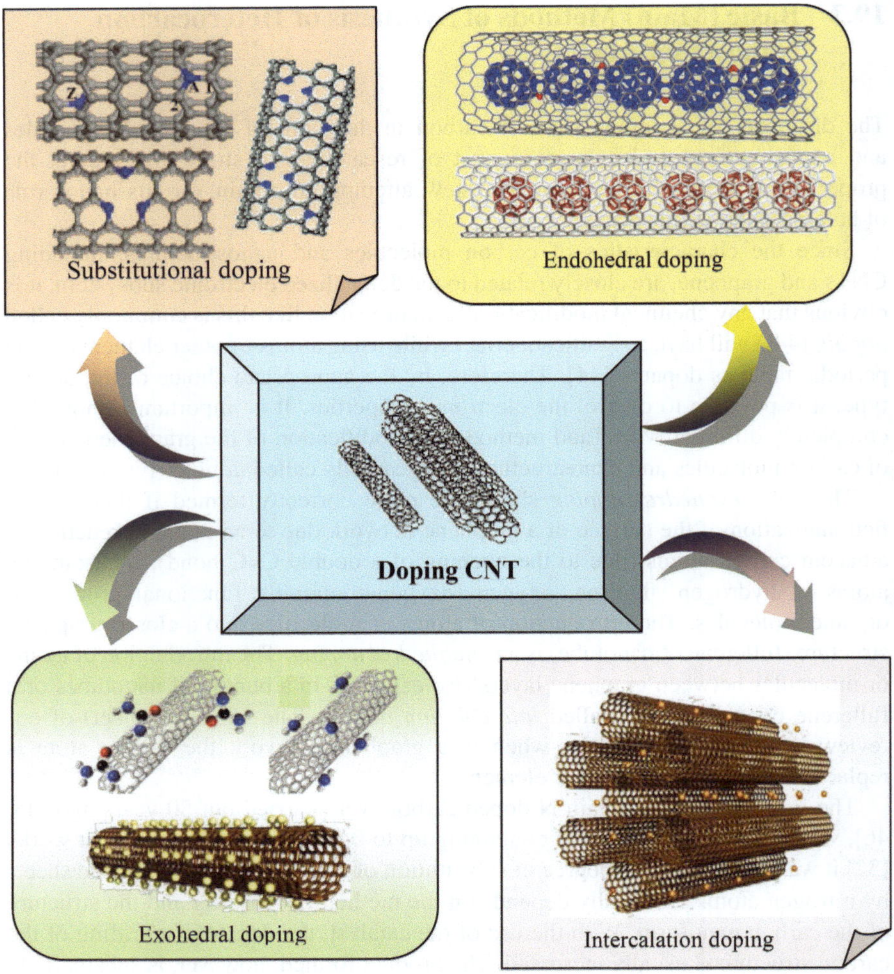

Fig. 19.2 Different approaches to chemical modification of carbon nanotubes: substitutional, endohedral, exohedral and intercalation doping

high-resolution electron microscopes [48]. However, it was later found that the discoverers of CNTs are not Iijima in 1991, as most researchers believe, but Soviet scientists who published clear TEM images of their CNTs almost 40 years earlier, in 1952 in the article in the Journal of Physical Chemistry [49]. Of course, the name "CNT]] did not exist then, so the primacy of Iijima can only refer to the definition (appropriation of the name, definition) of a "carbon nanotube". According to the authors (editorial article) [50] in journal Carbon, the long obscurity of the priority of the CNT discovery by Russian scientists was due to the cold war, when access to Russian scientific publications for Western scientists was complicated. In addition, many written in Russian language articles of Russian scientists were not mentioned in the literary databases, such as Current Contents. In [49], clear TEM microphotographs of carbon filaments with a continuous internal cavity are shown, thereby forming tubes. Despite the absence of a scale (in nm), the magnification indicated in the photographs makes it possible to calculate that the diameters of the first carbon tubes obtained are $\sim$50 nm, i.e. definitely nanosize. In addition, these nanotubes are obviously multi-walled, with walls of 15–20 graphene layers [50].

Depending on the synthesis conditions, the tubes can have single-, double- and multilayer walls. The diameter of binary and single-walled carbon nanotubes (SWCNTs) varies from 0.4 to 4 nm, and the diameter of multilayer (or multi-walled) nanotubes (MWNTs) can reach 100 nm [51]. Depending on the method of synthesis and purification, nanotubes are formed from 100 nm to tens and even hundreds of micrometers. The unique mechanical, optical, electrical properties of CNTs give rise to a variety of possibilities for their application [51].

It is known that, depending on the chirality indices, carbon nanotubes can be both metallic and semiconducting. However, substituting carbon atoms by other elements (for example, nitrogen or boron), it is possible to vary their electronic properties within certain limits. In addition, the doping additionally affects the reactivity of the samples and their mechanical strength. Therefore the task of finding simple methods for the synthesis of doped carbon nanotubes is in the extensive interest.

Due to the proximity of the atomic sizes of boron, nitrogen and carbon, boron and nitrogen are ideal elements capable to substitute carbon atoms in CNT walls. Firstly, in 1994, the doping of CNTs with nitrogen and boron atoms was carried out by an arc method [52]. Due to the proximity of the atomic radii boron (B) and nitrogen (N) are the most popular doping elements in comparison with other elements such as sulfur (S), phosphorus (P) and silicon (Si), because of the large discrepancy of their atom size in comparison to the size of carbon atoms [53]. It is important to note that significant changes in the initial tubular structure can occur at the place of heteroatomic substitution on the CNT wall. For example, doping with nitrogen leads to bamboo-like structures [53]; the implantation of sulfur causes branching [54]; and doping with boron leads to knee-like bending [55]. Note that doping with nitrogen often produces short CNTs which have several bamboo-like compartments [53], that may be an undesirable effect. At the present time, special N-doping methods have been developed that make it possible to obtain long CNTs with a high yield [56].

Doping of CNTs is one of the most effective ways of increasing their conductivity. Since its quasiperiodicity and an increase in the fraction of point defects that result from the replacement of carbon lattice atoms by impurities, lead to a decrease in the percolation threshold [57]. (In physics and chemistry, the phenomenon of percolation (from Latin, percôlâre – leaking, leaking) is the phenomenon of the flow or leakage of liquids through porous materials, electricity through a mixture of conductive and nonconducting particles and other similar processes). Carbon nanotubes doped with boron and nitrogen atoms attract attention of electronic nanodevices designers in connection with the prospects of its usage as nanostructured p- and n-type semiconductors, respectively [57–60]. The number of $2p$-electrons in nitrogen atoms is one more than that of carbon. So the excess electrons appear in nitrogen-doped CNTs. Nitrogen-doped CNTs have improved autoemission properties [61]. The incorporation of boron atoms into the nanotube structure leads to an increase in the hole concentration at the Fermi level.

Heteroatoms can be included in the carbon network of CNT either during synthesis or during subsequent processing of previously synthesized undoped CNTs (post-synthesis treatment). Among the nitrogen-containing reagents, NH_3 is the most preferred for doping pre-synthesized CNTs [56, 62]. For doping with boron, phosphorus and sulfur the effective reagents are boron or B_2O_3, triphenyl-phosphine [63] and elemental sulfur or thiophene [64, 65], respectively.

Methods based on the chemical vapor deposition (CVD) of nitrogen and carbon containing gas phase reagents are most effective for the synthesis of nitrogen-doped CNTs and allow the production of multi-walled N-CNTs (N-MWNT) (Fig. 19.3) with the highest (up to 15–20%) content of nitrogen [66]. At the same time, single-walled N-CNTs (N-SWNTs) (Fig. 19.4), as a rule, contain no more than 1%

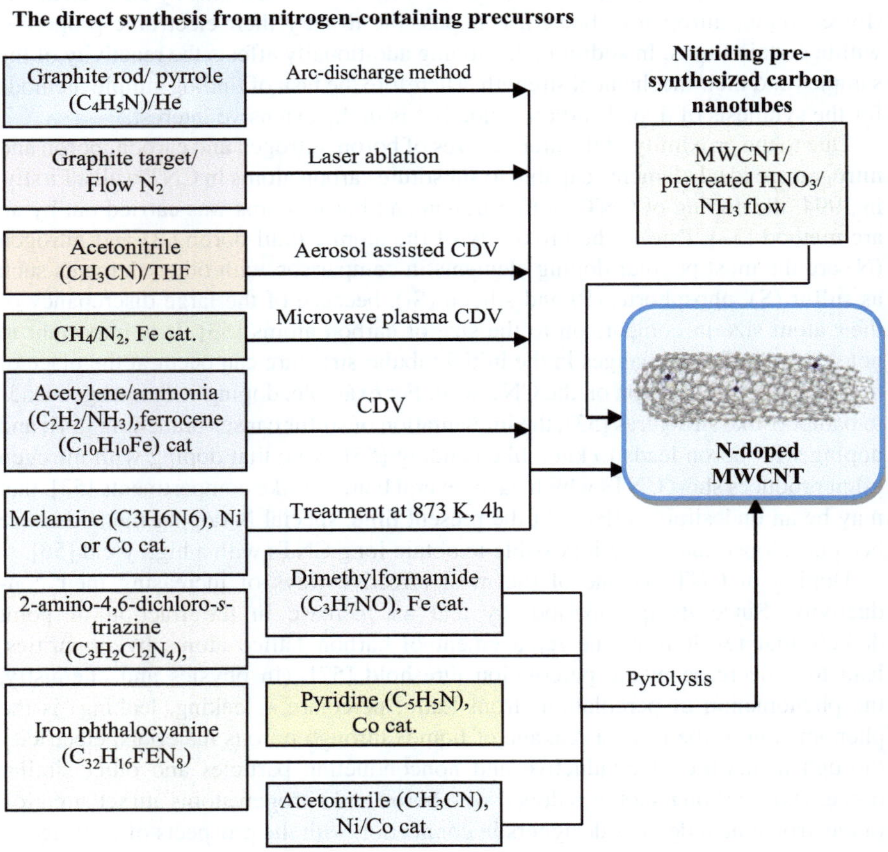

Fig. 19.3 Basic methods of the synthesis of N-doped multi-walled carbon nanotubes

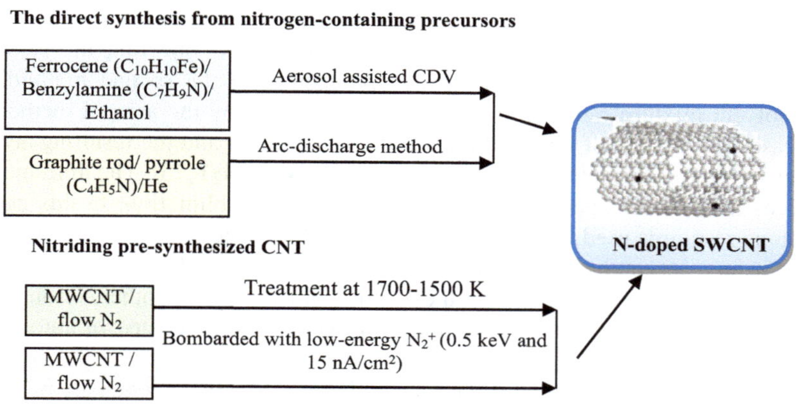

Fig. 19.4 Basic methods of the synthesis of N-doped single-walled carbon nanotubes

nitrogen, regardless of the synthesis method [67, 68]. Nitrogen-doped nanotubes have been synthesized by both classical CVD and CVD methods using aerosol-assisted chemical vapor deposition. Later, nitrogen doping with double walled nanotubes was demonstrated [69].

The content of nitrogen in doped CNTs depends significantly on the choice of nitrogen/carbon-containing precursors. Thus, CN nanotubes and nanofibers are synthesized either from compounds containing both nitrogen and carbon. For example, acetonitrile (CH_3CN), melamine ($C_3H_6N_6$), phthalocyanine ($C_{32}H_{18}N_8$) iron, dimethylformamide (C_3H_7NO), pyridine, or mixtures such as acetylene/ammonia, CH_4/N_2 [2, 6, 70]. It is noteworthy that when N-containing hydrocarbons are pyrolyzed, N-heterocarbon is usually obtained with higher nitrogen content than while heat treatment of mixtures involving ammonia or nitrogen. Carbon nanotubes obtained by pyrolysis of acetonitrile at 900 °C and 5 bar contain up to 3 at.% of nitrogen [71]. Multi-walled N-carbon nanotubes with 5 and 2% nitrogen content are formed from melamine (66.7 mass% nitrogen) and pyridine (17.1 mass% nitrogen), respectively [72]. In a number of studies, N-CNTs were successfully synthesized by CVD using the organometallic complex, mainly ferrocene $Fe(C_5H_5)_2$, ferrocene derivatives (substituted ferrocene, eg ferrocenylaniline) and iron pentacarbonyl $Fe(CO)_5$ [73].

It is noteworthy that in the CVD process, the organometallic complex serves simultaneously as a source of a metal catalyst and an additional source of carbon atoms for the construction of CNTs. Thus, the ferrocene already completely decomposes at a temperature of 500 °C with the formation of a number of fragments: $Fe(C_5H_5)_2 \rightarrow Fe + H_2 + CH_4 + C_5H_5 +$ other hydrocarbons [74]. Liquid catalysts, for example iron pentacarbonyl ($Fe(CO)_5$), can be introduced into the reaction space by a syringe injection procedure. However, most organometallic complexes are solids at the room temperature, so alternative strategies are needed to introduce them into the reactor. The introduction of ferrocene (FcH), for example, can be achieved in a process called injection CVD.

The injection CVD is carried out by dissolving FcH (which is usually the carbon source used to prepare CNTs) in a suitable solvent and injecting the resulting solution into the reactor using a syringe procedure. In the second method, the FcH/solvent mixture can be introduced into the reactor by the aerosol method. In this case, the mixture of FcH and the solvent is sonicated and the resulting aerosol is transferred to the reactor using a carrier gas (Ar, N_2 or H_2) [73]. The third method involves the introduction of FcH into the reaction by sublimation. In this case, a two-zone reactor is used. Passing over FcH in the first zone ($T \sim 250$ °C), the inert gas transfers the catalyst to the second zone ($T \sim 700$–1000 °C) where N-CNTs are formed. The fourth method uses the location of FcH and any other N-containing reagent in a closed environment (autoclave), which is then heated in autogenous conditions.

Thus, in [75], single-walled carbon nanotubes doped with nitrogen are produced by a method of aerosol-assisted chemical vapor deposition using ferrocene. The researchers [75] pyrolyzed ferrocene connected with aerosols of ethanol and benzylamine in the temperature range from 950 to 1100 °C and resulted in N-SWCNTs with a nitrogen content of about 1 at.%. With the help of transmission electron and scanning electron microscopy and Raman spectroscopy, the quality and morphology of nanotubes were monitored. The presence of nitrogen in the samples was determined by spectroscopy of the characteristic energy losses of electrons and X-ray photoelectron spectroscopy. The synthesized samples of N-SWCNTs are sufficiently thick "beams", the surface of which is covered with amorphous carbon and, in addition, containing a significant amount of large particles of Fe catalyst. Correction of parameters such as the temperature and composition of the carrier gas made it possible to vary the nanotube diameter in the range from 0.9 to 1.8 nm, and also to increase the reaction rate (>10 mg/s) and the percentage yield of N-SWCNTs. The maximum yield of N-SWCNT was obtained at 1000 °C in an argon-hydrogen mixture in a ratio of 9 : 1 [75].

Attempts to increase the number of dopant atoms in the samples were unsuccessful. According to the authors of [75], the reason was in the use of precursors with an increased nitrogen content inhibiting the growth of nanotubes. The further development of the CVD technique using organometallic complexes was the selection of new precursors for the production of N-CNTs. In a number of studies, the following precursors (or precursor mixtures) were deposited together with ferrocene in order to successfully synthesize N-CNTs containing up to 2% nitrogen: ethanol/benzylamine [76], xylene/ammonia/pyridine [77], 4-*t*-butylpyridine [78], toluene/benzylamine [79], ethanol/ethylenediamine [80], pyridine/pyrimidine [81]. When deposition together with ferrocene the monethanolamine [82] or melamine [83], and with pentacarbonyl iron mixtures of C_2H_2/NH_3 [84], NH_3/pyridine [85] or methylpyridine/pyridine/triazine [86] N-doped nanotubes with a higher (2–7%) nitrogen content were obtained. The highest nitrogen content in multi-walled N-MWNTs (the authors of [87] report the presence of 18–24% nitrogen in the product) was obtained by deposition of ethylenediamine with ferrocene in the presence of a Co catalyst.

The solvothermal synthesis is another technology developed for the synthesis of multi-walled N-doped carbon nanotubes [2]. The authors of [88] state that stable nanotubes of large diameter with a very high nitrogen content are obtained using cyanuric chloride as a starting material in an autoclave reaction (high pressure).

Two high-temperature methods of evaporation of a graphite target in a nitrogen atmosphere: an arc discharge [89–91] or laser ablation [92, 93] make it possible to obtain multi-walled nitrogen-doped carbon nanotubes (N-MWNTs). The authors of [94] reported the production of single-walled N-doped CNTs (N-SWNTs) using an arc discharge to vaporize a graphite target containing the catalyst in a nitrogen-helium atmosphere. It was noted [94] that doping with nitrogen leads to an increase in the content of more fibrous structures. Also, the arc discharge method was used to create heterostructures of single-walled nanotubes [95] by evaporation of composite anodes containing nitrogen-rich precursors in a helium atmosphere. However, the N-SWNTs synthesized by the arc method contain no more than 1 at.% nitrogen, and there are no reports of the successful use of the laser ablation method to create nitrogen-doped SWCNTs [2].

In addition to the incorporation of nitrogen into the walls of nanotubes during their growth, it is also possible N-doping the CNTs by post-treatment of the undoped (pristine) carbon nanotubes. Thus, in [96], heat treatment of undoped CNTs at 1500–1700 K in N_2 atmosphere leads to the appearance of uneven damaged nanotube walls with substitution of less than 1% N (comparable with arc discharge methods).

On the basis of theoretical studies in [97–101], ion implantation was used as a method to selectively doping carbon nanotubes with nitrogen. Undoped CNTs were deposited on silicon substrates and bombarded with low-energy N_2^+ (0.5 keV and 15 nA/cm^2). The concentration of nitrogen, equal to 0.47 at.%, was quantitatively determined by the XPS method [97].

Nitriding of pre-synthesized carbon nanostructures produces N-heterocarbon with low nitrogen content, and when a catalyst is used to increase it, N-nanostructures of carbon are contaminated with catalyst particles [6].

Boron-doped carbon nanotubes (B-CNTs), as well as other heterocarbon nanostructures, can be obtained both by direct synthesis from boron and carbon-containing precursors, and by post-synthesis treatment of previously synthesized undoped nanotubes. For the production of multi-walled B-doped CNTs (B-MWNTs), the following methods are most often used: decomposition of acetylene with boron-doped anode by an arc discharge [102], chemical vapor deposition of acetylene and BF_3 [103], microwave plasma induced chemical deposition from the vapor phase of methane and trimethylborate $B(OCH_3)_3$ [104] and pyrolysis of ferrocene ($C_{10}H_{10}Fe$) with diborane (B_2H_6) [105].

It is believed that among the known methods of B-CNT synthesis the chemical deposition from the gas phase is the most promising, since CVD method does not produce by-products [106]. The use of a metal-containing catalyst does not lead to a significant increase in boron content in B-CNTs, but promotes the formation of undesirable by-products. Thus, B-CNTs were prepared by catalytic decomposition of hydrocarbons (CH_4, C_6H_{14}) and boric acid esters ($B(OMe)_3$) on metal-supported mesoporous oxide catalysts $Met_{0.03}Mo_{0.1}MgO$, where Met = Fe, Co, Ni. According to the XPS data, the maximum boron content (~1 at.%) corresponds to the sample obtained at 750 °C on an iron-containing catalyst using a precursor mix of trimethyl borate and methanol ($B(OMe)_3$ – 75 vol%, MeOH – 25 vol%) and hexane [106]. B-CNTs of relatively high purity were obtained by chemical deposition of ethanol vapors and a water-alcohol solution of B_2O_3 in a H_2/Ar atmosphere at a temperature of 800 K on a copper substrate [106].

Multi-walled boron-doped CNTs were obtained by arc-discharge method of graphite evaporation [107]. As a result of evaporation of the graphite cathode in about 10 wt.% of amorphous boron, the dense build-up containing B-MWNTs was formed in a helium atmosphere under low pressure. It is established that the B-MWNT electric arc discharge is very short. Its average length does not exceed 1 micron [107]. In addition to the B-MWNT, there are also onion-like structures (multilayer fullerenes) present in the sample obtained by the electric arc method, the number of the latter being quite significant.

B-doped multilayer carbon nanotubes (B-MWNTs) with a boron content of 1–2 at.% were synthesized by CVD using a solution of boric acid in methanol as a precursor (Fig. 19.5) [108]. Boron-doped MWCNTs were grown on a Si substrate that was preliminarily coated with iron acetate as a metal-containing catalyst to activate the growth of B-MWNT.

The silicon substrate pre-coated with the iron acetate was placed inside a quartz tube and heated to 950 °C at a rate of 46 °C/min under a pressure of 10 torr. The temperature was maintained at 950 °C for 15 min to obtain pure iron from iron oxide. The temperature was then lowered to 730 °C at a rate of 14.7 °C/minute. Then vapors of boric acid solution in methanol was introduced into the reactor at a constant pressure of 200 torr for 4 h, after which the substrate containing the B-MWNT was cooled to room temperature [108].

Single-walled B-doped carbon nanotubes (B-SWCNTs) (Fig. 19.6) were synthesized by pulsed laser evaporation of targets containing 0.5 to 10 at.% boron [109].

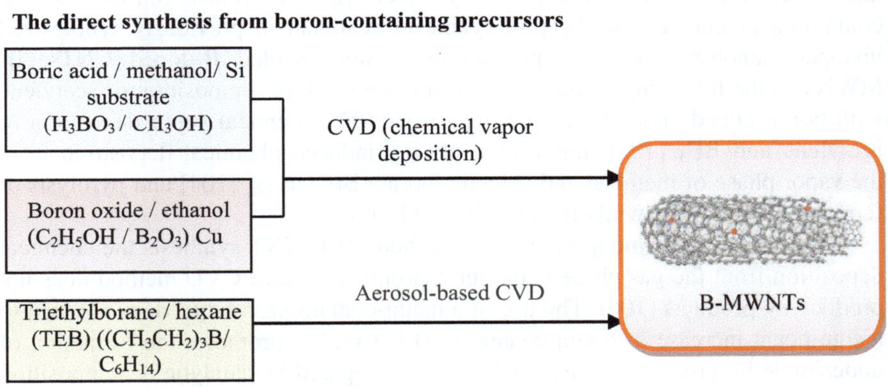

Fig. 19.5 Basic methods of the synthesis of B-doped multi-walled carbon nanotubes

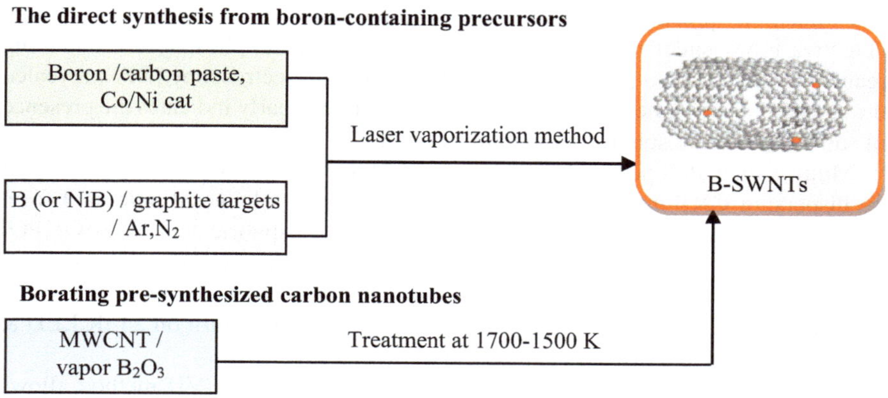

Fig. 19.6 Basic methods of the synthesis of B-doped single-walled carbon nanotubes

The targets were obtained by mixing elemental boron with a carbon paste with the addition of Co/Ni catalyst. The products of laser evaporation of five targets with different boron content were certified by high-resolution transmission electron microscopy (TEM) and electron energy loss characteristic spectroscopy (EPRS). It is established that beams of single-walled B-doped CNTs are present only in the evaporation products of targets containing not more than 3 at.% boron. The degree of doping of the obtained B-SWNTs is ~0.05–0.1 at.% boron [109]. With a boron content in the 4.5 at.% target, a small fraction of the products contained bilayer B-CNT. At a boron concentration in the more than 4.5 at.% target, the presence of single-layer and double-layer B-CNTs in the products was not detected.

The production of single-layer B-SWCNTs by the method of post-synthesis high-temperature (1523–1803 °C) treatment of previously synthesized SWCNTs in the presence of boron-containing compounds (B_2O_3) was reported [110]. However, studies aimed at controlled doping and the characteristics of boron-doped nanotube bundles are not available in [110].

Reports of the successful synthesis of phosphorus- and sulfur-doped carbon nanotubes are very few. Phosphorus-doped CNTs (P-CNTs) with the high phosphorus content (1.66%) were obtained by a low-temperature hydrothermal method by treating a mixture of CNT-H_3PO_4 at 170 °C [111]. Hydrothermal P-CNT synthesized by the hydrothermal method showed: improved thermal stability in the oxidation process in pure oxygen medium (with a weight loss peak at 694 °C); good electrocatalytic activity for the oxygen reduction reaction, and improved cyclic stability compared with undoped CNTs [111].

The phosphorous-doped multi-walled carbon nanotubes (P-MWNTs) were successfully synthesized by spray pyrolysis (pyrolysis of aerosols) using a solution of ferrocene ($C_{10}H_{10}Fe$) and triphenylphosphine ($C_{18}H_{15}P$) in toluene [112]. Electron microscopy images show P-MWNT corrugated tubes with a special morphology that is shorter than undoped tubes grown under the same conditions using ferrocene and toluene as precursors. Raman spectroscopy indicates the formation of more defective tubes under the increasing of the content of the phosphorus-containing reagent in the precursor solution. X-ray photoelectron spectroscopy (XPS) revealed the chemical environment of phosphorus atoms, which clearly indicates the presence of substitutional phosphorus in nanotubes [112].

Multi-walled P-doped carbon nanotubes obtained by post-synthesis treatment by dispersing CNT in a mixture of monohydrate of *p*-toluenesulfonic acid form (pTsOH·H_2O) and triethyl phosphate (triethyl orthophosphoric acid $(C_2H_5O)_3PO$) in a molar ratio of 1 : 1 were used as the electrodes in the supercapacitor cell [113]. Due to the high surface area and high electrical conductivity of P-MWNT electrodes, such a supercapacitor provides enough energy to turn on an IR LED at 30 mW and emit light for a certain period of time [113].

Doping of carbon nanotubes with an elemental sulfur by CVD method allows obtaining S-CNTs with distorted sidewalls of extremely diverse morphology [114]. The sulfur powder and ferrocene ($C_{10}H_{10}Fe$) were dissolved in acetonitrile. The resulting solution (20 mg/ml) was injected into the CVD reactor with a syringe pump at a constant rate of 0.4 ml/min for 0.5–1 h in the temperature range 940–1020 °C [114]. A mixture of Ar and H_2 was used as the carrier gas. Transmission electron microscopy of the obtained S-CNTs revealed a variety of the morphology of the nanotubes obtained. Twisted, bamboo-shaped, tree-like and bead-like S-CNTs were found in the product [114]. It has also been reported that an attempt was made to obtain S-CNT by pyrolysis of sucrose which carbonized by sulfuric acid [115]. However, the results of the investigation of the obtained MWNTs indicate the presence of sulfur only in the catalyst particles trapped inside the nanotubes, and sulfur was not detected in the side walls of the synthesized nanotubes [115].

19.3.2 Heterographene

After obtaining a monolayer of graphene for the first time with the help of Scotch tape method, later called "the method of mechanical peeling", the team of Nobel laureates in 2010, Novoselov and Geim [116, 117], its unusual electronic, optical and mechanical properties have attracted great interest of researchers from the fundamental physics and technological points of view [118, 119]. The most important task for creating nanodevices based on graphene is to precisely control its electronic properties by appropriate functionalization method, resulting in the inclusion of different atoms, in particular nitrogen, into the structure of graphene.

Two main approaches are used to synthesize N-heterographene (or N-graphene), as well as other forms of N-heterocarbon (with the exception of azafullerenes): direct synthesis, which involves the use of nitrogen-containing reagents in the synthesis process, and post synthesis treatment, which is the introduction of nitrogen atoms into the lattice of pre-synthesized undoped graphene (Fig. 19.7).

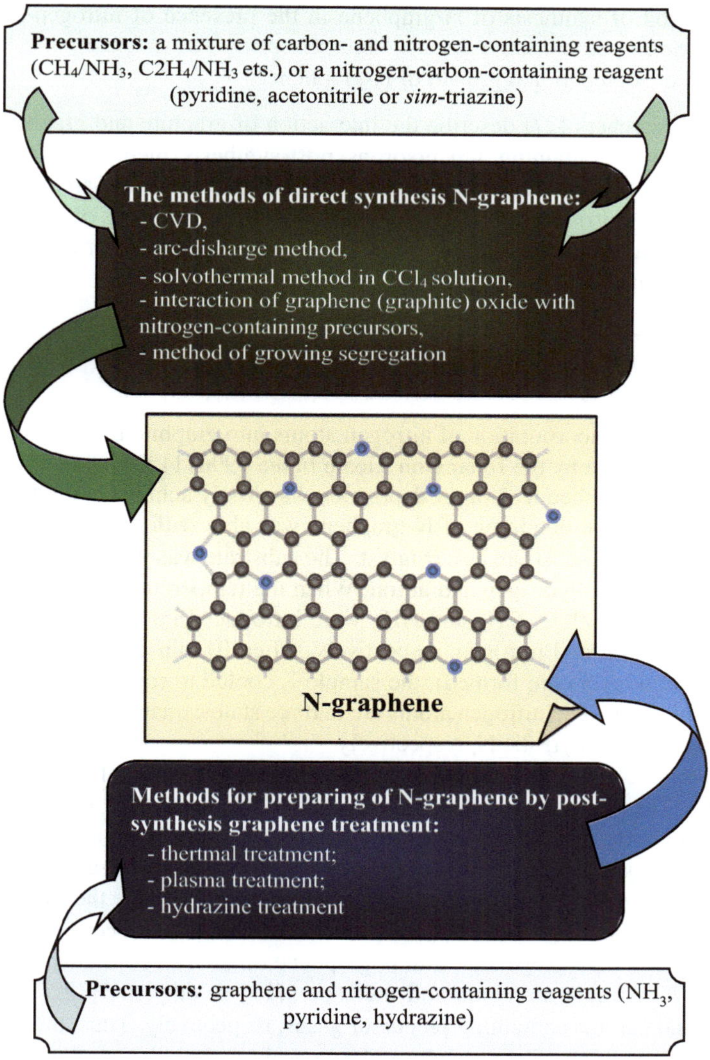

Fig. 19.7 Basic methods of the synthesis of N-doped graphene

Direct synthesis from precursors is carried out in the following ways:

- deposition of nitrogen and carbon-containing precursors (CVD) from the gas phase;
- chemical deposition of CCl_4 with a nitrogen precursor previously dissolved in it (solvothermal method);
- interaction of graphene oxide with nitrogen-containing precursors;
- interaction of graphite oxide with nitrogen-containing precursors;

- arc method of synthesis of N-graphene in the presence of nitrogen-containing precursors;
- method of segregational growth of N-graphene.

Some researchers [27] describe the interaction of graphite and graphene oxides with a nitrogen-containing precursor as post-synthesis methods. However, we believe that in this case both graphene oxide and graphite oxide are precursors of graphene, so its nitriding refers to the methods of direct synthesis of N-graphene.

In post-synthesis methods, the following types of treatment of pre-synthesized graphene are used:

- heat treatment;
- plasma treatment;
- treatment with hydrazine.

The chemical incorporation of nitrogen atoms into graphite by the CVD method was first carried out by the Johansson's team in the 1990s [120], while the synthesis of N-graphene by chemical vapor deposition was firstly achieved in 2009 [121]. In the study [121], the obtaining of N-graphene was able with the usage of a copper film grown on a Si substrate as a catalyst. The substrate was placed in a quartz tube blown by a flow of hydrogen and argon. When the temperature in the center of the furnace reached 800 °C, CH_4 and NH_3 were introduced into the inert gas flow as sources of carbon and nitrogen, respectively. After 10 min growth of N-graphene layers (several layers are formed), the sample is cooled to room temperature in the H_2 atmosphere. Doped nitrogen atoms are in three states: quaternary (substitutional) N, pyridinic N and pyrrolic N, respectively.

Currently CVD has become one of the main methods of synthesis of N-graphene. The method of chemical vapor deposition (CVD) of nitrogen- and carbon-containing precursors makes it possible to obtain N-doped graphene with the high nitrogen content. Single-layer N-heterographene (N-graphene) with a maximum nitrogen content of up to 16% was synthesized by the thermal deposition of ethylene and hydrogen vapor in the presence of ammonia on copper foil [122]. It is assumed that the nitrogen atom in N-graphene on the foil is in the pyridine state. In the CVD method, ammonia and methane [121] or ethylene [122] are used as nitrogen and carbon-containing precursor gases, respectively. Thus, the application of the CVD method using a mixture of ammonia and methane allows to implant 8.9% of nitrogen in the graphene structure [123, 124].

Pyridine, acetonitrile or *sim*-triazine (*s*-triazine $(HCN)_3$) can be used as precursors containing both elements [125]. Precursors are deposited on a substrate of a copper or nickel catalyst.

N-graphene consisting of several layers is produced by nitriding of CH_4. Single-layer N-graphene is produced by nitriding of ethylene, pyridine or acetonitrile. The nitrogen content of N-doped graphene depends on the ratio of nitrogen/carbon-containing precursors and the rate of its flow over Cu or Ni catalyst substrate. The maximum nitrogen content in N-graphene obtained by deposition CH_4/NH_3 mixture is 8.9%, acetonitrile – 9%, pyridine – 2.4%. Single-layer N-graphene containing up to 2% nitrogen was obtained by CVD by the method of deposition the *s*-triazine on a nickel substrate at a vapor pressure of triazine of 1×10^{-6} mbar at a substrate temperature of 630 °C for 30 min [125].

The solvotermal method is one of the few ways to obtain N-doped graphene in gram quantities. While a lithium nitride or a mixture of cyanuric chloride and lithium nitride is dissolved in carbon tetrachloride and while the resulting solutions are heated to ~300 °C, the reaction proceeds forming 1–6 layers of N-graphene containing 4.5 and 16.4% of nitrogen, respectively.

The interaction of graphene oxide (GO) with ammonia [126], a mixture of ammonia and hydrazine N_2H_4 [127], melamine [27] or plasma H_2/N_2 [128] produces N-doped graphene, which contains up to 10.78, 5.21, 10.1 and 2.5% nitrogen, respectively. N-graphene with a nitrogen content of 10.13% was synthesized using a simple hydrothermal reaction of graphene oxide (GO) and urea [129]. N-doping and reduction of GO were achieved simultaneously by a hydrothermal reaction. In the manufacture of N-graphene, nitrogen enriched in urea plays a key role in the formation of N-graphene nanosheets with a high nitrogen containing.

During the hydrothermal process, urea releases NH_3 on a continuous basis, which reacts with the oxygen of the functional groups of the GO, and then the nitrogen atoms are embedded in the graphene skeleton, which leads to the formation of N-graphene.

Treating thermally expanded graphite oxide with ammonia in an argon flow (NH_3/Ar) [130], plasma N_2 [131] or in a solution of hydrazine [132], N-graphene is obtained with a low, 1–3%, nitrogen content.

Arc method allows to synthesize N-doped multilayer (2–6 layers) graphene in gram quantities [133]. In [133], the arc discharge (120 A) was conducted in a water-cooled steel chamber of stainless steel. Two graphite electrodes with a diameter of 8 mm were evaporated in the atmosphere of He and NH_3 (760 torr). A high proportion of NH_3 (more than 50%, by volume) is a prerequisite for the production of N-doped multilayer graphene. In [134], this method was modified using the atmosphere H_2 + pyridine or H_2 + ammonia. By using the arc method, N-graphene is obtained. It contains only about 1 at.% N. It should be noted that the main drawback of using an arc discharge for the synthesis of carbon materials (including N-graphene) is the formation of a large number of undesirable by-products. As a result, the process requires complex and well-controlled stages of purification [135].

The synthesis of two-three-layer N-doped graphene was carried out by an arc method in the presence of a mixture of H_2/pyridine (NG1) or H_2/ammonia (NG2) [27]. N-doped graphene is also obtained by transformation of nanodiamond in the presence of pyridine (NG3) [27]. According to photoelectron spectroscopy (XPS), all three methods do not provide a high level of doping of graphene with nitrogen. The nitrogen content in the samples NG1, NG2 and NG3 is 0.6, 0.9 and 1.4 at.%, respectively. The signal N 1s consists of two peaks at 398.3 and 400 eV, the first of which is characteristic for pyridinic nitrogen (sp^2-hybridization), the second – for nitrogen in the graphene layer.

The essence of the method of segregation growth of graphene is the dissolution of carbon in the metal at high temperatures, followed by segregation (release) on the surface upon cooling. The solubility of carbon in the metal, the crystal lattice of the surface and the conditions of the growth process determine the morphology and thickness (number of layers) of the graphene film. Growth on a hexagonal lattice is often called epitaxial, even if there is no significant coincidence between the lattice and the substrate.

The method of segregation growth of N-graphene is sequentially depositing layers of nitrogen-containing boron and carbon-containing nickel on a SiO_2/Si substrate by means of electron-beam evaporation followed by annealing in a vacuum. During annealing, boron atoms are captured by nickel and migrate into the metal layer. Carbon atoms are segregated on the nickel surface and interact with nitrogen on the Ni substrate to form N-graphene containing 0.3–2.9% nitrogen [136].

Thermal post-synthesis treatment refers to high-temperature methods for the production of N-graphene. The nitrogen content of N-graphene synthesized by this method is relatively low. It is reported that the highest nitrogen content of N-graphene obtained by heat treatment of graphene in ammonia was 2.8 at.% at an annealing temperature of 800–900 °C [137]. N-graphene obtained by heat treatment of graphene in NH_3 at a higher temperature (1100 °C) contains less (1.1%) amount of nitrogen. The low level of doping of N-graphene obtained by the thermal post-synthesis method is explained by the authors [130, 138, 139] by two reasons. One reason is the insufficient number of defects in the original graphene of high quality, as a result of which nitrogen in the pyridine and pyrrole configuration [139] is located only along the edges of the graphene plane [138]. Another reason is the high annealing temperature, which causes the C-N bonds break in N-graphene [138].

In [140], N-graphene was produced using an arc discharge: graphene in an ammonia atmosphere was placed between graphite electrodes. In [141], graphene was placed in an arc discharge in the presence of hydrogen and pyridine. In both cases, the nitrogen content in the obtained samples of heterographene did not exceed 1%.

The nitriding of graphene with intense electric heating in ammonia [27] leads to the formation of C-N bonds at the most reactive edges of graphene. N-heterographene (up to 1.1% nitrogen) with nitrogen-containing radicals (N) and ions (N^+) was obtained by annealing graphene samples subjected to preliminary N ($^+$) ion bombardment in NH_3 [27]. It is assumed [27] that (N) and (N^+) easily form covalent bonds with the carbon lattice and are, therefore, in a stable state.

In the processing of graphene synthesized by chemical method, an N_2-plasma produces N-doped graphene with a low nitrogen content of $\sim$1.3%. It is also reported in [11, 131, 142] that N-graphene can be obtained from graphene by exposing it to nitrogen plasma. It is important to note that the prolonged action of plasma N_2 (40 min) causes destruction (split) of the graphene plane [142]. NH_3-plasma was also used to obtain N-graphene from graphene that was obtained by mechanical peeling [143].

Using ultrasonification, N-doped graphene can be synthesized by treating graphene in hydrazine (N_2H_4) [132]. The nitrogen content of the resulting N-graphene is about 1 at. %. Such N-graphene contains only pyridinic and pyrrolic nitrogen, so the authors of [132] come to the conclusion that when using the method of processing graphene in hydrazine, nitrogen atoms can be introduced only on graphene edges.

Compared to N-doping, doping graphene by boron is a more difficult process. So far, only a few research groups have performed effective synthesis of B-doped graphene (B-graphene). In the study [134], two methods of synthesis were used. In the first, B-graphene was obtained by arc method by evaporation of graphite electrodes in the presence of hydrogen, helium and diborane (B_2H_6). B_2H_6 vapors were generated by adding BF_3-diethylamine etherate to sodium borohydride in tetraglyme (tetraethylene glycol dimethyl ether ($(CH_3OCH_2CH_2OCH_2CH_2)_2O$). The B_2H_6 vapors were introduced into the arc chamber in a stream of hydrogen. In the second synthesis method, B-graphene was produced by an arc method using boron-doped graphite electrodes in a flow of hydrogen and helium. The results of the XPS show that B content in the products is up to 3.1 at%. In [144], a solvothermal method for the synthesis of boron-doped graphene is presented. B-graphene was prepared by reacting CCl_4, K and BBr_3 at a temperature of 160 °C for 20 h in a sealed Teflon autoclave made of stainless steel. The content of boron (1.1%) was determined on the basis of the measurement of the intensity of the XPS.

Currently, only two studies report the successful synthesis of S-doped graphene (S-graphene) [120]. In [145], S-doping was performed by heat treatment of graphene oxide (GO) and dibenzyl disulphide ($C_6H_5CH_2SSCH_2C_6H_5$) in Ar. The mixture of GO and dibenzyl disulphide was ultrasonically dispersed in ethanol for about 30 min. The resulting suspension was evaporated and dried in a dish until a uniform solid mixture was formed. The dried mixture was placed in a quartz tube and annealed in an argon atmosphere at 600–1050 °C. By varying the ratio of the masses of the GO and the dibenzyl disulphide, as well as the annealing temperature, the sulfur content in S-graphene reaches $\sim$1.53% by weight. Another method of synthesis involves the pyrolysis of a sulfur-containing liquid precursor [64]. The sulfur powder was dissolved in hexane under ultrasound to form a clear liquid, which was used as a precursor for the production of S-doped graphene. Copper foil was placed in a quartz tube, and the whole system was evacuated to 10^{-2} Torr, and then heated in a flow of a mixture of hydrogen and argon. When the temperature in the reactor reached 950 °C, the H_2/Ar flow was turned off and the precursor S/hexane mixture from the vapor phase was introduced into the reaction chamber. 2.5 min after the growth of the S-graphene layer on the Cu substrate, the sample was quickly cooled to 800 °C, and then the whole system was cooled to room temperature in a H_2/Ar atmosphere. According to XPS, the degree of doping of graphene with sulfur is about 0.6 at.%.

Some et al. [63] reported the production of phosphorus-doped graphene (P-graphene) by treating graphene with triphenylphosphine ($C_{18}H_{15}P$). Field effect transistors based on double-layer P-graphene show much stronger behavior of *n*-type than double-layer N-graphene, and, due to strong nucleophilicity, the authors of the study [63] believe that they are more stable not only in air, but also in the atmosphere oxygen.

19.3.3 N-Onions

After the discovery of C_{60} and C_{70} fullerenes, other substances of the fullerene series were discovered behind them: the higher fullerenes C_{2n} ($2n > 70$) and onion-like carbon, which are classified as giant fullerenes. The onion-like carbon consists of fullerene-like layers enclosed in each other. For the first time, the formation of the onion-like carbon in a transmission electron microscope was discovered by Ugarte in 1992 after electron beam exposure on carbon nanoparticles and nanotubes [146].

While N-doping of carbon-onion, nitrogen atoms contribute to the curvature of layers of the graphene-like layered structure, forming an onion or onion-like structure, since it facilitate the formation of pentagons (namely, the pentagons provide the warping of graphene-like layers of carbon atoms) and create active centers for the formation of additional interlayer bonds between the shells of onion [31]. Therefore, the N-doping of the onion structure significantly improves the characteristic of the onion-like structure of carbon combination of such mechanical properties as high microhardness and superelastic reduction [30]. N-doping of the carbon-onion provides significantly improved catalytic activity and stability (compared with expensive Pt or Ir) in alkali and acid electrolytes, in particular, in the oxygen reduction reaction (ORR) [35–37, 39].

The nitrogen-doped onion-carbon has excellently proved to be a new efficient electrocatalyst for the Li-O_2 battery [34], a stable catalyst for dye-sensitized solar cells [43]. Therefore, the task of obtaining N-onion-like carbon (N-OLC) is extremely relevant. However, reports on the developed methods for its obtaining are very few. The nitrogen content in the synthesized samples of N-OLC is usually small (no more than ~5%). A higher nitrogen content was recorded only in samples with an onion-like structure containing azafullerene $C_{48}N_{12}$ composition in the core [147]. At the center of such structure, the nitrogen concentration reaches 15–30%. The material is obtained as thin films on a substrate by magnetron sputtering of a graphite sample in a gaseous medium consisting of Ar/10% N_2. Initially, an ultrahigh vacuum ($<210^{-7}$ Pa) is created in the device in the temperature range 200–450 °C. Si (001) and NaCl [147] are used as substrates. However, the magnetron method of N-doped onions synthesizing, developed by the authors of [147], provides only thin films and does not allow obtaining of compact samples of a material with a high nitrogen content.

A compact carbon-nitrogen material with an onion structure is obtained by high-temperature treatment of fullerite, for example, C_{60} and C_{70} [32, 33]. The initial carbon material (fullerite, pre-crushed to a particle size of less than 1 μm) is placed in the working volume and nitrogen is injected to a working pressure of 220–250 MPa. At the first stage, the fullerite is heated to 1670–2020 K and the exposure is not less than 1 min. After that, the temperature is lowered to room temperature, and the pressure is stabilized to atmospheric pressure. In a second step, the resulting carbon-nitrogen material is treated in a high pressure device at a pressure of 7–15 GPa and a temperature of 1620–1770 K for at least 1 min. The resulting onion-like carbon contains 2.2–5.0 at.% N, has a Young's modulus of 43–67 GPa, a

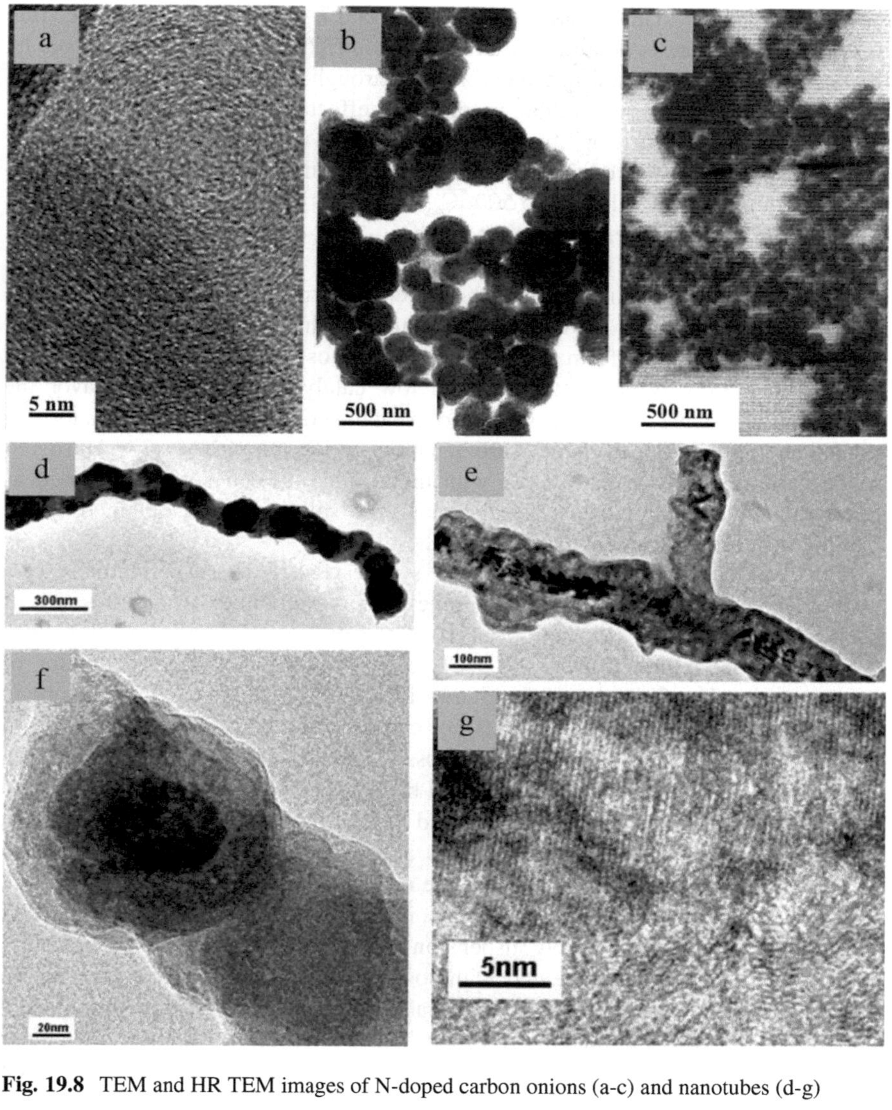

Fig. 19.8 TEM and HR TEM images of N-doped carbon onions (a-c) and nanotubes (d-g)

microhardness of 5.4–8.2 GPa, and a superelastic reduction parameter of 92–97%. The bulk N-doped onion-like carbon with a high (up to 15%) nitrogen content was obtained only in Kharlamov's laboratory under special pyrolysis conditions of pyridine (Fig. 19.8) [3, 6, 12]. High-nitrogen containing N-onion carbon due to polymerization and polycondensation of pyridine molecules is formed in the reaction flow and is carried to the low-temperature zones of the reaction space.

Fullerene soot with an onion-like structure with 0.7–4.5 at.% N incorporated into its carbon framework is obtained by an arc method [32]. Synthesis is carried out by feeding N_2 gas to the center of the carbon arc through a hollow graphite electrode. The formation of fullerenes soluble in toluene is effectively suppressed under these conditions. Temperature programmable oxidation with mass spectrometric detection of gaseous products showed that the nitrogen content did not decrease when the N-doped fullerene soot was heated to 500 °C.

However, the main methods for the synthesis of N-doped onion-like carbon are associated with the catalytic thermal decomposition of C or CN-containing gas phase precursors, often in an atmosphere rich in nitrogen [73]. When ferrocene (FcH) was used together with source C and source N, the Fe/C ratio had a significant effect on the preferential formation of N-carbon nanospheres (N-CNS) or N-doped N-CNT [73]. Low N concentrations and low catalyst concentrations favor the production of N-CNT, while the absence of a catalyst only leads to the formation of N-CNS and other N-doped shaped carbon nanomaterials (N-SCNM) [73]. Nitrogen-doped carbon onions, encapsulated with metal alloys ($FeNi_3$ or FeCo), used as an effective and stable catalyst for dye-sensitized solar cells, are obtained by the method of light and scalable pyrolysis [43].

Nitrogen-doped onion-like carbon was synthesized by thermal treatment of a "hybrid" containing hexamethylenediamine precursor complex in the presence of Co and Fe to produce an electrocatalyst of the oxygen reduction reaction [39]. Hexamethylenediamine (HDA, ($H_2N(CH_2)_6NH_2$)) was used as a precursor of nitrogen and carbon for a complex with transition metals (Fe and Co) [39]. To obtain nitrogen-doped onion-like carbon, 5.0 g of carbon black was pre-treated in 500 ml of 6.0 M HCl solution for 24 h to remove possible metal impurities. The surface area of the treated HCl carbon black, measured by BET, is $780 m^2/g$. A mixture of two transition metal salts [$Co(NO_3)_2/6H_2O$ and $FeCl_3$] in a total amount of 0.1 moles with different ratios of Co : Fe atoms was added to a 0.1 M solution of HDA in ethanol. When the transition metal salts are mixed with HDA at room temperature, a dark yellow suspension gradually appears, indicating the formation of the HDA-M complex (M: Fe and Co) [39]. A suspension of the HDA complex with a transition metal is continuously impregnated with soot for 24 h by continuous soaking (in a ratio by weight of carbon to HDA of about 1 : 10). Porous soot impregnated with HDA is dried in a vacuum and then heat treated at 900 °C using a tubular furnace in a flow of nitrogen for one hour. These heat-treated samples were then chemically leached in 0.5 M H_2SO_4 solution at 80 °C for 8 h to remove unstable and inactive substances from the catalysts to meet the stability requirements for the acidic Nafion-based fuel cell environment. The obtained samples of N-doped onion-like carbon, thoroughly washed with deionized water and dried in a vacuum oven overnight at 80 °C, show excellent electrocatalyst performance of the oxygen reduction reaction [39].

19.4 Toxicology of Nanocarbon and Nanoheterocarbon Materials

In recent years, toxicological features of carbon nanostructures are studied in deep details. The accumulated array of toxicological studies of carbon nanostructures allows us to conclude that some results of studies of the degree of toxicity of nanoparticles on biomaterial (*in vitro*) and in living organisms (*in vivo*) are quite contradictory. Undoubtedly, the toxicity of carbon nanomaterials depends significantly on the size, surface state and the method of preparation of the nanochemical object under study [148–150]. Equally important are methods of studying toxicity, primarily associated with the definition of exposure standards for various uses, including inhalation, injection, ingestion and skin contact. For example, carbon nanotubes, depending on their intake in the body of animals, exhibit various toxic activities. So, the inhalation of rats and mice with carbon nanotubes causes inflammation and fibrosis, the accumulation of neutrophils and protein in the lung tissue [151]. And the injections of large amounts of MWNT into the lungs of mice can cause asbestos-like pathogenicity [152].

At the oral administration the hydroxylated (water-soluble) CNTs are distributed to tissues and organs, excluding the brain. Multi-walled CNTs decrease the cell viability and increase the content of the inflammatory marker of interleukin-8 [151]. However, the toxicity of carbon molecules and nanostructures is primarily due to its ability to induce active oxygen species (oxygen ions, free radicals and various kinds of peroxide) that induce apoptosis and necrosis of the lung cells [153] and kidneys [154]. It was established [155] that the concentration limit of the cytotoxicity of a carbon nanotubes suspension is about 0.01 mg/ml. Carbon nanotubes are carcinogenic agents, it damage DNA and the nucleus of cells, cause fibrosis and granulomas in mice, and also destroy stem cells of mouse embryos [149]. Single-walled and multi-walled carbon nanotubes varies in the degree of toxicity and the ability to induce oxidative stress: single-walled carbon nanotubes are more toxic than multilayered CNTs and fullerenes [156]. SWNTs significantly break phagocytosis of alveolar macrophage at a low doze 0.38 mkg/sm^2, while MWNT and C_{60} cause an infringement only at a high doze 3.06 mkg/sm^2. Fullerenes, graphene and graphene oxide also generate reactive oxygen species and induce oxidative damage to the membrane and nucleus of liver and human blood cells [148, 149].

Thus, fullerene C_{60} is toxic at a very low concentration (0.02 ppb). Intravenous inward of doses (15 and 25 mg/kg) of fullerene just in 5 min resulted in the death of two of twenty rats. While at the oral administration, the lethal dose for rats was 600 mg/kg. At the sublethal doses, the nephropathy developed in animals. Graphene and MWNT strongly damage the cell membrane. It is shown that single-walled carbon nanotubes damage the cell membrane due to their "needle-like" shape. The cytotoxicity of graphene oxide was explained by the physical damage to the cell membrane by the edge of the carbon nanolayers [149]. The long-term effect of graphene oxide results in a hyperpermeable state of the intestinal barrier and changes in genes that are necessary for normal bowel function [149]. In [157], the cytotoxicity of graphene oxide (GO) was studied *in vitro* and *in vivo* depending on the size of nano flakes (diameter 89 and 277 nm) and on the time of exposure to cells. The impact of GO particles of any size leads to cell destruction within 48 h. GO flakes of larger nanosize rapidly reduce cell viability.

It is important to note that the toxicity of the most promising materials for nanosensors (in particular, for CBRNE defense) and others nanodevice based on hetero-carbon, such as heterofullerenes, heteronanotubes, heterographene, onion-like N-doped carbon (Fig. 19.8) [6], O-doped carbon nitride (O-C_3N [12, 13], O-C_3N_4 [3, 14]) has not been studied in practice. A few toxicological studies of N-doped carbon structures indicate an important feature: one of the properties that may be most important for the future prospects of the use of heterocarbon nanomaterials is its ***biocompatibility*** [2]. N-doped nanotubes, functionalized to render it being soluble in water, were directly excreted from the body of lab rats in their urine [158]. Whereas studies of pristine nanotubes demonstrated its absorption and storage in the body (notably within the organs) [159]. Toxicological studies of nitrogen-doped nanotubes showed a significantly lower toxic reaction in rats than their pristine MWNT analogs [160]. Thus, the chemical activation of the surface of carbon nanostructures, caused by doping nitrogen and the possible presence of an amide surface, makes it more biocompatible, and therefore more promising for the usage in biology and medicine than their pristine counterparts. Consequently, further toxicological studies of heterocarbon nanostructures are an extremely promising topical task.

19.5 Conclusions

The growing examples of international terrorist threats against citizens and principal infrastructures lead to increased security measures and the complication of sensing technologies. The tendency to complicate sensing technologies, in turn, dictates the need to develop innovative and new heterocarbon nanomaterials based on carbon molecules and nanostructures modified (doped) by atoms of other elements. Nanoheterocarbon materials are widely used in the field of national security (in particular, in the field of chemical, biological, radiological, nuclear and explosive (CBRNE) defense). However, the toxicity of the most promising nanomaterials for nanosensors, catalysts and other nanodevices based on heterocarbon has not yet been studied. Therefore, along with the need to develop new methods for the synthesis of heterocarbon nanostructures, the most important task is to study their potential toxicological effects and biocompatibility.

References

1. Zaporotskova IV, Boroznina NP, Parkhomenko YN et al (2016) Carbon nanotubes: sensor properties. A review. Mod Electron Mater 2:95–105
2. Ewels CP, Glerup M, Krstic V (2008) Nitrogen and boron doping in carbon nanotubes. In: Basiuk VA, Basiuk EV (eds) Chemistry of carbon nanotubes, Part 2. American Scientific Publishers, Stevenson Ranch, pp 2–65
3. Kharlamova G, Kharlamov O, Bondarenko M, Khyzhun O (2016) Hetero-carbon nanostructures as the effective sensors in security systems. In: Bonca J, Kruchinin S (eds) Nanomaterials for security, nato science for peace and security series a: chemistry and biology, chapter 19. Springer, Dordrecht, pp 239–258
4. Kharlamova G, Kharlamov O, Bondarenko M et al (2013) Hetero-carbon: heteroatomic molecules and nano-structures of carbon. In: Vaseashta A, Khudaverdyan S (eds) Advanced sensors for safety and security. Nato science for peace and security series B: physics and biophysics, Part 7. Springer, Dordrecht, pp 339–357
5. Kharlamova G, Kharlamov O, Bondarenko M (2015) Nanosensors in systems of ecological security. In: Bonca J, Kruchinin S (eds) Nanotechnology in the security systems. Nato science for peace and security series C, chapter 20. Environmental Security, Springer, Dordrecht, pp 231–242
6. Kharlamov AI, Kharlamova GA, Bondarenko ME (2013) Preparation of onion-like carbon with high nitrogen content (~15%) from pyridine. Russ J Appl Chem 86(10):1493–1503
7. Sandoval LM, Martinez H, Terrones M (2004) Fabrication of vapor and gas sensors using films of aligned CN_x nanotubes. Chem Phys Lett 386:137–143
8. Kharlamov AI, Kharlamova GA, Bondarenko ME (2013) New products of a new method for pyrolysis of pyridine. Russ J Appl Chem 86(2):167–175
9. Zhang J, Lei J, Pan R et al (2011) In situ assembly of gold nanoparticles on nitrogen-doped carbon nanotubes for sensitive immunosensing of microcystin-LR. Chem Commun 47: 668–670
10. Lv R, Li Q, Botello-Méndez AR et al (2012) Nitrogen-doped graphene: beyond single substitution and enhanced molecular sensing. Sci Rep 2(586):1–8
11. Wang Y, Shao Y, Matson DW et al (2010) Nitrogen-doped graphene and its application in electrochemical biosensing. Electrochem Biosensing ACS Nano 4(4):1790–1798
12. Kharlamov O, Bondarenko M, Kharlamova G (2015) O-Doped carbon nitride (O-g-C_3N) with high oxygen content (11.1 mass%) synthesized by pyrolysis of pyridine. In: Camesano TA (ed) Nanotechnology to aid chemical and biological defense. Nato science for peace and security series A: chemistry and biology, chapter 9. Springer, Dordrecht, pp 129–145
13. Kharlamov AI, Bondarenko ME, Kharlamova GA (2014) New method for synthesis of oxygen-doped graphite-like carbon nitride from pyridine. Russ J Appl Chem 87:1284–1293
14. Kharlamov A, Bondarenko M, Kharlamova G (2016) Method for the synthesis of water-soluble oxide of graphite-like carbon nitride. Diamond Relat Mater 61:46–55
15. Kharlamov A, Bondarenko M, Kharlamova G et al (2016) Features of the synthesis of carbon nitride oxide (g-C_3N_4)O at urea pyrolysis. Diamond Relat Mater 66:16–22
16. Kharlamov A, Bondarenko M, Kharlamova G et al (2016) Synthesis of reduced carbon nitride at the reduction by hydroquinone of water-soluble carbon nitride oxide (g-C_3N_4)O. J Solid State Chem 241:115–120
17. Nafise G (2011) CVD synthesis of nitrogen doped carbon nanotubes using iron pentacarbonyl as catalyst. Ph.D. Thesis. University of the Witwatersrand, Johannesburg, p 97
18. Plaza MG, Pevida C, Arenillas A et al (2007) CO_2 capture by adsorption with nitrogen enriched carbons. Fuel 86(14):2204–2212
19. Stavropoulos GG, Samaras P, Sakellaropoulos GP (2008) Effect of activated carbons modification on porosity, surface structure and phenol adsorption. J Hazard Mater 151(2–3):414–421
20. Hulicova D, Kodama M, Hatori H (2006) Electrochemical performance of nitrogen-enriched carbons in aqueous and non-aqueous supercapacitors. Chem Mater 18(9):2318–2326

21. Lee YF, Chang KH, Hu CC et al (2011) Synthesis of N-doped carbon nanosheets from collagen for electrochemical energy storage/conversion systems. Electrochem Commun 13:50–53
22. Shao YY, Sui JH, Yin GP et al (2008) Nitrogen-doped carbon nanostructures and their composites as catalytic materials for proton exchange membrane fuel cell. Appl Catal B-Environ 79(1–2):89–90
23. Ao Z, Li S (2011) Hydrogenation of graphene and hydrogen diffusion behavior on graphene/graphane interface. In: Gong JR (ed) Graphene simulation. InTech, Croatia, pp 53–74
24. Simon F, Kuzmany H, Fülöp F (2006) Encapsulating C_{59}Nazafullerenes inside single-wall carbon nanotubes. Phys Status Solidi B 243(13):3263–3267
25. Cuong NT, Otani M, Iizumi Y et al (2011) Origin of the n-type transport behavior of azafullerene encapsulated single-walled carbon nanotubes. Appl Phys Lett 99(5):053105–053108
26. Simon F, Kuzmany H, Bernardi J et al (2006) Encapsulating C_{59}Nazafullerene derivatives inside single-wall carbon nanotubes. Carbon 44:1958–1962
27. Wang H, Maiyalagan T, Wang X (2012) Review on recent progress in nitrogen-doped graphene: synthesis, characterization, and its potential applications. ACS Catal 2(5):781–794
28. Hirata A, Igarashi M, Kaito T (2004) Study on solid lubricant properties of carbon onions produced by heat treatment of diamond clusters or particles. Tribology Int 37(11–12): 899–905
29. Su DS, Maksimova N, Delgado JJ et al (2005) Nanocarbons in selective oxidative dehydrogenation reaction. Catal Today 102–103:110–114
30. Neidhardt J, Czigany Z, Hultman L (2003) Superelastic fullerene-like carbon nitride coatings synthesised by reactive unbalanced magnetron sputtering. Surf Eng 19(4):299–303
31. Czigany Zs, Brunell IF, Neidhardt J et al (2001) Growth of fullerene-like carbon nitride thin solid films consisting of cross-linked nano-onions. Appl Phys Lett 79(16):2639–2641
32. Belz T, Baue A, Find J et al (1998) Structural and chemical characterization of N-doped nanocarbons. Carbon 36(5–6):731–741
33. Gurevich AM, Terekhov AV, Kondrashev DS, Dolbin AV (2006) Low-temperature heat capacity of fullerite C_{60} doped with nitrogen. Low Temp Phys 32(10):967–969
34. Shu C, Lin Y, Su D (2016) N-doped onion-like carbon as an efficient oxygen electrode for long-life Li–O_2 battery. J Mater Chem A 4:2128
35. Wu G, Santandreu A, Kellogg W et al (2016) Carbon nanocomposite catalysts for oxygen reduction and evolution reactions: from nitrogen doping to transition-metal addition. Nano Energy 29:83–110
36. Wu G, Mack NH, Gao W et al (2012) Nitrogen-doped graphene-rich catalysts derived from heteroatom polymers for oxygen reduction in nonaqueous lithium-O_2 battery cathodes. ACS Nano 6:9764–9776
37. Li Q, Pan H, Higgins D et al (2015) Metal–organic framework-derived bamboo-like nitrogen-doped graphene tubes as an active matrix for hybrid oxygen-reduction electrocatalysts. Small 11:1443–1452
38. Matter PH, Zhang L, Ozkan US et al (2006) The role of nanostructure in nitrogen-containing carbon catalysts for the oxygen reduction reaction. J Catal 239:83–96
39. Wu G, Nelson M, Ma S et al (2011) Synthesis of nitrogen-doped onion-like carbon and its use in carbon-based CoFe binary non-precious-metal catalysts for oxygen-reduction. Carbon 49:3972–3982
40. Lahaye J, Nansé G, Bagreev A et al (1999) Porous structure and surface chemistry of nitrogen containing carbons from polymers. Carbon 37:585–590
41. Wu G, Li D, Dai C et al (2008) Well-dispersed high-loading Pt nanoparticles supported by shell-core nanostructured carbon for methanol electrooxidation. Langmuir 24:3566–3575
42. Wu G, Swaidan R, Li D et al (2008) Enhanced methanol electro-oxidation activity of PtRu catalysts supported on heteroatom-doped carbon. Electrochim Acta 53:7622–7629

43. Zhu C, Xu F, Chen J et al (2016) Nitrogen-doped carbon onions encapsulating metal alloys as efficient and stable catalysts for dye-sensitized solar cells. J Power Sources 303:159–167
44. Glerup M, Krstić V, Ewels C et al (2008) Doping of carbon nanotubes. In: Chen W (ed) Doped nanomaterials and nanodevices. American Scientific Publishers, New York, pp 169–242
45. Zanchetta J, Marchand A (1965) Electronic properties of nitrogen doped carbons. Carbon 3:332–332
46. Marchand A, Zanchetta JV (1966) Proprietes electroniques d'un carbone dope a l'azote. Carbon 3:483–491
47. Iijima S (1991) Helical microtubules of graphite carbon. Nature 354:56–58
48. Kharlamov AI, Kirillova NV, Zytheva ZA, Golovkova ME (2007) New state of carbon: transparent thread-like anisotropic crystals. Rep Acad Sci Ukraine 5:101–106
49. Radushkevich LV, Lukyanovich VM, (1952) O strukture ugleroda, obrazujucegosja pri termiceskom razlozenii okisi ugleroda na zeleznom kontakte. Zurn Fisic Chim 26:88–95
50. Monthioux M, Kuznetsov VL (2006) Who should be given the credit for the discovery of carbon nanotubes? Carbon 44:1621–1623
51. Khabashesku VN (2011) Covalent functionalization of carbon nanotubes: synthesis, properties and applications of fluorinated derivatives. Russ Chem Rev 80(8):705–725
52. Stephen O, Ajayan PM, Colliex C et al (1994) Doping graphitic and carbon nanotube structures with boron and nitrogen. Science 266:1683–1685
53. Sumpter BG, Meunier V, Romo-Herrera JM et al (2007) Nitrogen-mediated carbon nanotube growth: diameter reduction, metallicity, bundle dispersability, and bamboo-like structure formation. ACS Nano 1:369–375
54. Romo-Herrera JM, Sumpter BG, Cullen DA et al (2008) An atomistic branching mechanism for carbon nanotubes: sulfur as the triggering agent. Angew Chem Int Ed 47:2948–2953
55. Hashim DP, Narayanan NT, Romo-Herrera JM et al (2012) Covalently bonded three dimensional carbon nanotube solids via boron induced nanojunctions. Sci Rep 2: 363-1–363-8
56. Lee DH, Lee WJ, Kim SO (2009) Highly efficient vertical growth of wall-number-selected, n-doped carbon nanotube arrays. Nano Lett 9:1427–1432
57. Wilder WG, Venema LC, Rinzler AG et al (1998) Electronic structure of atomically resolved carbon nanotubes. Nature 391:59–62
58. Chan LH, Hong KH, Xiao DQ et al (2004) Resolution of the binding configuration in nitrogen-doped carbon nanotubes. Phys Rev B 70:125408–125415
59. Pan H, Feng YP, Lin J (2006) Ab initio study of single-wall BC_2N nanotubes. Phys Rev B 74:045409–045413
60. Schutz D, Droppa R Jr, Alvarezet F et al (2003) Stability of small carbon-nitride heterofullerenes. Phys Rev Lett 90:015501–015504
61. Doytcheva M, Kaisera M, Verheijen MA et al (2004) Electron emission from individual nitrogen-doped multi-walled carbon nanotubes. Chem Phys Lett 396:126–130
62. Lee JM, Park JS, Lee SH et al (2011) Selective electron- or hole-transport enhancement in bulk-heterojunction organic solar cells with N- or B- doped carbon nanotubes. Adv Mater 23:629–633
63. Some S, Kim J, Lee K et al (2012) Highly air stable phosphorus-doped n-type graphene field-effect transistors. Adv Mater 24:5481–5486
64. Gao H, Liu Z, Song L et al (2012) Synthesis of S-doped graphene by liquid precursor. Nanotechnology 23:275605-1–275605-7
65. Xu J, Dong G, Jin C et al (2013) Sulfur and nitrogen co-doped, few layered graphene oxide as a highly efficient electrocatalyst for the oxygen reduction reaction. Chem Sus Chem 6: 493–499
66. Tang C, Bando Y, Golberg D et al (2004) Structure and nitrogen incorporation of carbon nanotubes synthesized by catalytic pyrolysis of dimethylformamide. Carbon 42:2625–2633
67. Terrones M, Terrones H, Grobert N et al (1999) Efficient route to large arrays of CNx nanofibers by pyrolysis of ferrocene/melamine mixtures. Appl Phys Lett 75:3932–3934
68. Yudasaka M, Kikuchi R, Ohki Y et al (1997) Nitrogen-containing carbon nanotube growth from Ni phthalocyanine by chemical vapor deposition. Carbon 35:195–201

69. Kim SY, Lee J, Na CW et al (2005) N-doped double-walled carbon nanotubes synthesized by chemical vapor deposition. Chem Phys Lett 413:300–305
70. Dommele S (2008) Nitrogen doped carbon nanotubes: synthesis, characterization and catalysis. Ph.D. Thesis Utrecht University, Utrecht
71. Khavrus VO, Ibrahim EMM, Leonhardt A et al (2008) Simultaneous synthesis and separation of single- and multi-walled CN_x nanotubes. CarboCat 9(12):17–18
72. Teddy J (2009) CVD synthesis of carbon nanostructures and their applications as supports in catalysis. Ph.D. Thesis Toulouse University, Toulouse
73. Nxumalo EN, Coville NJ (2010) Nitrogen doped carbon nanotubes from organometallic compounds. Rev Mater 3:2141–2171
74. Barreiro A, Hampel S, Rümmeli MH (2006) Thermal decomposition of ferrocene as a method for production of single-walled carbon nanotubes without additional carbon sources. J Phys Chem B 110:20973–20977
75. Koós AA, Dillon F, Nicholls RJ et al (2012) N-SWCNTs production by aerosol-assisted CVD method. Chem Phys Lett 538:108–111
76. Villalpando-Paez F, Zamudio A, Elias AL et al (2006) Synthesis and characterization of long strands of nitrogen-doped single-walled carbon nanotubes. Chem Phys Lett 424:345–352
77. Maldonado S, Morin S, Stevenson KJ (2006) Structure, composition, and chemical reactivity of carbon nanotubes by selective nitrogen doping. Carbon 44:1429–1437
78. Ghosh P, Soga T, Ghosh K et al (2008) Vertically aligned N-doped carbon nanotubes by spray pyrolysis of turpentine oil and pyridine derivative with dissolved ferrocene. J Non-Cryst Solids 354:4101–4106
79. Koo's AA, Dowling M, Jurkschat K et al (2009) Effect of the experimental parameters on the structure of nitrogen-doped carbon nanotubes produced by aerosol chemical vapour deposition. Carbon 47:30–37
80. Li YL, Hou F, Yang ZT et al (2009) The growth of N-doped carbon nanotube arrays on sintered $A_{l2}O_3$ substrates. Mat Sci Eng: B 158:69–74
81. Liu J, Czerw R, Carroll DL (2005) Large-scale synthesis of highly aligned nitrogen doped carbon nanotubes by injection chemical vapor deposition methods. J Mater Res 20:538–543
82. Ghosh P, Tanemura M, Soga T et al (2008) Field emission property of N-doped aligned carbon nanotubes grown by pyrolysis of monoethanolamine. Solid State Commun 147:15–19
83. Jiang K, Eitan A, Schadler LS et al (2003) Selective attachment of gold nanoparticles to nitrogen-doped carbon nanotubes. Nano Lett 3:275–277
84. Lee CJ, Lyu SC, Kim HW et al (2002) Synthesis of bamboo-shaped carbon–nitrogen nanotubes using C_2H_2–NH_3–$Fe(CO)_5$ system. Chem Phys Lett 359:115–120
85. Liu J, Webster S, Carroll DL (2005) Temperature and flow rate of NH_3 effects on nitrogen content and doping environments of carbon nanotubes grown by injection CVD method. J Phys Chem B 109:15769–15774
86. Sen R, Satishkumar BC, GovindaraJ A et al (1997) Nitrogen-containing carbon nanotubes. J Mater Chem 7:2335–2337
87. Liang EJ, Ding P, Zhang HR et al (2004) Synthesis and correlation study on the morphology and Raman spectra of CNx nanotubes by thermal decomposition of ferrocene/ethylenediamine. Diamond Relat Mater 13:69–73
88. Cao C, Huang F, Cao C et al (2004) Synthesis of carbon nitride nanotubes via a catalytic-assembly solvothermal route. Chem Mater 16:5213–5215
89. Bill J, Riedel R (1992) Boron carbide nitride derived from amine-boranes. Mater Res Soc Symp Proc 271:839–844
90. Suenaga K, Colliex C, Demoncy N et al (1997) Synthesis of nanoparticles and nanotubes with well-separated layers of boron nitride and carbon. Science 278:653–655
91. Zhang Y, Gu H, Suenaga K et al (1997) Heterogeneous growth of BCN nanotubes by laser ablation. Chem Phys Lett 279:264–269
92. Glenis S, Cooke S, Chen X et al (1994) Photophysical properties of fullerenes prepared in an atmosphere of pyrrole. Chem Mater 6:1850–1853

93. Pradeep T, Vijayakrishnan V, Santa AK et al (1991) Interaction of nitrogen with fullerenes: nitrogen derivatives of C_{60} and C_{70}. J Phys Chem 95:10564–10565
94. Droppa R Jr, Hammer P, Carvalho ACM et al (2002) Incorporation of nitrogen in carbon nanotubes. J Non-Cryst Solids 299:874–879
95. Glerup M, Steinmetz J, Samaille D et al (2004) Synthesis of N-doped SWNT using the arc-discharge procedure. Chem Phys Lett 387:193–197
96. Goldberg D, Bando Y, Bourgeois L et al (2000) Large-scale synthesis and HRTEM analysis of single-walled B- and N-doped carbon nanotube bundles. Carbon 38:2017–2027
97. Morant C, Andrey J, Prieto P et al (2006) XPS characterization of nitrogen-doped carbon nanotubes. Physica Status Solidi A 203:1069–1075
98. Kruchinin SP, Repetsky SP, Vyshyvana IG (2016) Spin-depent transport of carbon nanotubes with chromium atoms. In: Bonca J, Kruchinin S (eds) Nanomaterials for Security. Springer, pp 65–97
99. Ermakov V, Kruchinin S, Hori H, Fujiwara A (2007) Phenomena of strong electron correlastion in the resonant tunneling. Int J Mod Phys B 11:827–835
100. Kruchinin S, Pruschke T (2014) Thermopower for a molecule with vibrational degrees of freedom. Phys Lett A 378:157–161
101. Ermakov V, Kruchinin S, Pruschke T, Freericks J (2015) Thermoelectricity in tunneling nanostructures. Phys Rev B 92:115531
102. Maultzsch J, Reich S, Thomsen C, Webster S et al (2002) Raman characterization of boron-doped multiwalled carbon nanotubes. Appl Phys Lett 81(14):2647–2649
103. Mondal KC, Coville NJ, Witcomb MJ et al (2007) Boron mediated synthesis of multiwalled carbon nanotubes by chemical vapor deposition. Chem Phys Lett 437:87–91
104. Chen CF, Tsai CL, Lin CL (2003) The characterization of boron-doped carbon nanotube arrays. Diam Rel Mater 12(9):1500–1504
105. Sharma RB, Late DJ, Joag DS et al (2006) Field emission properties of boron and nitrogen doped carbon nanotubes. Chem Phys Lett 428(1–3):102–108
106. Ceragioli HJ, Peterlevitz AC, Quispe JC et al (2008) Synthesis and characterization of boron-doped carbon nanotubes. J Phys Conf Ser 100(5):1–4
107. Okotrub AV, Bulusheva LG, Kudashov AG et al (2008) Arrays of carbon nanotubes aligned perpendicular to the substrate surface: anisotropy of structure and properties. Nanotechnol Russ 3:191–200
108. Ishii S, Watanabe T, Ueda S et al (2008) Resistivity reduction of boron-doped multi-walled carbon nanotubes synthesized from a methanol solution containing a boric acid. Appl Phys Lett 92(20):202116–202116-3
109. McGuire K, Gothard N, Gai PL et al (2005) Synthesis and raman characterization of boron-doped single-walled carbon nanotubes. Carbon 43:219–227
110. Goldberg D, Bando Y, Han W et al (1999) Single-walled B-doped carbon, B/N-doped carbon and BN nanotubes synthesized from single-walled carbon nanotubes through a substitution reaction. Chem Phys Lett 308:337–342
111. Guo M, Huang J, Kong X et al (2016) Hydrothermal synthesis of porous phosphorus-doped carbon nanotubes and their use in the oxygen reduction reaction and lithium-sulfur batteries. New Carbon Mater 31(3):352–362
112. Larrude DG, Maia da Costa MEH, Monteiro FH et al (2012) Characterization of phosphorus-doped multiwalled carbon nanotubes. J Appl Phys 111(6):064315-064315-6
113. Patiño J, López-Salas N, Gutiérrez MC et al (2016) Phosphorus-doped carbon–carbon nanotube hierarchical monoliths as true three-dimensional electrodes in supercapacitor cells. J Mater Chem A 4:1251–1263
114. Cui T, Lv R, Huang Z et al (2011) Effect of sulfur on enhancing nitrogen-doping and magnetic properties of carbon nanotubes. Nanoscale Res Lett 6(77):1–6
115. Kucukayan G, Ovali R, Ilday S et al (2011) An experimental and theoretical examination of the effect of sulfur on the pyrolytically grown carbon nanotubes from sucrose-based solid state precursors. Carbon 49:508–517

116. Novoselov KS, Geim AK, Morozov SV et al (2004) Electric field effect in atomically thin carbon films. Science 306(5696):666–669
117. Novoselov KS, Jiang D, Schedin F et al (2005) Two-dimensional atomic crystals. Proc Natl Acad Sci USA 102(30):10451–10453
118. Geim AK, Novoselov KS (2007) The rise of grapheme. Nat Mater 6:183–191
119. Repetsky SP, Vyshyvana IG, Kruchinin SP, Molodkin VB, Lizunov VV (2017) Influence of the adsorbed atoms of potassium on an energy spectrum of grapheme. Metallofiz Noveishie Tekhnol 39:1017–1022
120. Hu Y, Sun X (2013) Chemically functionalized graphene and their applications in electrochemical energy conversion and storage. In: Aliofkhazraei M (Ed.) Advances in Graphene Science, chapter 7. InTech, pp 161–189
121. Wei D, Liu Y, Wang Y (2009) Synthesis of N-doped graphene by chemical vapor deposition and its electrical properties. Nano Lett 9(5):1752–1758
122. Luo Z, Lim S, Tian Z et al (2011) Pyridinic N doped graphene: synthesis, electronic structure, and electrocatalytic property. J Mater Chem 21:8038–8044
123. Chernozatonskii LA, Sorokin PB, Artukh AA (2014) Novel graphene-based nanostructures: physicochemical properties and applications. Russ Chem Rev 83(3):251–279
124. Wei D, Liu Y, Wang Y et al (2009) Gold nanoparticles can induce the formation of protein-based aggregates at physiological pH. Nano Lett 9(2):666–671
125. Usachov D, Vilkov O, Gruneis A et al (2011) Nitrogen-doped graphene: efficient growth, structure, and electronic properties. Nano Lett 11(12):5401–5407
126. Zhang LS, Liang XQ, Song W et al (2010) Identification of the nitrogen species on N-doped graphene layers and Pt/NG composite catalyst for direct methanol fuel cell. Chem Chem Phys 12:12055–12059
127. Long D, Li W, Ling L et al (2010) Preparation of nitrogen-doped graphene sheets by a combined chemical and hydrothermal reduction of graphene oxide. Langmuir 26:16096–16102
128. Jeong HM, Lee JW, Shin WH et al (2011) Nitrogen-doped graphene for high performance ultracapacitors and the importance of nitrogen-doped sites at basal-planes. Nano Lett 11:2472–2477
129. Sun L, Tian C, Tan T (2012) Nitrogen-doped graphene with high nitrogen level via a one step hydrothermal reaction of graphene oxide with urea for superior capacitive energy storage. RSC Adv 2:4498–4506
130. Geng D, Chen Y, Chen Y et al (2011) High oxygen-reduction activity and durability of nitrogen-doped grapheme. Energy Environ Sci 4:760–764
131. Imran JR, Rajalakshmi N, Ramaprabhu S (2010) Nitrogen doped graphene nanoplatelets as catalyst support for oxygen reduction reaction in proton exchange membrane fuel cell. J Mater Chem 20:7114–7117
132. Wang DW, Gentle IR, Lu GQ (2010) Enhanced electrochemical sensitivity of PtRh electrodes coated with nitrogen-doped graphene. Electrochem Commun 12:1423–1427
133. Li N, Wang Z, Zhao K (2010) Large scale synthesis of N-doped multi-layered graphene sheets by simple arc-discharge method. Carbon 48(1):255–259
134. Panchokarla L, Subrahmanyam K, Saha S (2009) Synthesis, structure and properties of boron and nitrogen doped grapheme. Adv Mater 21(46):4726
135. Kim H, Kim H (2006) Preparation of carbon nanotubes by DC arc discharge process under reduced pressure in an air atmosphere. Mater Sci Eng B 133(1–3):241–244
136. Zhang C, Fu L, Liu N et al (2011) Synthesis of nitrogen-doped graphene using embedded carbon and nitrogen sources. Adv Mater 23:1020–1024
137. Guo B, Liu Q, Chen E et al (2010) Controllable N-doping of grapheme. Nano Lett 10(12):4975–4980
138. Kinoshita K (1988)Carbon: electrochemical and physicochemical properties. Wiley, New York
139. Wang X, Li X, Zhang L et al (2009) N-doping of graphene through electrothermal reactions with ammonia. Science 324:768–771

140. Lia N, Wang Z, Zhao K et al (2010) Large scale synthesis of N-doped multi-layered graphene sheets by simple arc-discharge method. Carbon 48(1):255–259
141. Panchakarla LS, Subrahmanyam KS, Saha SK et al (2009) Synthesis, structure, and properties of boron- and nitrogen-doped grapheme. Adv Mater 21(46):994726–994730
142. Shao Y, Zhang S, Engelhard MH et al (2010) Nitrogen-doped graphene and its electrochemical applications. Mater Chem 20:7491–7494
143. Lin YC, Lin CY, Chiu PW (2010) Controllable graphene N-doping with ammonia plasma. Appl Phys Lett 96:133110–133113
144. Lin T, Huang F, Jiang J (2012) A facile preparation route for boron-doped graphene, and its CdTe solar cell application. Energy Environ Sci 4(3):862–865
145. Yang Z, Yao Z, Li G (2012) Sulfur-doped graphene as an efficient metal-free cathode catalyst for oxygen reduction. ACS Nano 6(1):205–211
146. Ugarte D (1992) Curling and closure of graphitic networks under electron-beam irradiation. Lett Nature 359:707–709
147. Hultman L, Stafström S, Czigány Z et al (2001) Cross-linked nano-onions of carbon nitride in the solid phase: aza-fullerene. Phys Rev Lett 8722(22):225503-1–225503-4
148. Kharlamov O, Bondarenko M, Khyzhun O, Kharlamova G (2016) Anthology and genesis of nanodimensional objects and GM food as the threats for human security. In: Bonca J, Kruchinin S (eds) Nanomaterials for security. NATO science for peace and security series A: chemistry and biology, chapter 24. Springer, Dordrecht, pp 297–310
149. Kharlamov O, Bondarenko M, Kharlamova G et al (2015) Nanoecological security of foodstuffs and human. In: Bonca J, Kruchinin S (eds) Nanotechnology in the security systems. NATO science for peace and security series C: environmental security, chapter 19. Springer, Dordrecht, pp 215–229
150. Ermakov V, Kruchinin S, Fujiwara A (2008) Electronic nanosensors based on nanotransistor with bistability behaviour. In: Bonca J, Kruchinin S (eds) Proceedings of NATO ARW "Electron Transport in Nanosystems". Springer, pp 341–349
151. Donaldson K, Aitken R, Tran L et al (2006) Carbon nanotubes: review of their properties in relation to pulmonary toxicology and workplace safety. Toxicol Sci 92(1):5–22
152. Volder MFL, Tawfick SH, Baughman RH et al (2013) Carbon nanotubes: present and future commercial applications. Science 339:535–539
153. Wadhwa S, Rea C, O'Hare P et al (2011) Comparative in vitro cytotoxicity study of carbon nanotubes and titania nanostructures on human lung epithelial cell. J Hazardous Matter 191(1–3):56–61
154. Cui SD, Tian F, Ozkan CS et al (2005) Effect of single wall carbon nanotubes on human HEK293 cells. Toxicol Lett 155:73–85
155. Kartel MT, Ivanov LV, Kovalenko SN, Tereschenko VP (2011) Carbon nanotrubes: biorisks and biodefence. In: Mikhalovsky S, Khajibaev A (eds) Biodefence. NATO science for peace and security series A: chemistry and biology. Springer, Dordrecht, pp 11–22
156. Jia G, Wang HF, Yan L et al (2005) Cytotoxicity of carbon nanomaterials: single-wall nanotube, multi-wall nanotube and fullerene. Env Sci Technol 39:1378–1383
157. Mendes RG, Koch B, Bachmatiuk A et al (2015) A size dependent evaluation of the cytotoxicity and uptake of nanographene oxide. J Mater Chem B 12(3):2522–2529
158. Singh R, Pantarotto D, Lacerda L et al (2006) Tissue biodistribution and blood clearance rates of intravenously administered carbon nanotube radiotracers. PNAS 103:3357–3362
159. Muller FHJ, Moreau N, Missonet P et al (2005) Respiratory toxicity of multi-wall carbon nanotubes. Toxicol Appl Pharmacol 207:221–231
160. Carrero-Sanchez JC, Mancilla R, Arrellin G et al (2006) Biocompatibility and toxicological studies of carbon nanotubes doped with nitrogen. Nano Lett 6:1609–1616

Chapter 20
Carbon Nitride Oxide (g-C_3N_4)O and Heteroatomic N-Graphene (Azagraphene) as Perspective New Materials in CBRN Defense

O. Kharlamov, M. Bondarenko, G. Kharlamova, P. Silenko, O. Khyzhun, and N. Gubareni

Abstract In the twenty first century, there was an abrupt growth in the number of terrorist attacks and local military conflicts. Weapons of mass destruction, in particular, chemical weapons, are now being used again in the Syrian conflict. Therefore, the development of modern means of protection against weapons of mass destruction is becoming extremely relevant. The most important modern tool for the timely detection of substances that are deadly dangerous to humans in CBRN defense is a new generation of sensors – nanosensors based on nanomaterials. As a promising material for the creation of a new generation of nanosensors, graphite-like carbon nitride and its nanostructured and doped derivatives attract special attention. The water-soluble carbon nitride oxide (g-C_3N_4)O was synthesized by the gas phase method under special reaction conditions of pyrolysis of melamine and urea. Reduction by the hydroquinone of carbon nitride oxide (g-C_3N_4)O yields nanostructured reduced carbon nitride (or reduced multilayer azagraphene).

Keywords Azagraphene · Carbon nitride oxide · CBRN defense · Nanosensors

O. Kharlamov (✉)
Frantsevich Institute for Problems of Materials Science of NASU, Kiev, Ukraine

Taras Shevchenko National University of Kyiv, Kiev, Ukraine
e-mail: akharlamova@ukr.net

M. Bondarenko · P. Silenko · O. Khyzhun · N. Gubareni
Frantsevich Institute for Problems of Materials Science of NASU, Kiev, Ukraine

G. Kharlamova
Taras Shevchenko National University of Kyiv, Kyiv, Ukraine

J. Bonča, S. Kruchinin (eds.), *Nanostructured Materials for the Detection of CBRN*, NATO Science for Peace and Security Series A: Chemistry and Biology,
https://doi.org/10.1007/978-94-024-1304-5_20

20.1 Introduction

Since the beginning of the twenty first century, the number and intensity of local military conflicts and terrorist attacks have increased enormously. The optimistic hopes of the last century, arising after the collapse of the Soviet system, that the unconditionally predetermined harmonious liberal-democratic future of human civilization (deeply depicted by Francis Fukuyama in the essay "The End of History" and in the book "The End of History and the Last Man") are went in the dust. At least they are put into question. Nowadays, the balance of power in international relations has shaken, the unipolar system may be degraded and the ghost of a global military conflict arose in front of the world.

So, quite recently, it seemed that the use of chemical weapons remained in the horrible past of humanity. But recently it has been once again used in the Syrian conflict. Therefore, the development of modern means of protection against weapons of mass destruction becomes extremely relevant. One of crucial tasks of protection against weapons of mass destruction (nuclear, chemical and biological weapons, etc. (CBRN defense)) is the in time detection of deadly dangerous substances for humans. The newest generation of sensors – nanosensors created on the basis of nanomaterials – became the most important modern tool for CBRN defense. Nanotechnology is the foundation of scientific-technical revolution in the twenty first century. However there is the back side of the coin – the high probability that the development of nanotechnologies will inevitably lead to the creation of a number of fundamentally new threats to human life. For example, in the near future it seems likely to create fundamentally new types of weapons of mass destruction – self-developing hybrid bio-like nanosystems and nanostructures based on technologies that integrate the latest achievements in the field of microrobotics, nanobiotechnology and genetics.

The proper response to the growing variety of threats in the twenty first century can only be in the development of new, modern nanomaterials for the efficient and timely detection of lethal chemical, biological, radioactive, explosive substances (CBRNE) at the lowest concentrations. A new generation of nanosensors should detect and distinguish the chemical components of complex systems. Nanosensors are very important tools for identifying toxic chemicals that can be found in industrial products and in the environment in small concentrations, and for the detection of dangerous biological systems (bacteria, viruses). Recently, as a basis for creating of a new generation of nanosensors, one of the most promising materials – graphite-like carbon nitride [1–8] – attracts special attention of researchers.

20.2 Carbon Nitride as Perspective Material for Nanosensors for New Security Systems in CBRN Defense and Others Devices

The rapid development of technologies in the twenty first century determines not only the increased variety of threats, but also the complexity of the nature of their interconnection and interdependence (Fig. 20.1). (To illustrate the interconnection and interdependence of threats, it is sufficient to mention only the actual danger of terroristic cyber-attacks, which can provoke anthropogenic catastrophe at nuclear facilities, biological and chemical industries, accompanied by the release of relevant damaging substances.)

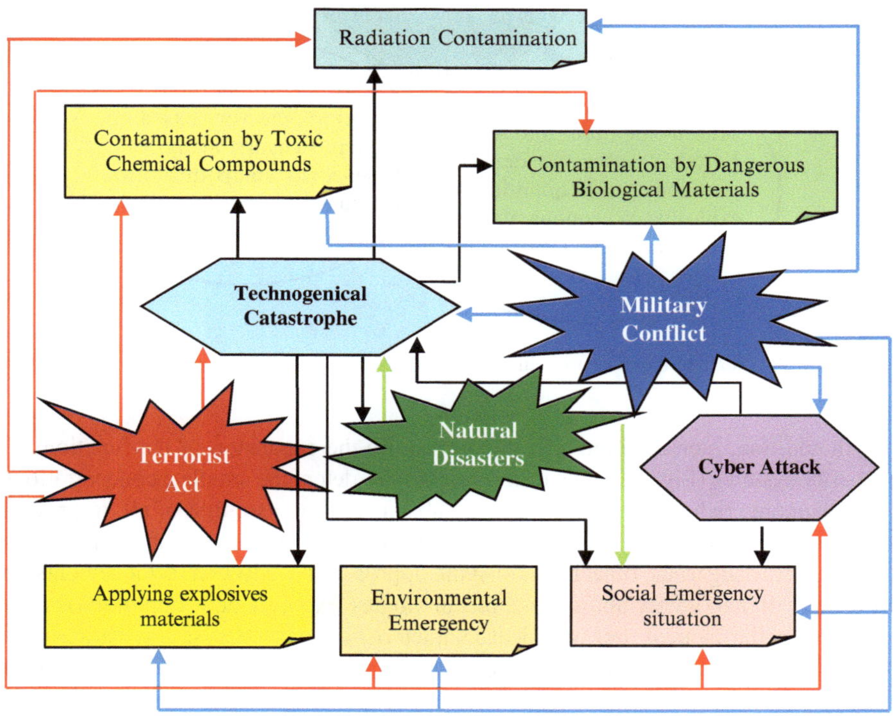

Fig. 20.1 Interrelation and interdependence of threats arising from the impacts of natural, human-made (terroristic acts, military conflicts etc.) and artificial characters

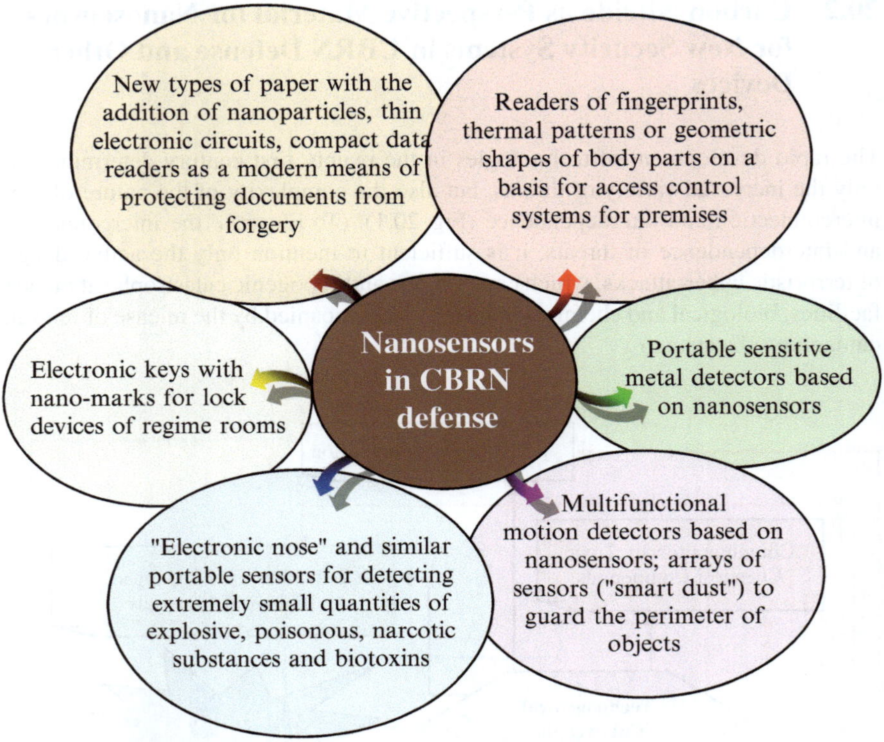

Fig. 20.2 Application of nanosensors in modern defense systems

In addition, increased terrorist threats require the expansion of the functions of security systems (Fig. 20.2). Its task is not only to detect the smallest concentrations of poisonous, radioactive and explosive substances, but also, first of all, to prevent the possibility of its usage, for example, in time to detect an attempt to penetrate potentially dangerous object “on distant approaches”. Modern achievements in the field of nanomaterials and nanotechnologies exposed new opportunities for improving the tactical and technical characteristics of such systems, antiterrorist means. For example, one of the most promising areas is the development of nanosensors for self-organizing sensor networks. To perform such a task, it is required to create miniature sensors capable to operate autonomously in any conditions and using radio waves to be integrated into local networks to transmit the collected information to a central computer. Ideally, such a micro-sensor should be, in fact, a chemical laboratory, located on the plate no more than 1 cm^2, which detects the presence of CBRN particles. The task of the sensor systems creating at bording checkpoints and in the airports, to protect critical infrastructure (metro and other public areas) can only be solved if a new generation of nanosensors is created. Therefore, to solve such a complex problem, it is necessary to use a new generation of nanomaterials, one of the most promising materials being graphite-like carbon nitride and its derivatives [1–8].

For the first time, a product similar in composition to C_3N_4 (however containing 0.6 to 2.4% hydrogen) as a porous bulk amorphous mass was obtained as early as 1922 by decomposition of mercury thiocyanate ($Hg(SCN)_2$) [9]. However, the interest of researchers, in addition to demonstrating the spectacular chemical experience of the "Pharaoh's snake", had not caused a new substance.

The first surge of interest in the search for new ways of synthesizing g-C_3N_4 led to the hypothesis that sp^2-carbon nitride could become an intermediate link in obtaining a hypothetical (and now it can be argued that mythical) superhard β-C_3N_4 with carbon in sp^3-hybridization. However, the unique electrophysical and physicochemical characteristics of g-C_3N_4 discovered in recent years have stimulated intensive searches for new ways of its synthesis. Though graphite-like carbon nitride is one of the oldest materials described in the chemical literature, only today, at the level of modern technology g-C_3N_4 experiences a renaissance as a highly active photocatalyst. Given that this polymer, not containing metal, is capable to generate hydrogen in visible light [1]. Thus, the renewed interest of researchers in graphite-like carbon nitride occurred in the twenty first century primarily as one of the most promising materials of green energy (as a photocatalyst for the production of hydrogen from renewable, natural sources (water, greenhouse gas) that can generate hydrogen with increased ($\sim$135 μmol$\cdot$h^{-1}) speed) [10].

The photocatalytic (g-C_3N_4) splitting of water with the formation of hydrogen is already being applied on an industrial scale, including the company KAUST in Saudi Arabia. Numerous experimental results [11, 12] show that the g-C_3N_4 is a unique non-metallic photocatalyst that differs from the much more toxic and less photocorrosion-resistant metal oxides, nitrides, sulfides, phosphides and organo-metal complexes [13]. Carbon nitride is widely used as a multifunctional non-metallic catalyst for environmental protection, conversion and storage of energy [12]. Carbon nitride, characterized as a non-toxic and chemically highly resistant material, is used for tribological coatings, biocompatible medical coatings, chemically inert coatings, insulators and for energy storage. Based on carbon nitride, applications have been developed in the field of medical sensing, bioimages [1–3].

The study of g-C_3N_4 is associated not only with catalysis (photocatalysis, electrocatalysis and photoelectrocatalysis), the wide future of carbon nitride is also due to the fact that this material is a metal-free and completely stable semiconductor [1]. Carbon nitride, as an organic polymer is widely used as a semiconductor (with a band gap 2.7 eV, corresponding to an optical wavelength of 460 nm) in photoelectric nanodevices, photovoltaic solar cells and chemical nanosensors, in particular in CBRN defense [2].

20.3 Carbon Nitride Oxide (g-C_3N_4)O and Heteroatomic N-graphene (Azagraphene) as New Perspective Materials for Nanosensors and Others Modern Technological Devices

In recent times, the object of increasing interest of researchers is the doped with atoms of other elements (in particular, oxygen) and nanostructured g-C_3N_4. The actual task for expanding the scope of application of carbon nitride is to increase and optimize its catalytic, electronic and optical properties both by increasing the surface area of g-C_3N_4 and by increasing the number of active centers of the carbon nitride network, due to an increase in the number of structure defects due to doping of carbon nitride.

Numerous attempts to solve the problem of increasing structural defects, increasing the number of active centers and, as a result, a significant change in the physicochemical characteristics of carbon nitride are carried out mainly in two main directions. First, the problem of obtaining a more developed surface is attempted to be solved by achieving the nanosized particles of synthesized carbon nitride and by searching for methods of its dispergation in various media (including the aquatic environment). However, partial exfoliation of particles of the pre-synthesized carbon nitride is possible only in aggressive media (concentrated acidic [3] and alkaline [14] solutions). Nevertheless, even nanoporous g-C_3N_4 with a high surface area (201–209 $m^2 \cdot g^{-1}$) and a large pore volume (0.50–0.52 $m^3 \cdot g^{-1}$) has a photocatalytic activity ten times accelerating the photodegradation of methyl orange (MO) under visible light ($\lambda > 420$ nm) [15].

Nanosheets dispersed in an aqueous medium (6–12 two-dimensional heteroatomic monolayers 2–4 nm thick) obtained by ultrasonic peeling of powdered g-C_3N_4 in an acid medium [3] have almost 6.4 times greater photocatalytic activity [16]. Carbon nitride nanosheets are distinguished by stronger photoluminescence [17], good biocompatibility and are more effective nonmetallic biosensors [3, 18] than bulk g-C_3N_4.

The current task of increasing and optimizing the catalytic, electronic and optical properties of g-C_3N_4 is solved by functionalization carbon nitride with fragments of organic molecules or atoms of various elements, predominantly oxygen [19, 20], silicon [13], sulfur [21] and phosphorus [21] atoms. For example, the functionalization of carbon nitride by heteroaromatic rings contributes to the improvement of separation and transmission of photoexcited charge carriers [22]. As a result, the resulting composite co-polymer demonstrates a significant improvement in the photocatalytic activity in the visible spectrum to generate H_2 as compared to the non-functionalized g-C_3N_4. As a new efficient sorbent, g-C_3N_4 functionalized with aromatic tri-*sim*-triazine fragments was used in analytical chemistry for pretreatment of samples with solid phase extraction (SPE), magnetic solid phase extraction (MSPE) and solid state microextraction (SPME) regimes [23].

A composite material based on doped graphite-like carbon nitride is presented as one of the best materials for hydrogen storage [24, 25]. And in oxygen-doped g-C_3N_4 obtained by hydrothermal treatment of carbon nitride synthesized from a mixture of melamine and thiourea, for example, at decomposition of the Acid Orange 7 dye ($C_{16}H_{11}N_2NaO_4S$, Orange II), the photocatalytic response rate is 10 times higher, while the photocatalytic activity in 7 times higher than untreated g-C_3N_4 [20]. Doping of carbon nitride with oxygen not only increases its photoactivity by 7 times [20], the boundary of its photoactivity in the visible spectrum expands from 460 to 498 nm [19]. These first results attach particular importance to the search of new efficient methods for synthesizing nanoscale particles of carbon nitride and the functionalization of its structure, in particular, by oxygen. We emphasize that several known methods of functionalization of carbon nitride are multistage (at least two-stage) methods, since post-synthesis treatment of previously synthesized g-C_3N_4 is assumed. For example, doped with oxygen (~7% O), carbon nitride is obtained as follows. First, undoped carbon nitride is synthesized by heating the dicyandiamide at 550 °C for 5 h in an argon flow [19]. The product is then oxidized by dispergation in a 30% H_2O_2 solution, treating the resulting suspension in a Teflon sealed autoclave at 140 °C for 10 h, followed by five times washing with water and centrifugation [19]. The Kharlamov's laboratory firstly proposed a one-stage method for the direct synthesis of oxygen-doped carbon nitride [4, 5, 26–30].

For the first time, a new substance – carbon nitride oxide (g-C_3N_4)O (Figs. 20.3c and 20.4a), which is an analogue of graphite oxide (GO) (Fig. 20.3d), was synthesized by the gas phase method under special reaction conditions of pyrolysis of melamine [5, 26] and urea [28]. The synthesis of carbon nitride oxide by thermal urea conversion [28] was carried out in accordance with the pyrolysis method described for the production of simultaneously carbon nitride oxide ((g-C_3N_4)O) and oxygen-doped carbon nitride (O-g-C_3N_4) from melamine [5, 26]. Studies were conducted under ambient pressure by varying the temperature, the amount of precursor in the crucible and the duration of its heat treatment. The experiments were carried out in the temperature range 320–450 °C. The precursor was localized in an open ceramic crucible placed in a tubular quartz reactor with one open end to exit a self-generated reactive vapor-gas mixture that bubbled through water. The rate of heating of a ceramic crucible with a precursor did not exceed 10 deg/min. Synthesized samples were analysed by X-ray diffraction analyses, XPS and IR spectroscopy, optical and scanning electron microscopy (SEM) as well as temperature-programmed desorption with mass spectrometry (TPDMS).

X-ray phase analysis of the samples was carried out on a DRON UM-1 diffractometer with Cu K_α-radiation and a nickel filter. The IR spectra in the reflection mode were recorded in the range from 4000 to 400 cm^{-1} with a spectral resolution of 8 cm^{-1} using a Nicolet Nexus FTIR spectrometer (Thermo Scientific).

At the heat treatment of urea (450 °C) a yellow product is formed (Fig. 20.5b, c), in the diffractogram (Fig. 20.4b) of which only three distinct reflexes are present (at $2\theta = 10.76°$ ($d = 0.818$ nm), 21.45° ($d = 0.414$ nm) and 27.49° ($d = 0.324$ nm)), which are characteristic for carbon nitride oxide (g-C_3N_4)O [5, 26]. The most intense reflex at $2\theta = 27.49°$ ($d = 0.324$ nm), characterizing the interlayer distance in g-C_3N_4, is preserved in the diffractogram of a carbon nitride oxide synthesized from urea. The appearance of an additional reflex at $2\theta = 21.45°$ ($d = 0.414$ nm)

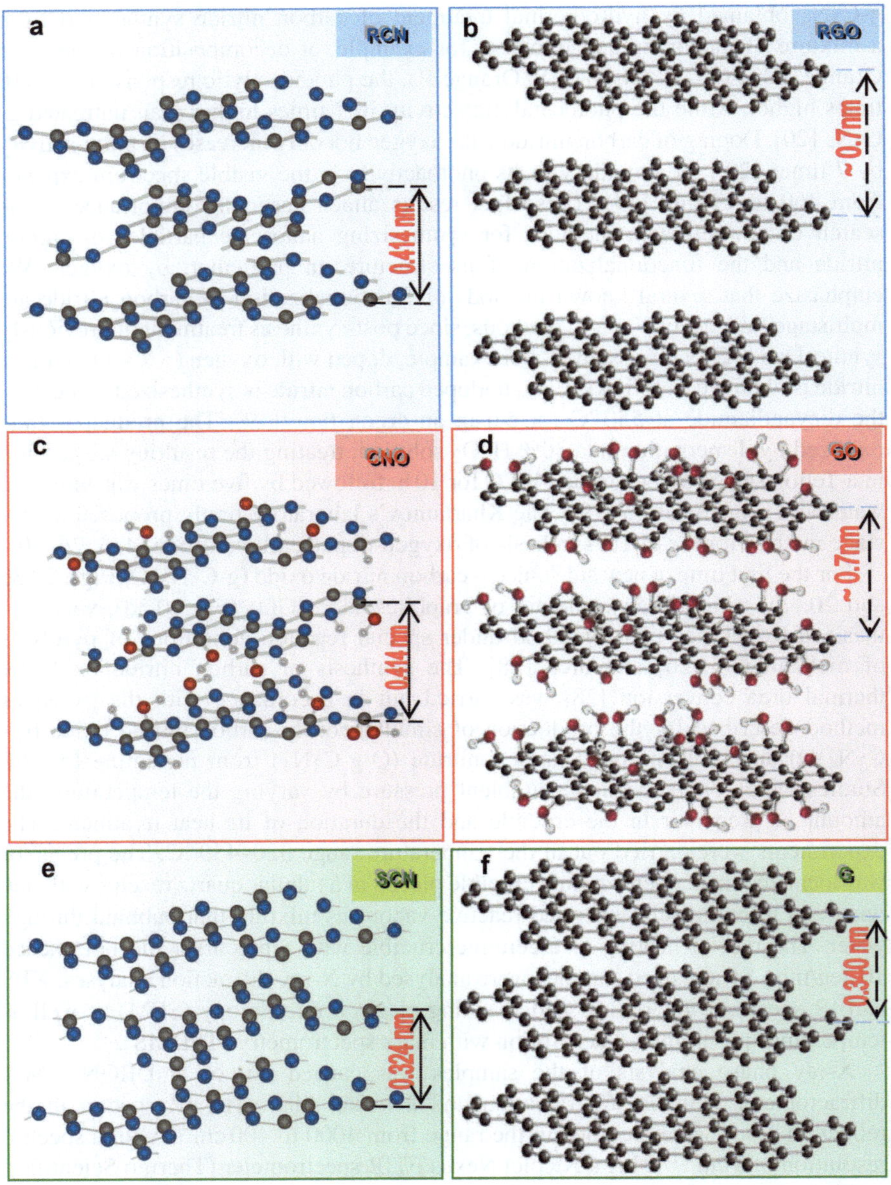

Fig. 20.3 Schematic atomic model of the layers and interplanar distances between adjacent layers of synthesized carbon nitride g-C_3N_4 (SCN) (**e**), carbon nitride oxide (g-C_3N_4)O (CNO) (**c**) and reduced carbon nitride g-C_3N_4 (RCN) (**a**); graphite (G) (**f**), graphene (graphite) oxide (GO) (**d**) and reduced graphene oxide (RGO) (**b**)

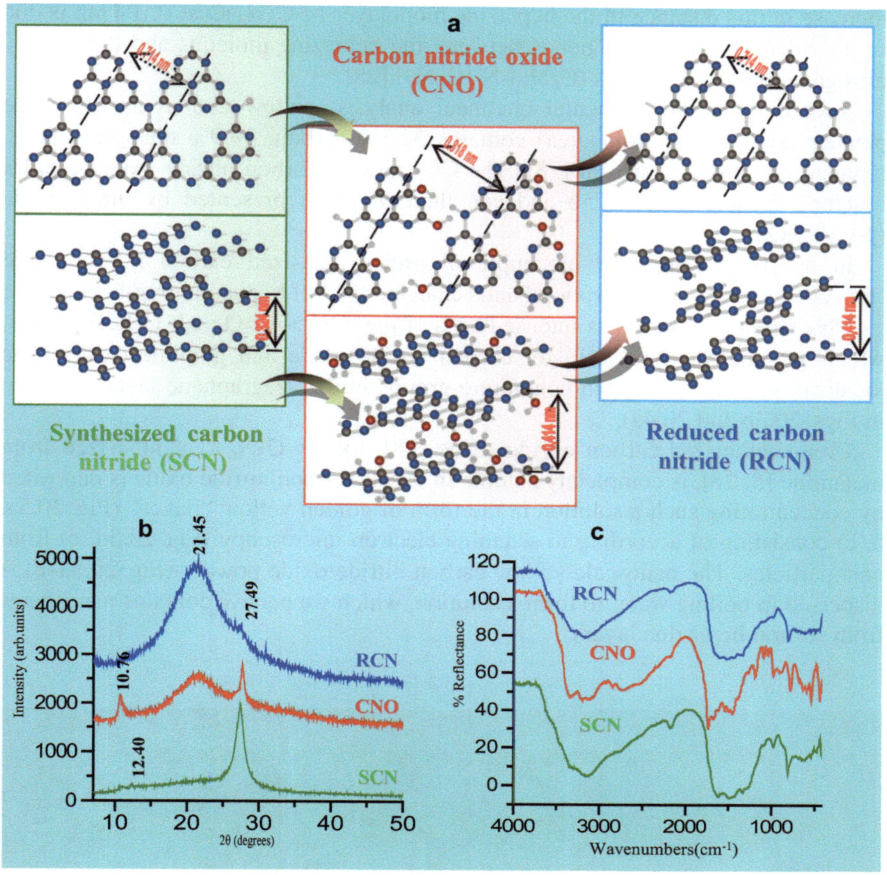

Fig. 20.4 Schematic atomic model of one layer (in plane) and interplanar distances between adjacent layers of synthesized carbon nitride g-C_3N_4 (SCN), carbon nitride oxide (g-C_3N_4)O (CNO) and reduced carbon nitride g-C_3N_4 (RCN) (**a**); X-ray diffraction (**b**) and IR (**c**) spectra of SCN, CNO and RCN

in the X-ray diffraction pattern of carbon nitride oxide is associated, as suggested in [26], with partial distortion of the planarity of its polymer network $((C_6N_7)\text{-}N)_n$ due to the dearomatization of some heterocycles during oxidation. With an increase in the number of oxygen-containing groups between adjacent planes, the peak $2\theta = 21.45°$ transforms into a wide halo, but the reflex at $2\theta = 27.49°$ also persists, since in the carbon nitride oxide the oxygen-containing groups are predominantly localized in the voids of the "openwork" plane (Figs. 20.3b and 20.4a). It is also noteworthy that the reflex at 12.40° (d = 0.714 nm), characterizing the distance between the fragments (C_6N_7) in the g-C_3N_4 plane, is shifted to 10.76° (d = 0.818 nm) in the diffractogram of carbon nitride oxide synthesized from urea. The increase in this distance in the heptazine monolayer ($g\text{-}C_3N_4$)O at 0.104 nm is due to the break of some C-N bonds between the heptazine moieties and the tertiary nitrogen atom in the dimers (C_6N_7)-N-(C_6N_7) [28].

According to the elemental chemical analysis, yellow carbon nitride oxide powder (synthesized from urea) contains 32.3% carbon, 49.9% nitrogen, 16.2% oxygen, and 1.6% hydrogen [28]. The C/N ratio in carbon nitride oxide samples obtained from urea is 0.756 and can therefore be represented by the formula ($g\text{-}C_3N_4$)O.

In the IR spectrum of a sample of urea-synthesized carbon nitride oxide (Fig. 20.4c), there are absorption bands characteristic of heptazine fragments of the $g\text{-}C_3N_4$ structure, as well as intense bands of the -OH,> C=O and -COOH groups, which are characteristic only for carbon nitride oxide and graphite oxide. (The localization of these oxygen-containing groups in the azagraphene layer is shown in Figs. 20.3b and 20.4a).

Urea-synthesized carbon nitride oxide [28], like ($g\text{-}C_3N_4$)O synthesized from melamine [5, 26], is completely soluble in water. Carbon nitride oxide is deposited by concentrating such a solution in the form of golden yellow "flakes" (Fig. 20.5a, b, c) consisting of according to scanning electron microscopy (Fig. 20.5d, e) from nanoparticles. The nanoscale yellow carbon nitride oxide powder (Fig. 20.5a–c) is dispersed in boiling water to form a solution, which we believe contains nanosheets from several heptazine layers.

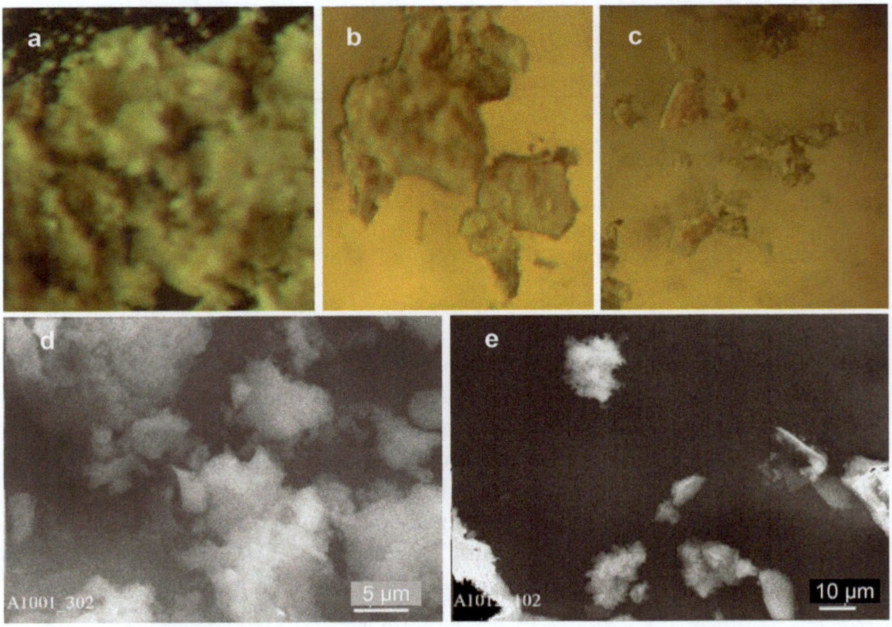

Fig. 20.5 Optical microscopy (**a**, **b**, **c**) and SEM (**d**, **e**) images of carbon nitride oxide ($g\text{-}C_3N_4$)O sample

Synthesize N-graphene (azagraphene [5, 26, 28, 29], in the monolayer of which the maximum possible number of carbon atoms is replaced by nitrogen atoms), from carbon nitride (as graphene was obtained from graphite) is the most interesting and topical task of the next stage of development of nanochemistry after successful synthesis of carbon monatomic molecules and nanostructures. In Kharlamov's laboratory, the first steps were taken to obtain an isostructural analog of graphene – azagraphene by a method analogous to the method of synthesis multilayer graphene by reduction graphite oxide (Fig. 20.3c). For the first time reduced carbon nitride (RCN) (or reduced multilayer azagraphene) (Fig. 20.3a) was obtained by reduction of carbon nitride oxide with hydroquinone [29]. In contrast to the light-brown carbon nitride powder g-C_3N_4, synthesized by pyrolysis of melamine, RCN has a dark, almost black color. RCN can be considered as a structural analogue of the reduced (from graphite oxide (GO) (Fig. 20.3d)) graphene (RG) (Fig. 20.3b).

According to the results of the chemical analysis, the XPS and IR (Fig. 20.4c) spectrometry, elemental composition and chemical bonds between atoms in the heteroatomic plane of reduced carbon nitride (RCN) correspond to the bonds of synthesized carbon nitride (SCN). The IR spectrum of the RCN (Fig. 20.4c) corresponds almost completely to the IR spectrum of the synthesized g-C_3N_4, since it contains all the absorption bands in the 1200–1650 cm^{-1} region and the band near $\sim$810 cm^{-1}, which are characteristic of g-C_3N_4. It is important to note that the absorption bands of oxygen-containing groups, characteristic both for (g-C_3N_4)O [26] and for graphite oxide, are absent in the IR spectrum of the reduced carbon nitride. However, according to the results of the X-ray phase analysis (Fig. 20.4b), the RCN has a significantly larger (at 0.09 nm) interplanar distance between adjacent planes $d = 0.414$ nm than the interplanar distance between the layers of synthesized carbon nitride ($d = 0.324$ nm). We believe that the reduced carbon nitride, as well as the reduced (from GO) graphene, consists of weakly connected heteroatomic azagraphene and monoatomic graphene layers, respectively. Further studies to obtain azagraphene by reduction of carbon nitride oxide are extremely promising (given the unique physicochemical properties of graphite-like nitride carbon nanostructures with planar graphene-like structure [2]). It has been established that the 2D carbon nitride nanostructures consisting of several layers have excellent stability, high quantum yield of fluorescence, low toxicity, excellent biocompatibility, unique electroluminescent and photoelectrochemical properties that make them a unique material for biosensing (with the prospect of extensive use in bioanalysis and diagnosis of diseases) [6].

The semiconductor properties of g-C_3N_4 stimulate research in the direction of the replacement in the future of silicon by 2D carbon nitride with a planar graphene-like structure in electronics [7, 31–34]. We note that carbon nitride as an organic (nonmetallic) polymer, which is a stable semiconductor [1] with a band gap of 2.7 eV [2], has a competitive advantage over graphene in this field of research. The semimetallic nature of graphene and the absence of an electron-band gap have hampered the development of a switch based on graphene [7].

Currently, the research aimed at developing photocatalysts based on 2D-C_3N_4 nanostructures occupies flagship positions in order to meet the global demand for renewable energy sources (hydrogen production by photocatalytic H_2O splitting) and to solve the problem of environmental pollution (photocatalytic conversion of carbon dioxide CO_2 into energy carriers) [8]. According to the author's expressive phrase [8], "photocatalysis appeared as one of the Sacred Grails of sustainable and environmentally friendly solar energy technologies for energy conversion, energy storage and environmental restoration", and graphite-like carbon nitride (g-C_3N_4) is one of the most promising photocatalysts.

And, in conclusion, we would like to mention the bright noteworthy fact given in the newest (April 2017) review [8], emphasizing the perspectives of materials based on g-C_3N_4 and, in particular, the importance of developing methods for obtaining 2D-C_3N_4. Since the first study, culminating in the successful use of g-C_3N_4 in the photocatalytic production of H_2 from H_2O in 2009 [35], there has been an exponential growth in scientific research on materials based on g-C_3N_4 – only in the previous (2016) year, according to the Web of Science, 800 publications [8]. Perspectives for the use of nanostructured and doped carbon nitride derivatives (O-doped carbon nitride O-C_3N [4, 27] and O-C_3N_4 [5, 26, 28, 29], carbon nitride oxide (g-C_3N_4)O [5, 26, 28, 29], reduced carbon nitride (azagraphene) [29]) in photovoltaic nanodevices, solar cells and chemical nanosensors of a new generation, in particular in CBRN defense, should stimulate further studies of these new compounds.

20.4 Conclusions

The number and intensity of military conflicts and terrorist attacks that have increased in the twenty first century demand extraordinary measures in the development of modern means of protection against weapons of mass destruction. A new generation of modern sensors – nanosensors based on nanomaterials – should become an important tool for the well-timed detection of substances that are lethal to humans and for CBRNE defense. Particular attention is attracted to promising materials for the creation of a new generation of nanosensors, – graphite-like carbon nitride and its nanostructured and doped derivatives, in particular water-soluble carbon nitride oxide (g-C_3N_4)O and reduced carbon nitride (or reduced multilayer azagraphene). Nanostructured reduced carbon nitride is obtained by reduction with hydroquinone of carbon nitride oxide (g-C_3N_4)O. Water-soluble carbon nitride oxide (g-C_3N_4)O is synthesized under special reaction conditions of pyrolysis of melamine and urea.

References

1. Liu J, Wang H, Antonietti M (2016) Graphitic carbon nitride "reloaded": emerging applications beyond (photo) catalysis. Chem Soc Rev 45:2308–2326
2. Zhang Y, Pan Q, Chai G et al (2013) Synthesis and luminescence mechanism of multicolor-emitting g-C_3N_4 nanopowders by low temperature thermal condensation of melamine. Sci Rep 3:1943–1950
3. Ma TY, Tang Y, Dai S et al (2014) Proton-functionalized two-dimensional graphitic carbon nitride nanosheet: an excellent metal-label-free biosensing platform. Small 10:2382–2389
4. Kharlamov O, Bondarenko M, Kharlamova G (2015) O-Doped carbon nitride (O-g-C_3N) with high oxygen content (11.1 mass%) synthesized by pyrolysis of pyridine, chapter 9. In: Camesano TA (ed) Nanotechnology to aid chemical and biological defense, NATO science for peace and security series A: chemistry and biology. Springer, Dordrecht, pp 129–145
5. Kharlamova G, Kharlamov O, Bondarenko M, Khyzhun O (2016) Hetero-Carbon nanostructures as the effective sensors in security systems, chapter 19. In: Bonca J, Kruchinin S (eds) Nanomaterials for security, NATO science for peace and security series A: chemistry and biology. Springer, Dordrecht, pp 239–258
6. Xiong M, Rong Q, Meng H et al (2017) Two-dimensional graphitic carbon nitride nanosheets for biosensing applications. Biosens Bioelectron 89:212–223
7. Cooper AI, Bojdys MJ (2014) Carbon nitride vs. graphene – now in 2D. Mater Today 17:468–469
8. Ong WJ (2017) 2D/2D graphitic carbon nitride (g-C_3N_4) heterojunction nanocomposites for photocatalysis: why does face-to-face interface matter? Front Mater 4:1–10
9. Franklin EC (1922) The ammono carbonic acids. J Am Chem Soc 44(3):486–509
10. Martha S, Nashima A, Parida KM (2013) Facile synthesis of highly active g-C_3N_4 for efficient hydrogen production under visible light. J Mater Chem A 1:7816–7824
11. Xiangqian F, Zheng X, Zhu S (2015) Improved photocatalytic activity of g-C_3N_4 derived from cyanamide–urea solution. RSC Adv 5:8323–8328
12. Zheng Y, Liu J, Liang J et al (2012) Graphitic carbon nitride materials: controllable synthesis and applications in fuel cells and photocatalysis. Energy Environ Sci 5:6717–6731
13. Wu P, Shi J, Chen J, Wang B et al (2012) Graphitic carbon nitride modified by silicon for improved visible light driven photocatalytic hydrogen production, chap 33. In: Mathur S, Ray SS (eds) Nanostructured materials and nanotechnology VI: ceramic engineering and science proceedings. Wiley, pp 137–148
14. Fan X, Peng W, Li Y et al (2008) Deoxygenation of exfoliated graphite oxide under alkaline conditions: a green route to graphene preparation. Adv Mater 20:4490–44932
15. Han K, Wang C, Li Y et al (2013) Facile template-free synthesis of porous g-C_3N_4 with high photocatalytic performance under visible light. RSC Adv 3:9465–9469
16. Zhang S, Li J, Zeng M et al (2013) In situ synthesis of magnetic graphitic carbon nitride photocatalyst and its synergistic catalytic performance. ACS Appl Mater Interfaces 5:12735–12743
17. Yuan B, Chu Z, Li G et al (2014) Ribbon-like graphitic carbon nitride (g-C_3N_4): green synthesis, self-assembly and unique optical properties. J Mater Chem C 2:8212–8215
18. Zhang X, Xie X, Wang H et al (2013) Enhanced photoresponsive ultrathin graphitic-phase C_3N_4 nanosheets for bioimaging. J Am Chem Soc 135:18–21
19. Li J, Shen B, Hong Z et al (2012) A facile approach to synthesize novel oxygen-doped g-C_3N_4 with superior visible-light photoreactivity. Chem Commun 48:12017–12019
20. Ming L, Yue H, Xua L et al (2014) Hydrothermal synthesis of oxidized g-C_3N_4 and its regulation of photocatalytic activity. J Mater Chem A 2:19145–19149

21. Ma X, Lv Y, Xu J et al (2012) A strategy of enhancing the photoactivity of g-C_3N_4 via doping of nonmetal elements: a first-principles study. J Phys Chem C 116:23485–23493
22. Tian J, Zhang L, Fan X et al (2016) A post-grafting strategy to modify g-C_3N_4 with aromatic heterocycles for enhanced photocatalytic activity. J Mater Chem A 4:13814–13821
23. Sun Y, Ha W, Chen J et al (2016) Advances and applications of graphitic carbon nitride as sorbent in analytical chemistry for sample pretreatment: a review. TrAC Trends Anal Chem 84:12–21
24. Nair AAS, Sundara R, Anitha N (2015) Hydrogen storage performance of palladium nanoparticles decorated graphitic carbon nitride. Int J Hydrogen Energy 40:3259–3267
25. Nair AAS, Sundara R (2016) Palladium cobalt alloy catalyst nanoparticles facilitated enhanced hydrogen storage performance of graphitic carbon nitride. J Phys Chem C 120:9612–9618
26. Kharlamov A, Bondarenko M, Kharlamova G (2016) Method for the synthesis of water-soluble oxide of graphite-like carbon nitride. Diamond Relat Mater 61:46–55
27. Kharlamov AI, Bondarenko ME, Kharlamova GA (2014) New method for synthesis of oxygen-doped graphite-like carbon nitride from pyridine. Russ J Appl Chem 87:1284–1293
28. Kharlamov A, Bondarenko M, Kharlamova G et al (2016) Features of the synthesis of carbon nitride oxide (g-C_3N_4)O at urea pyrolysis. Diam Relat Mater 66:16–22
29. Kharlamov A, Bondarenko M, Kharlamova G et al (2016) Synthesis of reduced carbon nitride at the reduction by hydroquinone of water-soluble carbon nitride oxide (g-C_3N_4)O. J Solid State Chem 241:115–120
30. Kharlamova G, Kharlamov O, Bondarenko M et al (2013) Hetero-carbon: heteroatomic molecules and nano-structures of carbon, Part VII. In: Vaseashta A, Khudaverdyan S (eds) Advanced sensors for safety and security. NATO science for peace and security series B: physics and biophysics. Springer, Netherlands, pp 339–357
31. Rodionov VE, Shmidko IN, Zolotovsky AA, Kruchinin SP (2013) Electroluminescence of Y_2O_3:Eu and Y_2O_3:Sm films. Mater Sci 2:210–220
32. Repetsky SP, Vyshyvana IG, Molodkin VB, Lizunov VV (2017) Influence of the adsorbed atoms of potassium on an energy spectrum of grapheme. Metallofiz Noveishie Tekhnol 39:1017–1022
33. Ermakov V, Kruchinin S, Hori H, Fujiwara A (2007) Phenomena of strong electron correlation in the resonant tunneling. Int J Mod Phys B 11:827–835
34. Ermakov V, Kruchinin S, Fujiwara A (2008) Electronic nanosensors based on nanotransistor with bistability behaviour. In: Bonca J, Kruchinin S (eds) Proceedings of the NATO ARW "Electron transport in nanosystems". Springer, pp 341–349
35. Wang X, Maeda K, Thomas A et al (2009) A metal-free polymeric photocatalyst for hydrogen production from water under visible light. Nat Mater 8:76–80

Chapter 21
Polymer Nanocomposites with Silver Nanoparticles Formed by Low-Energy Ion Implantation: Slow Positron Beam Spectroscopy Studies

T. S. Kavetskyy, M. O. Liedke, N. Srinivasan, A. Wagner, R. Krause-Rehberg, O. Šauša, T. Petkova, V. Boev, and A. L. Stepanov

Abstract Polymer nanocomposites formed by low-energy ion implantation were studied by means of positron annihilation spectroscopy with a variable-energy positron beam or slow positron beam spectroscopy. Silver ion implantation into polymethylmethacrylate (Ag:PMMA) and hybrid organic-inorganic ureasil (Ag:ureasil) was performed at different ion fluences with a constant energy of 30 keV and a current density of 1 μA/cm^2 in order to prepare Ag nanoparticles in the near-surface region of polymer matrix. Contribution of Doppler broadening slow positron beam spectroscopy technique for understanding Ag nanoparticles formation in Ag:PMMA and Ag:ureasil nanocomposite films is demonstrated.

Keywords Polymers · Nanocomposites · PMMA · Ureasil · Ion implantation · Silver nanoparticles · Positron annihilation · Slow positrons

---

T. S. Kavetskyy (✉)
Drohobych Ivan Franko State Pedagogical University, Drohobych, Ukraine

The John Paul II Catholic University of Lublin, Lublin, Poland

M. O. Liedke · N. Srinivasan · A. Wagner
Institute of Radiation Physics, Helmholtz-Zentrum Dresden-Rossendorf, Dresden, Germany

R. Krause-Rehberg
Department of Physics, University Halle, Halle, Germany

O. Šauša
Institute of Physics, Slovak Academy of Sciences, Bratislava, Slovakia

T. Petkova · V. Boev
Institute of Electrochemistry and Energy Systems, Bulgarian Academy of Sciences, Sofia, Bulgaria

A. L. Stepanov
Kazan Physical-Technical Institute, Russian Academy of Sciences, Kazan, Russia

Kazan Federal University, Kazan, Russia

J. Bonča, S. Kruchinin (eds.), *Nanostructured Materials for the Detection of CBRN*, NATO Science for Peace and Security Series A: Chemistry and Biology,
https://doi.org/10.1007/978-94-024-1304-5_21

21.1 Introduction

Polymer nanocomposites formed by low-energy (<100 keV) ion implantation are innovative materials to be used in advanced optoelectronic devices, including organic luminescent devices, solar cells, waveguides, diffractive elements, micro-components for integrated optical circuits [1–5], as well as for hard-materials [6–10] and anti-bacteria [11, 12] applications. Positron annihilation spectroscopy by means of a variable-energy positron beam or the so-called slow positron beam spectroscopy is a powerful experimental tool available to show a completely new way to understand better the ion-irradiation-induced processes in optoelectronic polymer materials on the nanoscale level [13–15]. This technique has also been successfully applied to make contribution in understanding nanoparticle-loaded polymer brushes [16, 17]. In the present work, polymer nanocomposite films fabricated by Ag ion implantation into polymethylmethacrylate (PMMA) and organic-inorganic ureasil were studied with Doppler broadening slow positron beam spectroscopy (DB-SPBS). The Doppler-broadening annihilation line-shape *S*-parameter as a function of incident positron energy dependencies have been discussed, which shed more light on evolution of ion implantation induced Ag nanoparticles (NPs) formation in Ag:PMMA and Ag:ureasil nanocomposites.

21.2 Experimental

21.2.1 Preparation of Polymer Nanocomposites

21.2.1.1 Preparation of Ag:PMMA Nanocomposites

As substrates for ion implantation, 1.2-mm-thick PMMA plates with a high optical quality and transparency in a wide spectral range 400–1000 nm were used as provided by the Kazan Physical-Technical Institute (KPTI, Russia). Ion implantation of the PMMA substrates was performed in vacuum under a pressure of 10^{-5} Torr at room temperature of the target with an ILU-3 ion accelerator at the KPTI [18, 19]. Ag:PMMA nanocomposites layers were fabricated with an energy of 30 keV, an ion current density of 1 μA/cm^2 and various fluences of 2.5×10^{16}, 1.0×10^{17}, and 1.5×10^{17} Ag^+/cm^2 sufficient to prepare Ag NPs in the near-surface region of polymer matrix [4, 5].

21.2.1.2 Preparation of Ag:ureasil Nanocomposites

As substrates for ion implantation, ureasil matrix was initially synthesized as follows [20–22]: (1) O,O'-bis(2-aminopropyl)-polypropylene glycol-block-polyethylene glycol-block-polypropylene glycol-500 (Jeffamine ED-600) was dried under vacuum for 30 min; (2) 3-isocyanatepropyltriethoxysilane (ICPTES), tetraethoxysilane (TEOS, 98%) and n-butyl amine were used as received; (3) Jeffamine and ICPTES were mixed in stoichiometric ratio of 1:2 in order to obtain a liquid ureasilicate monomer; (4) thereafter, TEOS (1.12 mmol) and n-butylamine were added to the mixture, which was kept under stirring for more than 20 min; (5) the mixture was transferred into a plastic Petri dish and jellified under appropriate conditions; (6) the obtained gels were heated in a vacuum furnace at 333 K at ambient conditions; and (7) a non-rigid, homogeneous and highly transparent xerogel in form of a disk with a diameter of 40 mm and a thickness of 0.25 mm was obtained within 1 day. After preparation, the ureasil sample was cut into some pieces for ion implantation. Ag ion implantation of the ureasil substrates was performed at the same conditions as described above for PMMA but at other ion fluences, namely, 2.5×10^{16}, 5.0×10^{16}, 7.5×10^{16}, and 1.0×10^{17} Ag^+/cm^2 sufficient to prepare Ag NPs in the near-surface region of polymer matrix [21].

21.2.2 Characterization of Polymer Nanocomposites

The investigated polymer nanocomposites were characterized by means of Doppler broadening slow positron beam spectroscopy technique developed at the Helmholtz-Zentrum Dresden-Rossendorf (HZDR, Germany). Doppler broadening of positron annihilation gamma-rays spectra as a function of incident positron energy were measured by a variable-energy positron beam "SPONSOR" with a radioactive ^{22}Na positron source and a semiconductor high purity Ge detector with an energy resolution of 1.09 ± 0.01 keV at 511 keV [23, 24]. The S-parameter was determined as a ratio of the counts in the very middle of the annihilation line: 511 ± 0.7 keV to the total counts contained in the peak 511 ± 17.4 keV. The positron incident energy E_p was changed in the range from 0.05 to 35 keV. Mean penetration depth of positrons (z_m) was estimated using the formula [25–27]:

$$z_m = (40/\rho)E_p^n, \tag{21.1}$$

where z_m is presented in nm, E_p the positron energy in keV, ρ is the density (1.18 g/cm^3 for both PMMA and ureasil) and $n = 1.62$. The density of PMMA sample was taken as a known commercial polymer [13]. The density of ureasil sample was measured based on the gravimetric method (Archimedean principle) [28].

21.3 Results and Discussion

Figure 21.1 shows the $S(E_p)$ curves obtained for the pristine PMMA and ion-implanted Ag:PMMA samples with various fluences. The S-parameter for the pristine sample is practically constant, while that for the irradiated samples shows

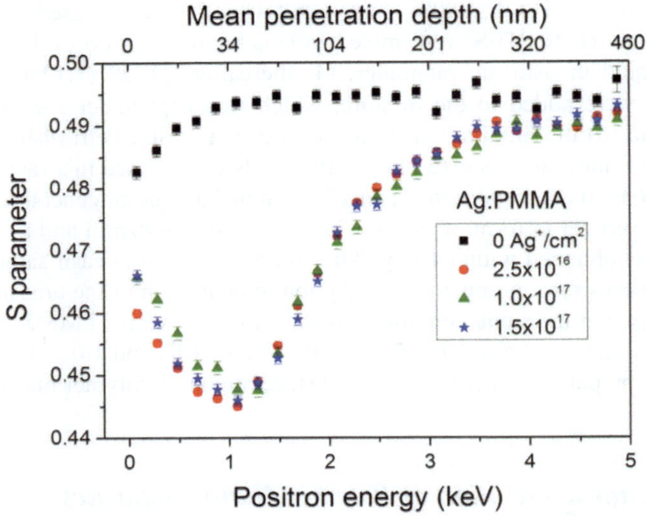

Fig. 21.1 Doppler broadening S-parameter for the pristine (non-irradiated) and irradiated Ag:PMMA samples with various fluences as a function of incident positron energy in the range of 0–5 keV. The error bars are within the size of the symbol

remarkable reduction below an incident positron energy $E_p \cong 2$ keV; the same as was previously detected for Ag:PMMA in [15]. Also, similar behaviour of the S-E_p dependence for ion-irradiated polymeric materials has been observed for 150 keV He^+-implanted poly-L-lactic acid (PLLA) polymer in [29, 30]. The decrease in the S-parameter could be interpreted by a reduction of the positronium (Ps) formation fraction [13, 30]. According to the analysis of S-E_p dependence performed in [30] and recovery of the S-parameter for E_p larger than 4 keV as seen in Fig. 21.1, it is suggested that $E_p \cong 2$ keV for Ag:PMMA, which correspond to the middle value of the S-parameter, could be explained as that the positrons implanted with such energy annihilate around the Bragg peak of the ion implantation [15]. Analysing the fluence dependence of S-parameter in the region up to 2 keV or 104 nm depth corresponding to the implanted layer, one can see that the S-parameter minimum at $E_p \cong 1$ keV demonstrates the lowest value for a fluence of $2.5 \times 10^{16} Ag^+/cm^2$, the highest value for a fluence of $1.0 \times 10^{17} Ag^+/cm^2$, and middle value for a fluence of $1.5 \times 10^{17} Ag^+/cm^2$.

The observed fluence dependence of S-parameter could be interpreted in terms of the Ps formation fraction in the nanocomposites studied within the model of different size carbon-shell Ag-core NPs of the quasi-spherical morphology [4, 31]. It must be noted that the approach of core-shell nanoparticles in polymer materials is widely used in literature as well [32–34] and here it is assumed for ion implantation process. It is also known that the physical and chemical properties of a core-shell nanoparticle dramatically depend on the sizes of its core and shell [35]. Because of in PMMA an unusual weak and very broad surface plasmon resonance (SPR) band with maximum at ~500 nm has been observed for Ag NPs, in contrast to the sharp and narrow SPR band at ~450 nm of Ag NPs in SiO_2, there are some difficulties to experimentally estimate the NPs sizes directly from optical absorption [1, 4, 12, 31] or cross-sectional transmission electron microscopy [12] measurements for Ag:PMMA. In order to analyze the possible changes of core-shell Ag NP size from the DB-SPBS data, additional experiment was performed for other type of polymer, namely, Ag:ureasil, which is characterized by a well resolved sharp and narrow SPR band at ~435 nm for Ag NPs formed by ion implantation with a fluence of $2.5 \times 10^{16} Ag^+/cm^2$; the SPR band becomes broadened and red shifted to ~500 nm as ion fluence rises to $5.0 \times 10^{16} Ag^+/cm^2$, indicating the changes of morphology and/or increasing size of NPs [21]. In other words, the expected fluence dependence of S-parameter for Ag:ureasil with foreseen behavior of Ag NPs size would be helpful to interpret the observed fluence dependence of S-parameter for Ag:PMMA with unforeseen behavior of Ag NPs size.

Figure 21.2 shows the $S(E_p)$ curves obtained for the pristine ureasil and ion-implanted Ag:ureasil samples with various fluences. Similarly as in case of Ag:PMMA, the S-parameter for the pristine sample is practically constant, while that for the irradiated samples shows remarkable reduction below $E_p \cong 2$ keV, probably due to the same density for both PMMA and ureasil.

The observed decrease in the S-parameter could be interpreted by a reduction of the Ps formation fraction as well [13, 30]. Also, it is suggested that $E_p \cong 2$ keV for Ag:ureasil, which correspond to the middle value of the S-parameter, could be explained as the positrons implanted with such energy annihilate around the Bragg

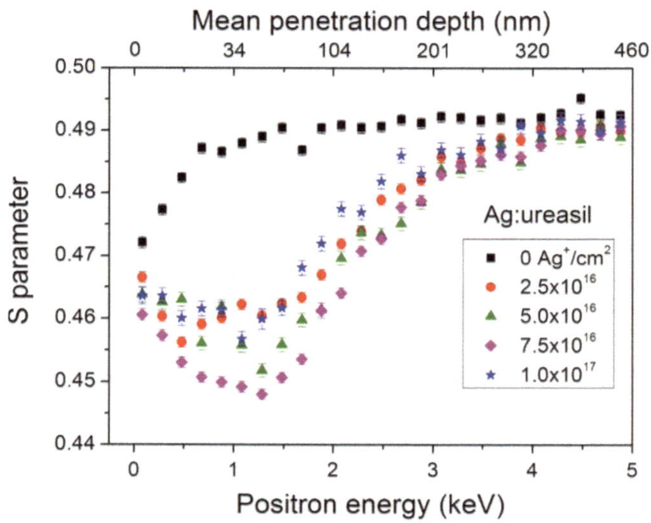

Fig. 21.2 Doppler broadening S-parameter for the pristine (non-irradiated) and irradiated Ag:ureasil samples with various fluences as a function of incident positron energy in the range of 0–5 keV. The error bars are within the size of the symbol

peak of the ion implantation. Analysing the fluence dependence of S-parameter in the region up to 2 keV or 104 nm depth corresponding to the ion-implanted layer, one can see that the S-parameter demonstrates a gradual decrease as ion fluence rises from 2.5×10^{16} to 7.5×10^{16} Ag^+/cm^2, while for the highest fluence of $1.0 \times 10^{17} Ag^+/cm^2$ examined the S-parameter shows an increase. The gradual decrease of S-parameter with ion fluence is possibly connected to preferential annihilation of positrons in Ag due to the positron affinity $A_+ = -5.36$ eV [36] and increased NPs size. At the same time, the increase of S-parameter for a fluence of $1.0 \times 10^{17} Ag^+/cm^2$ may be explained by the Ps formation in free-volume cavities created between NPs due to a growth and Ostwald ripening of Ag NPs as follows from the basic principles of metal nanoparticles formation by ion implantation [1, 4]. Thus, the detected fluence dependence of S-parameter in the region of ion-implanted layer for Ag:ureasil samples indicates the changes of morphology and/or size of NPs in a good agreement with optical measurements [21] and theoretical predictions [4].

Returning to Ag:PMMA (Fig. 21.1), one may suggest that the observed increase of S-parameter for a fluence of $1.0 \times 10^{17} Ag^+/cm^2$ should be also explained by the Ps formation in free-volume cavities created between core-shell Ag NPs due to a growth and Ostwald ripening. Middle values of S-parameter for the highest fluence of $1.5 \times 10^{17} Ag^+/cm^2$ would be connected with increasing of Ag filling factor in Ag:PMMA, that is based on comparative analysis of S-parameter for polymer samples with different Ag filling factor in the case of Ag NPs loaded poly(dimethylaminoethyl methacrylate) (PDMAEMA) brushes [16, 17]. Further research and more detail analysis to verify this conclusion are in progress.

21.4 Conclusion

Ag ion implantation into PMMA and ureasil has been performed at various ion fluences with a constant energy of 30 keV and a current density of 1 $\mu A/cm^2$ in order to prepare Ag NPs in the near-surface region of polymer matrix. The changes of morphology and/or size of NPs are in agreement with optical observations and basic principles of metal nanoparticles formation by ion implantation are possible to be identified by means of Doppler broadening slow positron beam spectroscopy technique applied, but further research and more detail analysis are still required.

Acknowledgements This work was financially supported in part by the Ministry of Education and Science of Ukraine (projects #0116U004737 and #0117U007143; to TK), by the Slovak Grant Agency VEGA (project #2/0157/17; to OŠ), and Slovak Research and Development Agency (project #APVV-16-0369; to OŠ), by the National Science Fund of the Bulgarian Ministry of Education (project #FNI-DN09/12-2016; to TK, TP and VB) and by the Russian Foundation for Basic Research (project #15-48-02525; to ALS).

References

1. Stepanov AL (2010) Synthesis of silver nanoparticles in dielectric matrix by ion implantation: a review. Rev Adv Mater Sci 26:1
2. Wang J et al (2012) Photoluminescence and reflectivity of polymethylmethacrylate implanted by low-energy carbon ions at high fluences. Appl Surf Sci 261:653
3. Gupta R et al (2012) Optical characterization of poly(methyl methacrylate) implanted with low energy ions. Appl Surf Sci 263:334
4. Stepanov AL (2004) Optical properties of metal nanoparticles synthesized in a polymer by ion implantation: a review. Tech Phys 49:143
5. Galyautdinov MF et al (2016) Formation of a periodic diffractive structure based on poly(methyl methacrylate) with ion-implanted silver nanoparticles. Tech Phys Lett 42:182
6. Lee EH et al (1992) Improved hardness and wear properties of B-ion implanted polycarbonate. J Mater Res 7:1900
7. Lee EH et al (1993) Ion beam application for improved polymer surface properties. Nucl Instrum Meth Phys Res B 74:326
8. Klepikov V, Kruchinin S, Novikov V, Sothikov A (2006) Composite materials with radioactive inclusions as artificial radiation covering. Rev Adv Mater Sci 12:127
9. Kruchinin S, Novikov V, Klepikov V (2008) Nonlinear current oscillations in a Josephson junction with fractal radioisotope composites. Metrol Meas Syst 15:281
10. Ermakov V, Kruchinin S, Fujiwara A (2008) In: Bonca J, Kruchinin S (eds) Proceeding of NATO ARW "Electron Transport in Nanosystems". Springer, pp 341–349
11. Boldyryeva H et al (2004) Surface modification and nanoparticle formation by negative ion implantation of polymers. Nucl Instr Meth Phys Res B 219–220:953
12. Boldyryeva H et al (2005) High-fluence implantation of negative metal ions into polymers for surface modification and nanoparticle formation. Surf Coat Technol 196:373
13. Kavetskyy T S et al (2014) Structural defects and positronium formation in 40 keV B+-implanted polymethylmethacrylate. J Phys Chem B 118:4194
14. Kavetskyy TS, Stepanov AL (2016) Ion-irradiation-induced carbon nanostructures in optoelectronic polymer materials. In: Monteiro WA (ed) Radiation effects in materials. InTech, Rijeka, pp 287–308
15. Kavetskyy T et al (2017) High-dose boron and silver ion implantation into PMMA probed by slow positrons: effects of carbonization and formation of metal nanoparticles. J Phys Conf Ser 791:012028
16. Panzarasa G et al (2016) Positron annihilation spectroscopy: a new frontier for understanding nanoparticle-loaded polymer brushes. Nanotechnology 27:02LT03
17. Panzarasa G et al (2017) Probing the impact of the initiator layer on grafted-from polymer brushes: a positron annihilation spectroscopy study. Macromolecules 50:5574
18. Stepanov AL et al (2000) Formation of metal-polymer composites by ion implantation. Phil Mag B 80:23
19. Stepanov AL et al (2015) Synthesis of porous silicon by ion implantation. Rev Adv Mater Sci 40:155
20. Kavetskyy T et al (2011) Nanovoids in glasses and polymers probed by positron annihilation lifetime spectroscopy. In: Riethmaier JP et al (eds) Nanotechnological basis for advanced sensors. Springer, Berlin, pp 103–110
21. Kavetskyy T et al (2012) New organic-inorganic hybrid ureasil-based polymer and glass-polymer composites with ion-implanted silver nanoparticles. Phys Status Solidi C 9:2444
22. Kavetskyy T et al (2013) New organic-inorganic hybrid ureasil-based polymer materials studied by PALS and SEM techniques. Mater Sci Forum 733:171
23. Anwand W et al (1995) A magnetically guided slow positron beam for defect studies. Acta Phys Pol A 88:7
24. Anwand W et al (2012) Design and construction of a slow positron beam for solid and surface investigations. Defect Diffus Forum 331:25

25. Makhov AF (1961) The penetration of electrons into solids. II. The distribution of electrons in depth. Sov Phys Solid State 2:1942
26. Vehanen A et al (1987) Profiling multilayer structures with monoenergetic positrons. Phys Rev B 35:4606
27. Kobayashi Y et al (2008) Application of positron beams to the study of positronium-forming solids. Appl Surf Sci 255:174
28. Kavetskyy T et al (2017) Acta Phys Pol A Network properties of ureasil-based polymer matrixes for construction of amperometric biosensors as probed by PALS and swelling experiments. 132:1515
29. Saito F et al (2004) Study of ion irradiated poly-lactic acid using slow positron beam. Mater Sci Forum 445–446:340
30. Saito F et al (2014) Characterization of ion-irradiated poly-L-lactic acid using nano-cutting. Phys Chem Chem Phys 16:26991
31. Kavetskyy TS et al (2018) Surface plasmon resonance band of ion-synthesized Ag nanoparticles in high dose Ag: PMMA nanocomposite films. In: Petkov P et al (eds) Advanced nanotechnologies for detection and defence against CBRN agents, Springer, Dordrecht. https://doi.org/10.1007/978-94-024-1298-7_5
32. Hanemann T, Szabo DV (2010) Polymer-nanoparticle composites: from synthesis to modern applications. Materials 3:3468
33. Wie S et al (2011) Multifunctional composite core-shell nanoparticles. Nanoscale 3:4474
34. Kumar KS et al (2013) Recent advancement in functional core-shell nanoparticles of polymers: synthesis, physical properties, and applications in medical biotechnology. J Nanoparticles 2013:672059
35. Gillet JN, Meunier M (2005) General equation for size nanocharacterization of the core-shell nanoparticles by X-ray photoelectron spectroscopy. J Phys Chem B 109:8733
36. Puska MJ, Lanki P, Nieminen RM (1989) Positron affinities for elemental metals. J Phys Condens Matter 1:6081

Chapter 22
Heterogeneous Systems with Ag Nanoparticles

V. Smyntyna and V. Skobeeva

Abstract Optical properties of hybrid nanosystem consisting of silver nanoparticles (Ag NPs) with a methylene blue dye (MB) investigated. The first detected the effect of Ag NPs on the structuring of the molecule MB and the effect of the dye structure on its optical and luminescent properties. This effect explained by the formation of nonluminescent aggregates of dye in the nanosystems with Ag NPs. Molecular aggregates are not involved in the radiative processes of the dye, resulting in reduced intensity of its luminescence.

Keywords Heterogeneous nanosystems · Silver nanoparticles · Methylene blue dye · Luminescence · Surface plasmon resonance · Aggregates of dyes

22.1 Introduction

Recently, studies of nanosystems that are formed from metallic nanoparticles, semiconductor quantum dots, luminescent organic molecules, biological objects are of great interest. The actuality of research on such nanosystems is due, firstly, to the possibility of creating generalized conception about the physical interaction between the components of heterocomposites, and secondly, to the development of new nanomaterials with a wide range of applications in photocatalysts, photovoltaic devices, biosensors, having the prospect of introduction in biomedicine and environmental environment.

This paper reports on the investigation of the optical properties of nanocomposites consisting of Ag NPs and methylene blue dye.

V. Smyntyna (✉)
Odessa I.I. Mechnikov National University, Odessa, Ukraine
e-mail: Smyntyna@onu.edu.ua

V. Skobeeva
Research Institute of Physics, I.I. Mechnikov National University, Odessa, Ukraine

J. Bonča, S. Kruchinin (eds.), *Nanostructured Materials for the Detection of CBRN*, NATO Science for Peace and Security Series A: Chemistry and Biology,
https://doi.org/10.1007/978-94-024-1304-5_22

Silver nanoparticles, and methylene blue dye, each individually have unique physico-chemical properties. Silver nanoparticles have unique optical properties that are caused by surface plasmon resonance (SPR) and have practical applications in plasmonics and nanophotonics. Silver nanoparticles are used in textile technology, in electronics, in optics, and especially in medicine [1–7]. One of the most promising materials for creating hybrid structures is methylene blue. Its properties make it possible to use it for various applications, including chemical sensors, biological markers and spectral-sensitization systems for photovoltaic cells and photocatalysis [8–11].

Cationic thiazine dyes under specific conditions form aggregates in aqueous solutions. Cationic thiazine dyes under specific conditions form aggregates in aqueous solutions. The processes of dye aggregation significantly influence the optical and luminescent properties of hybrid nanostructures of dyes with inorganic nanoparticles [12–16].

Studies show that Ag NPs can promote both to an increase in the luminescence intensity of dyes and to its quenching. The reason for magnification the luminescence intensity of a nearby luminescent dye is the interaction between the dipole moments of the dye and the surface plasmon field NPs [17, 18]. The authors of [19, 20] believe that the non-radiative transfer of energy from the dye to the nanoparticle leads to a decrease in the quantum yield of the luminescence of dye molecules.

Another mechanism that determines the photophysical properties, primarily the radiation characteristics of thiazine dyes, is the formation of aggregates [12]. The formation of aggregates from monomers of dyes depends on many factors (concentration, environment, chemical composition, etc.). It is of interest how the processes of structuring dyes affect the optical properties of dye composites with metal nanoparticles.

In this paper, we propose a mechanism for reducing the intensity of luminescent dyes, caused by the aggregation of dye molecules in the presence of Ag NP.

22.2 Experimental

Silver NPs synthesized by chemical reduction of silver salt by sodium citrate. Sodium citrate carry out functions as a reducing agent for silver ions and for the stabilizer of nanoparticles. Synthesis was carried out at the following technological parameters: concentrations of sodium nitrate ($AgNO_3$) and sodium citrate values correspond to 5×10^{-4} M; 1×10^{-3} M; 5×10^{-3} M; concentrations of methylene blue dye correspond to 5×10^{-6} M and 5×10^{-5} M; the temperature of synthesis $-100\,°C$. The synthesis procedure is described in the works [21, 22].

Optical absorption spectra were measured on a spectrophotometer SF-26 in the wave length region from 320 to 600 nm. To reduce the error associated with the effect of light scattering in the short-waveregion (320–360 nm) we used the UVLF-2 light filter. The measurement error did not exceed $\pm 1\%$. The luminescence was excited by a pulsed LCS-DTL-374QT laser with a wave length of light 355 nm (power is 35 mW) and by LED Gree-XB with a wave length of light 630 nm (power is 1 W).

22.3 Results and Discussion

Using these concentrations and synthesis conditions NPs Ag have a spherical shape (Fig. 22.1). In the absorption spectrum of nanoparticles, an intense SPR band appears, the maximum of which localized in the visible region of the spectrum and depends on the technological regimes. The positions of the maxima of the SPR are determined by the size of the nanoparticles and with an increase in the size of the NPs Ag the maximum shifts to the long-wave region of the spectrum.

The plasmon resonance band of nanoparticles grown in sodium citrate is localized in the region $\lambda_{max} = 390 \div 420$ nm (curve *1*, *2*, Fig. 22.2). When a tannin aqueous solution with a concentration of 5×10^{-4} M in a ratio of 1:1 is added to sodium citrate, the maximum of the band is shifted to the long-wavelength region and localized at $\lambda_{max} = 440$ nm (curve *3*, Fig. 22.2). The broadening of the band with $\lambda_{max} = 440$ nm indicates the formation of nanoparticles of different sizes. The average size of grown silver nanoparticles estimated by the method of dynamic light scattering and varied in the range of $20 \div 60$ nm.

The absorption spectra of heterogeneous systems of NP with MS illustrated in Fig. 22.3. The resulting silver nanoparticles show in the spectrum an absorption band caused by surface plasmon resonance. The maximum of Ag NPs SPR for NPs of size 30 nm is localized at $\lambda_{max} = 420$ nm. The absorption spectrum of MB dye contains a band $\lambda_{max} = 633$ nm (curve *1*). The absorption spectrum of Ag NPs – MB nanocomposite, in addition to the absorption bands of the silver nanoparticles ($\lambda_{max} = 420$ nm, for size NPs 30 nm and $\lambda_{max} = 440$ nm, for size NPs 60 nm) and the absorption band of the dye in the molecular form, contains a new band with $\lambda_{max} = 590$ nm. The nature of this absorption band is associated with the formation of molecular aggregates in the dye (curve *2* and *3*).

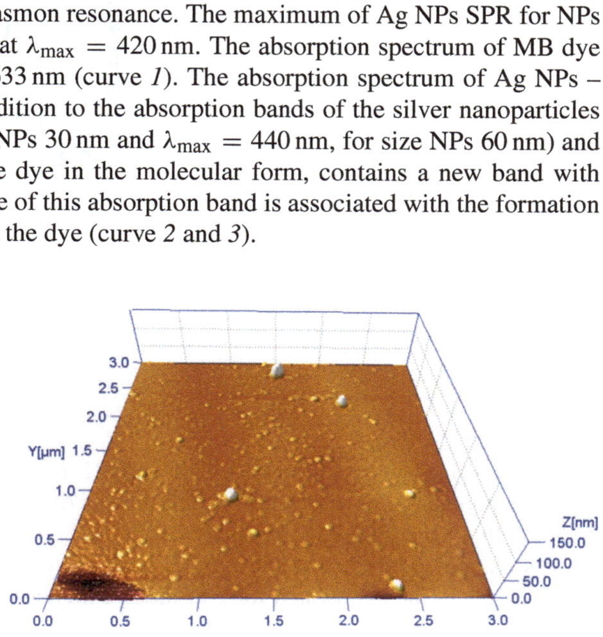

Fig. 22.1 3-D AFM image of a fragment of the surface of NP Ag deposited on a substrate of mica. The concentration of $AgNO_3 = 5 \times 10^{-4}$ M, sodium citrate $= 1 \times 10^{-3}$ M

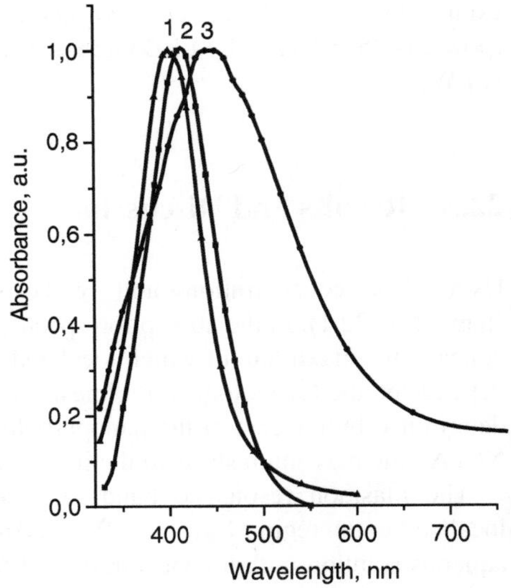

Fig. 22.2 Spectra of optical density of silver NP of different sizes, nm: 20 (*1*); 30 (*2*); 60 (*3*)

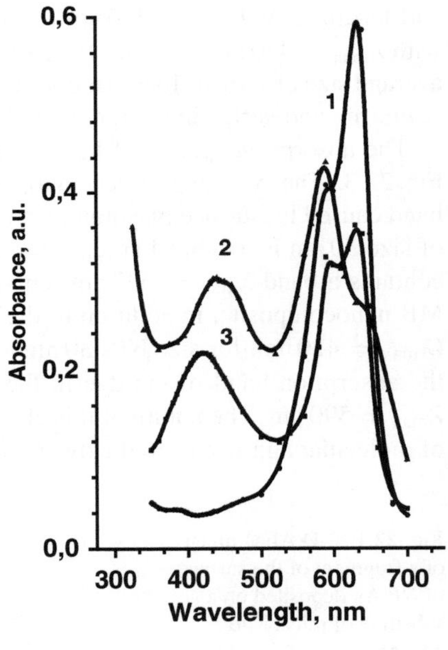

Fig. 22.3 Absorption spectra of MB dye (*1*) and NPs Ag with the addition of MB dye (*2*, *3*). Size NPs Ag, n̂: 60 (*2*), 30 (*3*)

Thus, a solution containing an Ag NP and MB dye is a plasmon absorption band of silver nanoparticles and a binary mixture of dye monomers and its associates (dimers) that have different absorption bands.

It is characteristic that as the Ag NPs size increases, the intensity of the band associated with the aggregated form of the dye increases and for nanoparticles with

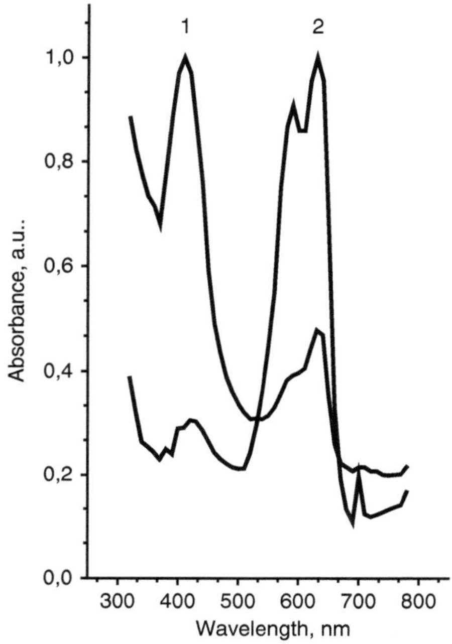

Fig. 22.4 Normalized absorption spectra of NPs Ag ($\lambda_{max} = 420$ nm) with the addition of MB dye, concentration: 5×10^{-6} Ì (*1*); 5×10^{-5} Ì (*2*)

a size of 60 nm becomes predominant (curve *2*, Fig. 22.2). This result confirms the fact of aggregation of MB molecules on the NPs surface. With increasing surface area, the number of dimer aggregates increases.

Another fact testifying to the effect of nanoparticles on the processes of structuring MS dye molecules is the dependence of the ratio of intensities of the absorption bands of nanoparticles and dye on the concentration of the dye. As can be seen from Fig. 22.4, the SPR band predominates in the absorption spectrum of the composite with a dye concentration of 5×10^{-6} Ì (curve *1*). With an increase in the concentration of the dye by an order of magnitude, the intensity of the nanoparticle absorption band decreases relatively to luminescence of MG (curve *2*). A decrease in the intensity of PPR may indicate a decrease in the total amount of silver particles in which conditions for the formation of plasmons are created. A possible reason for this may be the electrostatic interaction between the dye and NPs Ag.

Our studies show that, despite the increase in the concentration of methylene blue, aggregates of higher order do not appear in the composite, as is observed in the case of aqueous MB solution. It is obvious that the electrostatic interaction stabilizes the structural composition of the dye, preventing its desorption from the surface of the NPs.

In Fig. 22.5 represent the spectra the luminescence spectra of MS with silver NPs. As the results show, the formation of aggregates accompanied by a decrease in the intensity of the luminescence of MS.

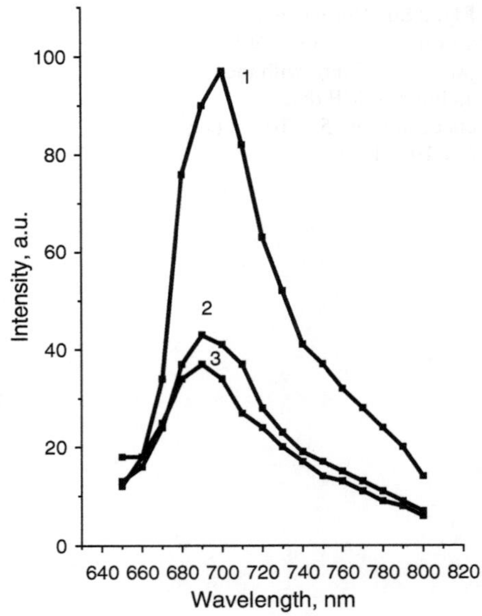

Fig. 22.5 The luminescence spectrum of the methylene blue dye, concentration of 10^{-5} M (*1*) and composites Ag NPs with MB at concentrations of $AgNO_3$: 2×10^{-4} M (*2*); 4×10^{-4} M (*3*)

This explained by the fact that dimers and aggregates of higher order do not luminesce, and the number of luminescing monomers decreases.

It is noted that as the Ag NPs concentration increases, the luminescence intensity of the composite decreases substantially. Here, we believe that the concentration of NP increases with increasing initial concentration of $AgNO_3$. We note that the absorption bands of Ag NPs and MC do not overlap and are in different regions of the spectrum, this indicates the absence of a mechanism for nonresonant energy transfer to the dye.

22.4 Conclusion

In the heterogeneous system of NPs Ag-MÂ electrostatic interaction of NPs with dye molecules occurs, which results in the formation of its aggregated forms. Since aggregates do not luminesce, the intensity of the dye undergoes quenching.

Studies have shown that NPs Ag promote the formation of aggregates in dimeric form and prevent the formation of aggregates of higher order. The degree of aggregation of molecules depends on the surface area and concentration of NP, the concentration of the dye.

References

1. Prabhu S, Poulose EK (2012) Silver nanoparticles: mechanism of antimicrobial action, synthesis, medical applications, and toxicity effects. Int Nano Lett 2(1):32–41. https://doi.org/10.1186/2228-5326-2-32
2. Sharma A, Tapadia K (2017) Simple and rapid quantitative determination of thiol-containing toxicants using silver nanoparticles as an affinity probe. J Appl Spectrosc 83(6):1068–1075. https://doi.org/10.1007/s10812-017-0409-3
3. Zhang F, Wu X, Chen Y, Lin H (2009) Application of silver nanoparticles to cotton fabric as an antibacterial textile finish. Fibers Polym 10(4):496–501. https://doi.org/10.1007/s12221-009-0496-8
4. Li Y, Guo M, Lin Z, Zhao M, Xiao M, Wang C, Xu T, Chen T, Zhu B (2016) Polyethylenimine-functionalized silver nanoparticle-based co-delivery of paclitaxel to induce HepG2 cell apoptosis. Inter J Nanomedicine 11:6693–6702. https://doi.org/10.2147/IJN.S122666
5. Kim SC, Kim SM, Yoon GJ, Nam SW, Lee SY, Kim JW (2014) Gelatin-based sponge with Ag nanoparticles prepared by solution plasma: fabrication, characteristics, and their bactericidal effect. Curr Appl Phys 14(2):S172–S179
6. Ermakov V, Kruchinin S, Fujiwara A (2008) Electronic nanosensors based on nanotransistor with bistability behaviour. In: Bonca J, Kruchinin S (eds) Proceedings of the NATO ARW "Electron transport in nanosystems". Springer, pp 341–349
7. Ermakov V, Kruchinin S, Hori H, Fujiwara A (2007) Phenomena in resonant tunneling through degenerated energy states with electron correlation. Int J Mod Phys B 21(11):1827–1836
8. Jockusch S, Lee D, Turro NJ, Leonard EF (1996) Photo-induced inactivation of viruses: adsorption of methyleneblue, thionine, and thiopyronine on Qf3 bacteriophage. Proc Natl Acad Sci USA 93:7446–7451
9. Gabrielli D, Belisle E, Severino D, Kowaltowski AJ, Baptista MS (2004) Binding, aggregation and photochemical properties of methylene blue in mitochondrial suspensions. Photochem Photobiol 79(3):227–232. https://doi.org/10.1111/j.1751-1097.2004.tb00389.x
10. Tardivo J, Giglio A, Oliveira C, Gabrielli D, Junqueira H, Tada D, Severino D, Turchiello R, Baptista M (2005) Methylene blue in photodynamic therapy: from basic mechanisms to clinical applications. Photodiagn Photodyn Ther 2(3):175–191. https://doi.org/10.1016/S1572-1000(05)00097-9
11. Tuite EM, Kelly JM (1993) Photochemical interactions of methylene blue and analogues with DNA and other biological substrates. J Photochem Photobiol B21(2–3):103–124. PMID: 8301408
12. Klochkov VK, Egorova AV, Sedykh OO, Efimova LE (2014) Aggregation of methylene blue and nile blue in the presence of nanoparticles $ReEuVO_4$ (Re = Cd, Y, La) with different form–factor. Quest Chem Chem Technol 1:106–109
13. Bujdak J, Janek M, Madejova J, Komadel P (2001) Methylene blue interaction with reduced-charge smectites. Clay Clay Miner 49(3):244–254
14. McArthur EA, Godbe JM, Tice DB, Weiss EA (2012) A study of the binding of cyanine dyes to colloidal quantum dots using spectral signatures of dye aggregation. J Phys Chem C 116(10):6136–6142. https://doi.org/10.1021/jp300478g
15. Rodionov VE, Shnidko IN, Zolotovsky A, Kruchinin SP (2013) Electroluminescence of Y_2O_3:Eu and Y_2O_3:Sm films. Mater Sci 31:232–239
16. Vlaskina S, Kruchinin S, Kuznetsova E, Rodionov V et al (2016) Nanostructures in silicon carbide crystals and films. Inter J Mod Phys B 30(13):1042015
17. Suvorov T, Balbekova A, Klyuev V, Latyshev A, Ovchinnikov O, Smirnov M, Rybalko A (2012) Magnification of luminescence of dye molecules in the presence of silver nanoparticles. Opt J 79:79–82

18. Martynenko I, Orlova A, Maslov V, Fedorov A, Berwick K, Baranov A (2016) The influence of phthalocyanine aggregation in complexes with CdSe/ZnS quantum dots on the photophysical properties of the complexes. Beilstein J Nanotechnol 7:1018–1027
19. Anger P, Bharadwaj P, Novotny L (2006) Enhancement and quenching of single-molecule fluorescence. Phys Rev Lett 96(11):113002. https://doi.org/10.1103/PhysRevLett.96.113002
20. Ritchie G, Burstein E (1981) Luminescence of dye molecules adsorbed at a Ag surface. Phys Rev B 24(8):4843. https://doi.org/10.1103/PhysRevB.24.4843
21. Smyntyn VA, Skobeeva VM (2012) Synthesis and optical properties of nanoparticles of silver, technical digest frontiers in optics (FiO) 2012 and laser science (LS) XXVIII meetings, FW3A.15. https://doi.org/10.1364/FIO.2012.FW3A.15
22. Smyntyna VA, Skobeeva VM, Vorobyov NK, Struz DA, Kogut IS, Sviridova OI (2012) The influence of external factors on the stability of the optical properties of silver nanoparticles. Sens Electron Microsyst Technol 3(9):34–38

Chapter 23
CBRN Risk Scenarios

F. Bruno, M. Carestia, M. Civica, P. Gaudio, A. Malizia, F. Troiani, R. Sciacqua, and U. Spezia

Abstract Events of dispersion of Chemical, Biological, Radiological and Nuclear (CBRN) materials due to natural, accidental or intentional events are considered risk situations and nowadays represent one of the most critical concerns for safety and security. Safe management of these materials, in particular the certainty of its location and control of their transport, is a key issue to be sure that no diversion to unwanted and unauthorized uses occurs. Particularly relevant in the control of the transportation is a stringent cross-border traffic control aimed at avoiding their illicit traffic. Starting from the processes on risk scenarios developed for radiological and nuclear material control, Sogin, in collaboration with the University of Rome "Tor Vergata", have defined a standard approach for analyzing the preparation of the border points in relation to both safety and security. The proposed approach take into consideration all the relevant aspects of the control such as instrumentation, control procedures and information exchange within the domestic infrastructure and with the other side of the border.

Keywords CBRN materials · Cross-border control · Illicit traffic · Threats · Adversary scenarios · Detection system · Standard approach · Best practice

23.1 Introduction

In the current world context the chemical, biological, radioactive and nuclear (CBRN) materials could be used for illicit purposes, for this reason it's necessary to implement a strategy to enhance safety and security and to ensure the protection of

F. Bruno · M. Civica · F. Troiani (✉) · R. Sciacqua · U. Spezia
SOGIN S.p.A., Rome, Italy
e-mail: Troiani@sogin.it

M. Carestia · P. Gaudio · A. Malizia
University of Rome "Tor Vergata", Rome, Italy

J. Bonča, S. Kruchinin (eds.), *Nanostructured Materials for the Detection of CBRN*, NATO Science for Peace and Security Series A: Chemistry and Biology, https://doi.org/10.1007/978-94-024-1304-5_23

citizens [1–5]. Countries sharing a border have to respond to this risk by engaging in a collective commitment to strengthen the protection and control of such material and to respond effectively to CBRN security events.

This commitment lead to analysis of borders and in particular to the external border of regional entities, such as the European Union, because the risks connected to management of CBRN material are a strategic issue. The final result of analysis of the situation on the external border shall lead to an improvement of the collaboration, for example, among the European Union countries and the neighbor countries [5–11].

This means to strengthen existing instruments and to implement them to enhance CBRN security, on the border areas. The detection and interception of the illicit traffic is necessary to eliminate or, at least, to reduce the risk of the use of CBRN agents for illegal purposes.

The CBRN defense is a complex challenge, it includes technical, tactical, operational and strategic level of decision-making. From a technical point of view, the framework of the CBRN threats is further complicated by the intrinsic differences between chemical, biological, radiological and nuclear agents.

This work aims to indicate a standard approach to implement a security system to control of cross border transfers of risk CBRN materials and to avoid illicit traffic. For this purpose Sogin and the University of Tor Vergata have collected best practices, recommendations and suggestions on the detection of CBRN materials, advancing the methodology explained in this document.

23.2 Objective

Objective of this proceeding is to describe the proposed standard approach based on best practices and applicable international references for the management of the threat related to the illicit traffic of CBRN material.

The start point is to identify the threats in terms of illicit traffic on the border and to identify the reference adversary responsible of malicious acts.

In particular the adversary are typically divided in three categories based on different motivation, intention and capabilities:

- The criminal motivated by economic gain. He wants to steal CBRN material because he (or they) believes it has a high value, may work as part of a well-organized team of criminals with funding and infrastructure support, may collude with an insider, and he does not want to die.
- The terrorist motivated by ideology. He wants to sabotage a facility or to steal nuclear, radioactive, chemical or biological material, he is part of a well-armed and trained group with a well-funded infrastructure, he is expected to conduct an attack on a facility, he is willing to die in the attempt, and this is the worst-case of adversary.
- The protestor motivated by an ideological goal to eliminate the use of nuclear power or chemical/biological industry. He wants to reveal the weaknesses of the

facility by demonstrating that the physical protection measures are ineffective or insufficient, may have several different subgroups within the organization, members may have peaceful intentions while a few individuals within the extremist core may have a violent agenda, could collude with an insider.

In order to provide a comprehensive analysis, it is necessary to develop a structured approach to identify credible scenarios. An adversary scenario is a sequence of successive events that an adversary would follow to achieve his undesirable objective. The scenarios for detection of CBRN material are defined and developed on statistical analysis on illicit traffic, in particular, it is needed to identify CBRN materials and the available detection technology.

The logistics, environmental and structural analysis of border points, identify the possible events and credible scenarios to prepare protection measures, operational and managerial procedures in order to provide an adequate level of security. The goal of this work is to define a standard approach for analyzing the preparation of the border points about safety and security aspects.

23.3 Threat Assessment

To perform the analysis is necessary a description of the border point; the possible events on the border derive from the information on CBRN events, defined by selected data bases. This information not refer only to events related to detection on the border of illicit traffic but also inside the territory of the country.

The scenarios are defined by means of the typology of crossing, typology of border point and related behavior.

From the combination of events and scenario, is possible to better describe the potential illicit act. The definition of the potential illicit act to be considered in the real situation, according to IAEA indications, should be defined by the appropriate State authorities, using various credible information sources.

Taking into consideration the scenarios and the need to detect all the possible materials, each side has to be equipped to detect all the materials in all the reference and credible crossing configurations. For this reason on both sides of the border detectors with the same characteristics should be available. It is better not to have the same detectors but detection systems covering the same areas of detection, this in order to reduce the risk to have shortfalls on detection capabilities.

23.4 Introducing a Standard Approach

The work aims to introduce a standard approach to reduce the illicit traffic on cross border. From a technical point of view, the framework of the CBRN threats is further complicated by the intrinsic differences between chemical, biological, radiological and nuclear agents that makes impossible to perform detection and identification with the same procedures and equipment. Moreover, the spread of such substances that can be the result of natural, industrial or malevolent evens may have a cross border impact, depending on the magnitude of the diffusion, therefore the controls are essential to reduce the risks.

The starting point of the proposed approach is the collection of basic inputs composed by best practice, threats and current situation of detection point; this elements allow to analyze the detection system.

The main international "best practices" in CBRN materials are:

- Radiological and nuclear materials from International Atomic Energy Agency (IAEA);
- standards on Safety and Security Chemical materials from Organization for the Prohibition of Chemical Weapons and Integrated Chemical Safety and Security Program (ICSSP);
- Biological materials from Biological and Toxin Weapons Convention.

The definition of possible threats is allowed by taking into account information on detection point, geopolitical situation and specific logistic configuration.

Other input are the current situation in detection points, defined by a collection of information and data from interviews and other specific documentation.

To implement a detection system is necessary to define standards of safety and security and detection point capabilities; the solution of identified gaps of differences between defines standards and the current situation allow to implement a reliable detection system. The process is represented by Fig. 23.1.

The detection and interception of CBRN materials are consequences of detection system and information system. The detection system is implemented by best practice, analysis of threats and current situation of detection point; the information services consist of national police, neighbor countries authorities and intelligence service.

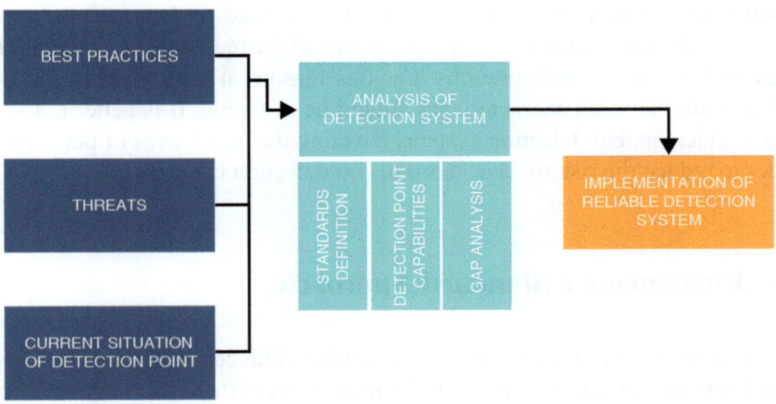

Fig. 23.1 Implementation of reliable detection system

During the detection, the alarm can be a real or false, the latter can be related to a false positive as statistical noise, background fluctuation and false information, interferences, etc. In this case it is necessary to record information and analyze the reason of false alarm.

The real detection can be related to a legal or an illicit traffic. A legal transportation means legal shipment of CBRN materials; shipment of material incidentally contaminated with CBRN materials, Naturally Occurring Radioactive Materials (NORM), etc. In this case, it is necessary to record information and analyze the situation.

There is an illicit traffic when a detected CBRN material is transported without necessary documentation or not respecting the norms and legal framework. In this case, it is necessary to isolate the CBRN source with the aim to protect operator, public and environment. Furthermore a response system need to be implemented.

The information services are domestic, international and intelligence services. The Domestic services are police, fire guards, and other Authorities and law enforcement agencies; the International services come from neighbor countries authorities, and the Intelligence services are involved for information related to possible arrival of materials from other countries.

About the measurement for detection, the implementation of reliable detection system comprise of standard equipment typologies, standard operational procedures, professional training, etc.

The approach allows to manage the dangerous events with the aim to protect operators, people and environment and by means of tactical and strategic response. The incidental command, included in the response, acts on scene, material management and incident investigations:

- on scene: follow standard procedures, establish off area; evacuate person from inner space of the off area and inform media;
- on material management: handle causalities and suspect; identify CBRN material; seize materials and temporarily store materials;
- to investigate: detain suspect, gather evidence and initiate legal process.

The described process is outlined on next Fig. 23.2.

23.5 International Best Practice

For what concerns the best practices on CBRN, is not still developed a standardize legal instrument for information exchange, border control and related cooperation; international references should be specified case by case taking into consideration local needs and constraints.

Therefore, the study on the best practices will include a comprehensive overview of the international obligations and non-binding guidelines concerning the authorized and illicit movement of CBRN materials at the border. It is necessary to highlight that a smooth cross-border cooperation is key to reduce illegal trafficking

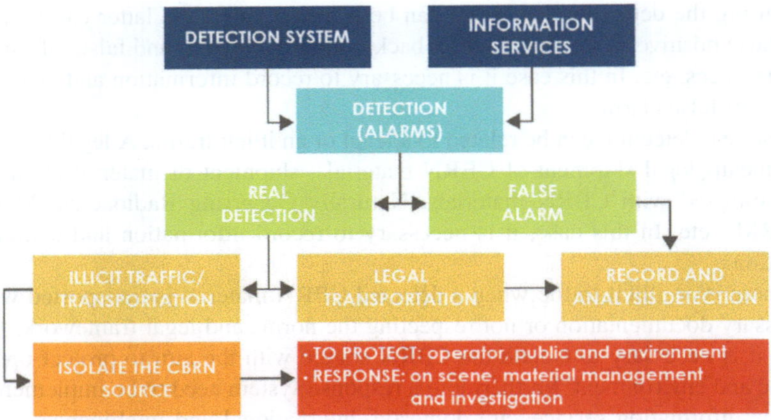

Fig. 23.2 Basic approach to detection point

and smuggling of CBRN materials; as a general principle the two different sides of the border should have the possibility to work in a complementary mode and/or doubling the controls [12–15].

23.5.1 International Agreements of CBRN Relevance

At the international level, a number of laudable initiatives have been promoted in order to enhance detection and interception of illicit CBRN materials, to face the challenges posed by the growing flux of people and goods moving worldwide. In the following a more detail on each of the main element constituting the best practices is given:

- UN resolution 1540;
- Security Council resolutions 1373 and 1377(2001);
- Resolutions by the UN General Assembly;

23.5.2 International Agreements and Standards Regarding Nuclear and Other Radioactive Materials

The best practices on radiological and nuclear materials are based on a correct legal framework both national and international, for the cross-border traffic it is more important the international framework than the national one and it is necessary to be sure the both parties correctly transfer the international legal framework on the national one. The international standards that for radiological and nuclear related issue are almost exclusively derived from IAEA. Such standard are organized in: fundamentals, recommendations, technical guidance, implementing guide and overview on the information of the events. The main references are listed below:

- International Convention for the suppression of Acts of Nuclear Terrorism (ICSANT);
- The Non Proliferation Treaty (NPT) and safeguards agreements with IAEA;
- Convention on Physical Protection of Nuclear Material and its amendment (CPPNM);
- Convention on Early Notification of a Nuclear Accident;
- Convention on Assistance in the Case of a Nuclear Accident or Radiological Emergency;
- IAEA Standards and guidance;
- EU integrated political crisis response arrangements (IPCR) June 2013 and EU civil protection legislation end of 2013;
- Council Directive 2013/59/EURATOM of 5 December 2013;
- Council Directive 2013/59 with effect from February 2018;
- Council Decision of 14 December 1987 on Community arrangements for the early exchange of information in the event of a radiological emergency (87/600/EURATOM).

23.5.3 International Agreements in the Area of Chemicals

On the international level, the UN Security Council Resolution 1540 (UNSCR 1540) gives some specific indications for the chemical and bacteriological materials. The Resolution calls upon States to "[...] renew and fulfil their commitment to multilateral cooperation, in particular within the framework of the International Atomic Energy Agency, the Organization for the Prohibition of Chemical Weapons and the Biological and Toxin Weapons Convention, as important means of pursuing and achieving their common objectives in the area of non-proliferation and of promoting international cooperation for peaceful purposes".

The major international legal reference for chemical material is provided by the "Convention on the Prohibition of the Development, Production, Stockpiling and Use of Chemical Weapons and on their Destruction", implemented by the Organization for the Prohibition of Chemical Weapons. The Convention aims to eliminate an entire category of weapons of mass destruction by prohibiting the development, production, acquisition, stockpiling, retention, transfer or use of chemical weapons by States Parties. States Parties must take the steps necessary to implement and enforce that prohibition while being recognized the right to develop and use chemicals for peaceful purposes.

In this context, the main references in chemical area are:

- The Convention on the Prohibition of the Development, Production, Stockpiling and Use of Chemical Weapons and on their Destruction Purpose, scope, status, implementation;

- The Chemical Weapons Convention entered into force 29 April 1997;
- Regulation (EC) No 1907/2006 of the European Parliament and of the Council of 18 December 2006 concerning the Registration, Evaluation, Authorization and Restriction of Chemicals (REACH), establishing a European Chemicals Agency;
- Directive 2006/121/EC of the European Parliament and of the Council of 18 December 2006 the Registration, Evaluation, Authorization and Restriction of Chemicals (REACH), establishing a European Chemicals Agency.

23.5.4 International Agreements in the Biological Area

In the international context of the UN Security Council Resolution 1540 (UNSCR 1540), recommendations have been also introduced about biological issues.

For what concern the biological material in the international legal framework, is necessary to cite the "Biological Weapons Convention", the result of a prolonged efforts by the international community to establish a new instrument to ban the production and use of biological agents for mass destruction.

The convention prohibits the development, production, acquisition, transfer, stockpiling and use of biological and toxin weapons to address the proliferation of weapons of mass destruction. The commitment to refrain from transferring any agent or weapon is quite general and thus may include terrorist organization as well as States. The reference for biological area is:

- Convention on the Prohibition of the Development, Production and Stockpiling of Bacteriological (Biological) and Toxin Weapons and on Their Destruction.

References

1. Center of Non-proliferation Studies (CNS) (2016) Global Incidents and Trafficking, tracking publicly reported incidents involving nuclear and other radioactive materials, 2015 Annual Report, Published Mar 2016. http://www.nti.org/media/documents/global_incidents_trafficking_report.pdf
2. CNS Global Incidents and Trafficking Database. http://www.nti.org/analysis/reports/cns-global-incidents-and-trafficking-database/
3. Convention on the Prohibition of the Development, Production, Stockpiling and Use of Chemical Weapons and on their Destruction. https://www.opcw.org/fileadmin/OPCW/CWC/CWC_en.pdf
4. Development, Use and Maintenance of the Design Basis Threat, Implementing Guide, Nuclear Security Series No. 10, Nuclear Security Culture. http://www-pub.iaea.org/MTCD/publications/PDF/Pub1386_web.pdf
5. EU CBRN Action Plan (2009) – Communication from the Commission to the European Parliament and the Council on Strengthening Chemical, Biological, Radiological and Nuclear Security in the European Union. https://ec.europa.eu/home-affairs/sites/homeaffairs/files/what-we-do/policies/pdf/com_2009_0273_en.pdf
6. Incident and Trafficking Database (ITDB), IAEA. http://www-ns.iaea.org/security/itdb.asp

7. International Atomic Energy Agency, Technical Guidance Reference Manual, Nuclear Security, Series No. 6, Combating Illicit Trafficking and other Radioactive Material. http://www-pub.iaea.org/MTCD/publications/PDF/pub1309_web.pdf
8. International Atomic Energy Agency, Technical Guidance Reference Manual, Nuclear Security Series No. 1, Technical and Functional Specifications for Border Monitoring Equipment. http://www-pub.iaea.org/books/IAEABooks/7400/Technical-and-Functional-Specifications-for-Border-Monitoring-Equipment
9. International Atomic Energy Agency, Recommendations, Nuclear Security Series No. 13, Nuclear Security Recommendations on Physical Protection of Nuclear Material and Nuclear Facility/INFCIRC/225/Revision 5. http://www-pub.iaea.org/MTCD/publications/PDF/Pub1481_web.pdf
10. International Atomic Energy Agency, Implementing Guide, Nuclear Security Series No. 7, Nuclear Security Culture. http://www-pub.iaea.org/MTCD/publications/PDF/Pub1347_web.pdf
11. International Atomic Energy Agency, Technical Guidance, Nuclear Security Series No. 12, Educational Program in Nuclear Security. http://www-pub.iaea.org/mtcd/publications/pdf/pub1439_web.pdf
12. OSCE Handbook of Best Practices at Border Crossings – A Trade and Transport Facilitation Perspective. http://www.osce.org/eea/88238?download=true
13. Web references concerning international best practice on CBRB materials. http://www.un.org/en/ga/search/view_doc.asp?symbol=S/RES/1540%20(2004); http://register.consilium.europa.eu/doc/srv?l=EN&f=ST%2015505%202009%20REV%201; https://ec.europa.eu/home-affairs/sites/homeaffairs/files/what-we-do/policies/crisis-and-terrorism/securing-dangerous-material/docs/cbrn_glossary_en.pdf
14. Web references concerning nuclear and radioactive materials. https://www.iaea.org/publications/documents/conventions/convention-physicalprotection-nuclear-material; https://www.iaea.org/publications/documents/treaties/npt; https://www.iaea.org/safeguards/safeguards-legal-framework/additional-protocol
15. Web references concerning chemical and biological materials. http://www.unog.ch/80256EE600585943/(httpPages)/77CF2516DDC5DCF5C1257E520032EF67?OpenDocument; http://www.wcoomd.org/en.aspx; http://www.oie.int/about-us/key-texts/cooperation-agreements/agreement-with-the-world-customs-organization/; http://www.who.int/csr/delibepidemics/biochemguide/en/; http://register.consilium.europa.eu/doc/srv?l=EN&f=ST%2015708%202003%20INIT; https://www.start.umd.edu/gtd/about/; http://www.factsonline.nl/browse-chemical-accidents-in-database/1; http://data.euro.who.int/cisid/?TabID=67; http://www.globalincidentmap.com/beta/

Chapter 24
Influence on the Energy Efficiency of Hydrocarbon Fuels

V. I. Morozov and I. V. Morozova

Abstract We consider some aspects of applications of the method of electrophysical influence on a hydrocarbon fuel with the purpose to improve the performance parameters of heat engines and the question of applications of the magnetic susceptibility and dielectric losses for the control over the quality of hydrocarbon fuels and oils.

Keywords Hydrocarbon fuel · Magnetic field · Magnetic moment · Spin of a nucleus · Resonance frequency · Nuclear polarization · Electrophysical method of influence · Magnetic susceptibility · Dielectric permeability · Dielectric losses

24.1 General Statement of the Problem

At the present time, the increase in the efficiency of the use of oil products and the economic efficiency of the power units of vehicles is of great importance in the realization of priority directions aimed at the energy conservation in the economy.

The efficiency of the use of oil products is determined, in many aspects, by a technological process of their fabrication and by the usage of various additives, which have positive properties but possess also a number of essential drawbacks such as the complexity of a technological production process, high prime cost, and, as a rule, narrow area of applications.

The problems of effectiveness and toxicity of the operation of heat engines are directly related to the quality of fuels. In this connection, the toughening of requirements to the quality and the improvement of the most crucial operation properties of a fuel and oils are the actual significant scientific-technical problems.

V. I. Morozov (✉) · I. V. Morozova
National Aviation University, Kiev, Ukraine
e-mail: vimoro@ukr.net

J. Bonča, S. Kruchinin (eds.), *Nanostructured Materials for the Detection of CBRN*,
NATO Science for Peace and Security Series A: Chemistry and Biology,
https://doi.org/10.1007/978-94-024-1304-5_24

24.2 Review of Publications and Analysis of Unsolved Problems

One of the ways to the improvement of operation properties of fuels and oils is the electrophysical influence (EPI) on hydrocarbon fluids. Its essence consists in that the technological fluid is passed with some velocity through an EPI unit, in which a constant magnetic field with known characteristics and a topography, as well as a resonance variable electromagnetic field, are created [1].

Earlier, the problem of EPI was reduced to the search and the rational usage of the magnetic treatment of water and aqueous solutions. But now, we can consider, on the firm foundation, the question about a targeted change of properties of hydrocarbon fluids and other various liquid systems with the help of external electrophysical influences.

The EPI method is based on the interaction of the molecular structure of a liquid system with external magnetic and electromagnetic fields.

The electrons of an atom and the nuclei of some elements (H^1, C^{13}, O^{17}, P^{31}, Li^7, F^{19}, and a number of other elements) possess magnetic and mechanical moments. The spin of an electron is manifested in the form of a fine structure in atomic spectra. For nuclei, this property is registered in the form of a superfine structure. The magnetic moment of an electron interacts with the magnetic moment of a nucleus. The interaction energy depends on the mutual orientation of spins or magnetic moments, and the number of possible orientations is determined by the spin of a nucleus. The total spin of a nucleus depends on whether the spins of components of a nucleus compensate one another or not [2].

In the magnetic field, the magnetic moment of a proton with the spin $I = \pm 1/2$ has only two orientations: along the field or in the opposite direction. Each nucleus with magnetic moment, being placed in a magnetic field, acquires the additional Zeeman energy $-\mu H_0$ [3–7]. In this case, the Hamiltonian has the form

$$\mathcal{H} = -\mu_n H_0, \tag{24.1}$$

where μ_n is the magnetic moment of a proton; H_0 is the strength of the magnetic field.

Under the customary conditions, the nuclear polarization P is insignificant due to the smallness of μ_n and is determined from the formula [8]

$$P = \frac{n_{(+\mu)} - n_{(-\mu)}}{n_{(+\mu)} + n_{(-\mu)}}. \tag{24.2}$$

The difference of the energies for the spin directed against the field and the spin directed along the field increases with the strength of an external magnetic field, which causes an increase in the polarization. However, this rearrangement of spins occurs not instantly. Prior to the switching-on of a magnetic field, the numbers of spins directed "up" and "down" were equal. But, in the field, a part of spins directed against the magnetic field turns over.

The reorientation of a nuclear spin is accompanied by a change of its energy. Therefore, this can occur only in the presence of some object, with which the nucleus can exchange the energy. Such objects are electrons, neighboring nuclei, paramagnetic elements (oxygen, etc.), ions, free radicals, or a radio-frequency circuit with high Q-factor that is tuned onto the precession frequency of nuclei in the external magnetic field.

In the bulk of a substance, the total magnetic moment of nuclei (magnetization), which is directed along the external field, is as follows [9]:

$$M = \frac{\mu_n^2 H_0 n_0}{kT}, \tag{24.3}$$

where n_0 is the number of nuclei in unit volume of a substance; k is the Boltzmann constant; T is the temperature.

A sharp decrease in the strength of a magnetic field leads to that the additional energy of a proton is absorbed, due to the superfine interaction, by atoms of elements with nonzero electron spin that are located near the hydrogen nucleus. The role of such elements can be played by paramagnetic oxygen, free radicals, etc., i.e., paramagnetic centers interact with the basic chemical compound, where they are placed.

An increase in the polarization of hydrogen nuclei under a change in the strength of a magnetic field induces inevitably the excitation of electron shells in molecules by means of the superfine interaction. This generates, in turn, induction currents, which influence the physical properties of fluids.

In order to enhance the efficiency, stability, and universality of the magnetic influence on a fuel, the electrophysical method based on the resonance absorption of the energy by the proton system of a fluid was proposed. The method consists in the simultaneous influence of an inhomogeneous constant magnetic field and a resonance high-frequency electromagnetic field H_1 on a fluid [10]. In this case, the frequency of oscillations coincides with the precession frequency of nuclei in the given magnetic field H_0, and the selective absorption of the energy of a generator is observed at the frequency

$$\omega = \gamma H_0, \tag{24.4}$$

where γ is the gyromagnetic ratio of a proton.

The variable electromagnetic field H_1 induces a perturbation of the system of nuclear spins, which is described by the Hamiltonian [2, 11–13]:

$$\mathcal{H}'(t) = \mathcal{H} \cos \omega t = q_n \beta (I_x H_{1x} + I_y H_{1y} + I_z H_{1z}) \cos \omega t. \tag{24.5}$$

The influence of an electromagnetic field H_1 causes the intense "upward" transitions, and, in this case, the Boltzmann occupation of levels is violated. The electromagnetic energy of the field H_1 is partially absorbed by the proton system of a fluid. So, the absorption signal is registered, and, hence, the energy of the medium increases [10]. The power P absorbed by spins from the variable electromagnetic field is

$$P = \hbar\,\omega\,W_h = \hbar\,\omega\,n_0 \frac{W}{1 + 2WT_1}. \tag{24.6}$$

Thus, the resonance absorption of the electromagnetic energy of an oscillatory circuit by the proton system of a fluid and then the exchange by this energy with the molecular system of a fluid lead to a change in the physical and physico-chemical parameters of a hydrocarbon fuel. Eventually, the electrophysical influence causes a change in the rate constant of a chemical reaction, which is determined by the Arrhenius law at a change in the activation energy E_a

$$K_f = K_0\sqrt{T} \cdot e^{-E_a/RT}, \tag{24.7}$$

which affects, in turn, the rate and the completeness of combustion and, as a result, the performance parameters of heat engines [11].

Thus, the considered theoretical aspects of the magnetic influence on hydrogen-containing liquid systems testify to a change in the energy of interaction of the magnetic moment of a nucleus with the spin system of electrons. In dependence on the method of influence, which reveals itself in the absorption signal in inhomogeneous magnetic fields or the emission signal with the use of the effect of double nuclear resonance, we get the possibility to control the proton system of a hydrogen-containing fluid, by using the electrophysical method of influence. This allows one to increase or decrease the energy of a hydrocarbon fluid and, in this way, to influence the physico-chemical and operation properties of a fuel.

24.3 Purpose of the Study

An enhancement of the energy efficiency of a fuel for heat engines by applying the electrophysical method to influence the proton system of a hydrocarbon fluid.

24.4 Results of Studies and Their Analysis

The study and substantiation of the method of EPI, as well as the determination of changes in the qualitative parameters of a fuel were executed step-by-step: the study of the results of theoretical and experimental investigations and methods of control over EPI; theoretical substantiation of the choice and optimization of a method of influence and control over the quality; its experimental verification by certifying methods of tests.

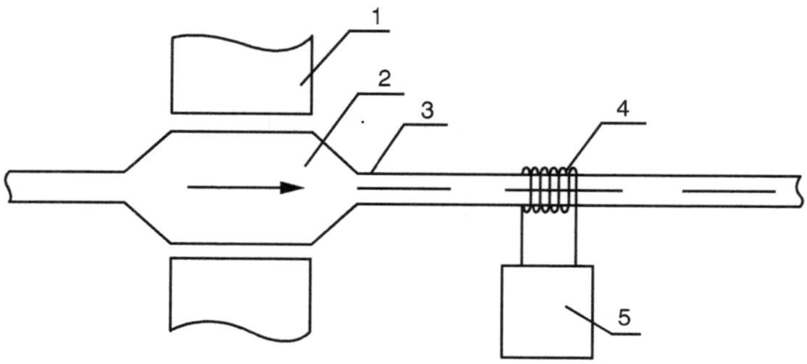

Fig. 24.1 Unit for EPI

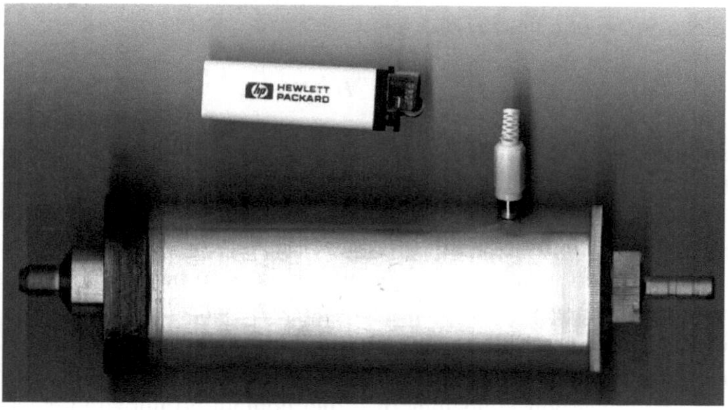

Fig. 24.2 General view of the unit for EPI

The block diagram of the experimental unit is shown in Fig. 24.1. The EPI unit includes constant magnet *1*. The field of a constant magnet acts on polarizing tank *2* filled in with a fluid. Then the fluid moves in pipeline *3*, by undergoing the action of the inhomogeneous scattering fields of constant magnet *1*, to high-frequency coil *4* of an LC oscillatory circuit connected with high-frequency generator *5*. The external view of the EPI unit is presented in Fig. 24.2.

To influence the proton system of a hydrocarbon fluid within the electrophysical method, the additional radio-frequency nutation coil (not shown in Fig. 24.1) fed from a generator of sinusoidal oscillations is placed between the polarizing magnet and high-frequency coil giving the absorption signal with the purpose to induce the double nuclear resonance.

The nutation coil, being in the inhomogeneous magnetic field, is oriented so that a fluid flows along the gradient, and the force lines of the oscillating field are directed normally to the external magnetic field. In this case, the establishment of the longitudinal component of the nuclear magnetization occurs as a result of the energy exchange between the system of nuclear spins and the molecular system. The rate of this exchange and, hence, of establishment of the longitudinal component is determined by the intensity of spin-lattice interaction.

The difference in the occupations of levels implies that the substance possesses the magnetic susceptibility. We write the complex susceptibility of a specimen as

$$\chi = \chi' - j\chi'', \tag{24.8}$$

where the imaginary part $j\chi''$ corresponds to the absorption of the energy of a variable electromagnetic field in a unit volume of the specimen [14]. In this case, the connection between microscopic and macroscopic properties can be determined, by calculating the mean value of the energy ω, which is absorbed per unit time, of the variable electromagnetic field $H_1 \cos \omega t$:

$$P_\omega = 2\omega\chi'' H_1^2 \cos \omega t. \tag{24.9}$$

This relation shows the simple interconnection between the absorbed power P, susceptibility χ'', and amplitude of the variable magnetic field H_1.

Near the resonance, we measure the strong resonance absorption of the energy from the oscillatory circuit of a high-frequency generator. In this case, we observe the signal of absorption of the energy, which is absorbed by a nucleus at the transition from one energy level to another one, $\Delta E = h\gamma H_0$.

A hydrocarbon fuel can be represented in the form of a diamagnetic chemical compound with added paramagnetic centers, which are organic radicals, dissolved oxygen, etc. Hydrocarbon molecules of a fuel contain no unpaired electrons, and their magnetic properties are caused by the diamagnetism of closed electron shells and the weak paramagnetism of atomic nuclei (in this case, nuclei of hydrogen) [15, 16].

When a diamagnetic substance enters a magnetic field, the energy of an electron shell is changed by ΔU, which leads to the appearance of a magnetic moment μ_z of the shell and to the magnetization of the substance:

$$M_z^d = -\frac{e}{2mC} N \sum_{i=1}^{z} \langle L_z \rangle - \frac{Ne^2}{6mC^2} H_0 \sum_{i=1}^{z} \langle r_i^2 \rangle. \tag{24.10}$$

Here, the first term is related to the orientation of the orbital moments of atoms without an external field and is equal to zero in our case. The second term is negative and determines the diamagnetic part of the magnetization

$$M_z^d = \chi_0^d H_0, \tag{24.11}$$

where χ_o^d is the diamagnetic susceptibility:

$$\chi_0^d = -\frac{Ne^2}{6mC^2}\sum_{i=1}^{z}\langle r_i^2\rangle, \tag{24.12}$$

$$\chi_0^d = -\left(10^{-5}\text{–}10^{-6}\right). \tag{24.13}$$

The magnetic moments of oxygen molecules, organic radicals, and other components of a fuel, which are characterized by the electron paramagnetism, are oriented under the action of an external magnetic field H_0, and their magnetization is proportional to its strength:

$$M_z^p = \chi_0^p H_0. \tag{24.14}$$

Since the moments are oriented in the direction of the field, we have $\chi_0^P > 0$.

For the considered effect, the magnetization of a substance that contains nuclear moments coupled through the superfine interaction with the magnetic moments of unpaired electrons is caused by the total magnetic moments of atoms and molecules, which are in the excited state. The interaction of nuclear spins of hydrogen, which received the energy from the variable electromagnetic field under EPI, with the electron spins of paramagnetic centers causes a change in the magnetization of a substance due to the electron-nucleus spin interaction. The interaction of the spins of electrons of the paramagnetic centers with the spins of nuclei of the basic chemical compound, into which these centers are placed, causes an increase in the diamagnetism of a substance. In other words, the collectivization of electrons of each of the atoms happens, by decreasing the paramagnetism of dissolved oxygen.

Of significant interest are the study of the electrophysical influence on the electrophysical and physico-chemical properties of hydrocarbon fluids such as the magnetic susceptibility, dielectric permeability, dielectric losses, content of resins, and acidity and the spectral and chromatographic studies [17, 18].

The experimental studies of the magnetic susceptibility of fuels on a radio-frequency installation, whose principle of action is based on the determination of a change in the resonance frequency of a high-frequency oscillatory circuit, indicate an increase in the diamagnetism of a fuel (Fig. 24.3).

In the process of measurement, the frequency stability was held to within 10^{-8}, which allowed us to find the diamagnetic susceptibility of fuels of order 10^{-6} with an accuracy of 0.01%.

It is seen from Fig. 24.3 that the maximum change in χ^d down to -0.7227×10^{-6} occurs in 60–70 min after EPI. Possibly, this occurs due to the relaxation processes, which characterize the energy exchange between the system of nuclear spins of hydrogen and the electrons of paramagnetic centers (dissolved O_2) with the formation of diamagnetic complexes.

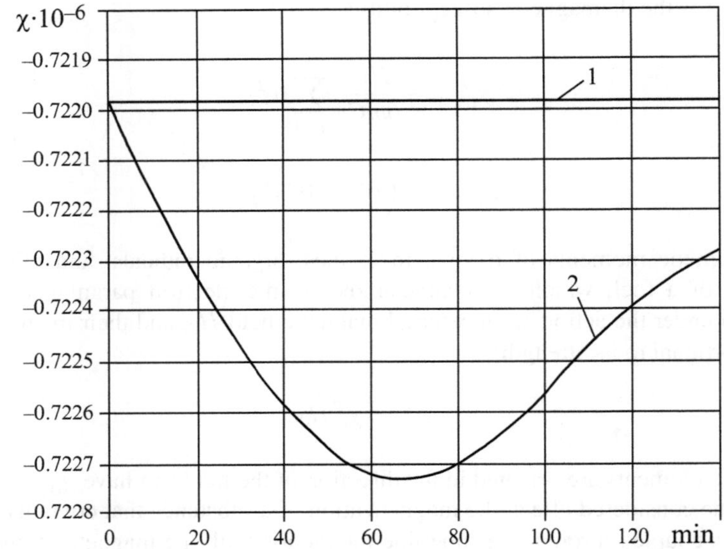

Fig. 24.3 Time variation of the magnetic susceptibility of a fuel: *1* – input fuel; *2* – fuel after EPI

If a fluid is present for a sufficiently long time in a constant magnetic field, it has a uniform magnetization of nuclei, which is directed in parallel to the field. A slow change in the direction of the magnetic field leads to a turn of the magnetization of nuclei together with the field. But if the turn of the magnetic field is realized for a time t much less than the period of precession of nuclei, then the magnetization of nuclei has no time to turn and deviates from the direction of the magnetic field. This effect was first measured by E.M. Purcell and R.V. Pound.

The deviation of the magnetization of nuclei from the direction of the magnetic field in a flowing fluid, which was preliminarily polarized in a strong magnetic field, happens, when the fluid flows inside the nutation coil. In this case, the nutation coil is located in inhomogeneous magnetic fields. Then the fluid passes a high-frequency influence coil, and we observe a change in the absorption signal amplitude on the output of the unit.

The influence of an electromagnetic field oscillating with a frequency equal to the precession frequency of nuclei in the given inhomogeneous magnetic field deflects the magnetization of nuclei from the field direction by some angle (the angle of nutation), which is determined by the formula

$$\theta = \gamma H_1 \Delta t, \tag{24.15}$$

where H_1 is the strength of a variable electromagnetic field, A/m.

The magnetization M of nuclei of a fluid, which enters a high-frequency influence coil, is proportional to the projection of the magnetization of nuclei of a fluid, which flows out of the nutation coil, and depends on the time of movement of the fluid between the coils

$$M = M_{z\,\text{output}} e^{-V_T/qT_1}. \quad (24.16)$$

The magnitude of M is proportional to the amplitude of an absorption (emission) signal in the high-frequency influence coil.

Under the resonance conditions, the energy of the electromagnetic field is absorbed in the nutation coil, which causes the transition of nuclear spins from the lower energy level onto an upper one. If the high-frequency influence coil (it is placed after the nutation coil) is passed by a fluid, in which the population of the upper energy level is higher than that of the lower, then the resulting absorption is negative, i.e., the system gives the energy more than receives. In this case, the absorption signal is changed in its amplitude and sign.

Thus, if we excite the oscillating field in the nutation coil, set the resonance frequency by the maximum effect of nutation, and select the amplitude of oscillations, then the negative absorption signal can appear in the high-frequency oscillatory circuit. This means that if the negatively polarized fluid enters the high-frequency influence coil, then the emission signal appears in it instead of the absorption signal. In other words, the spin system of nuclei emits the energy into the radio-frequency oscillatory circuit, and we register the emission signal, which indicates a decrease in the energy of the proton system and the fluid on the whole. Our task is the simplification of the means aimed at an increase in the octane number of benzines without a deterioration of their operation speed and quality. For its solution, we will apply the effect of double nuclear resonance.

The proposed method is realized in the following way. In order to create the conditions for the double nuclear resonance, a hydrocarbon fluid (benzine), which is used in the fuel systems of machines and mechanisms, is passed through a magnetic system with inhomogeneous constant magnetic field, in which two inductance coils are successively placed. They are connected with high-frequency generators that are tuned onto the precession frequencies of protons of hydrogen, which are determined by the strengths of the magnetic field at the places of location of the coils.

This is confirmed by the data of the executed experimental studies and the estimation of the detonation resistance of benzines with the help of a universal setup UTT-90 under the standard conditions of tests. It was established that the application of the method of double nuclear resonance allows one to increase the detonation resistance of benzines by 1.5–3.0 units [17, 19–22].

The studies carried out with a fuel indicate that the dielectric permeability of all specimens of fuels depends on the temperature and linearly decreases, as the temperature T increases. Under the heating, a fuel extends. As a result, a unit volume contains a less number of molecules, and the dielectric permeability of a fluid decreases. As for the EPI method, it has practically no effect on the dielectric properties of a fuel, which testifies to the invariance of the structure of a hydrocarbon medium in applications of the EPI method, which is also corroborated by spectral studies.

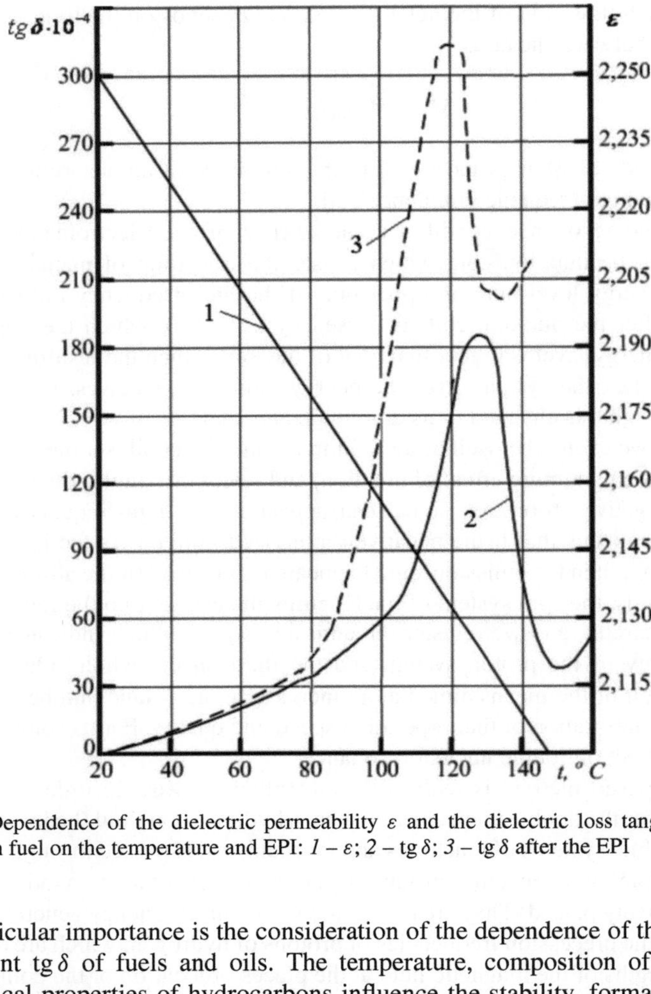

Fig. 24.4 Dependence of the dielectric permeability ε and the dielectric loss tangent tg δ of a hydrocarbon fuel on the temperature and EPI: *1* – ε; *2* – tg δ; *3* – tg δ after the EPI

Of particular importance is the consideration of the dependence of the dielectric loss tangent tg δ of fuels and oils. The temperature, composition of a mixture, and chemical properties of hydrocarbons influence the stability, formation of free radicals, and direction of the oxidative process on the whole. As is seen from Fig. 24.4, curve 2, the dependence of tg δ on the temperature of a fuel has extreme character, which is manifested in the interval of temperatures 373–423 K. It is clear that the process of oxidation of hydrocarbons is running differently for fuels with different compositions. It is also worth to note that, as the temperature increases higher than a certain value, a decrease in the rate of oxidation is observed, and, in a certain temperature interval for different fuels, the rate of change in the dielectric loss tangent has a negative value. Thus, as the temperature increases, the rate of formation of polar structures decreases in a certain temperature interval. As the temperature increases further, tg δ increases sharply, i.e., the mean rate of oxidation of a fuel takes a positive value, and the increase in tg δ for different hydrocarbon media is leveled. We may say that, as the temperature increases, the molecules receive the thermal energy and have possibility to orient themselves in the electric field, which leads to the appearance of the dipole-orientation polarization. As the molecules become free, the dipole-orientation polarization increases with the temperature. When all molecules can orient themselves in the electric field, the dipole-orientation polarization passes through the maximum. The further growth in the temperature causes a decrease in the dipole-orientation polarization, since the thermal motion hampers the molecules to orient themselves in the electric field. As a possible cause for such effect, we mention the extreme temperature dependence of the rate of the reaction of initiation of chains by the homogeneous mechanism.

After the EPI on a fuel (Fig. 24.4, curve *3*), the absorption of the electromagnetic energy occurs, and the favorable conditions for the transition of molecules in excited states are formed. In this case, the intermolecular associates and complexes favoring an increase in the rate of oxidation of hydrocarbons in the liquid phase volume are created. The dielectric losses of a fuel increase by several times, by remaining practically invariable at room temperature.

A sharp increase in the dielectric loss tangent with the temperature of liquid hydrocarbons can be explained, possibly, by the formation of ion-radical and charge-transfer complexes, which is accompanied by an increase in the relaxation losses. The latter is caused by an increase in the number of dipole molecules and weakly bound ions.

24.5 Conclusion

The established effect of a higher rate of oxidation of fuels after the EPI has a great practical meaning. In the engines, the rates of subsequent processes of oxidation of an evaporated fuel depend essentially on the "chemical prehistory" of a fuel. Moreover, the comparatively easily realized initiation of oxidation of drops of a fuel serves a method of regulation of the spontaneous ignition of a fuel in the engines.

References

1. Morozov VI, Usatenko ST, Umanets GN, A means of electrophysical treatment of hydrogen-containing fluids [in Russian], Auth. Certif. No. 693659, S 02 V 9/00, V 03 S 1/08, Applic. No. 2152729 on 4.07.1975, publ. 28.07.1979
2. Abragam A (1961) Principles of nuclear magnetism. Clarendon Press, Oxford
3. Jeffries CD (1963) Dynamic nuclear orientation. Interscience, New York
4. Kruchinin S, Nagao H, Aono S (2010) Modern aspect of superconductivity: theory of superconductivity. World Scientific, Hackensack, p 232
5. Kruchinin S, Dzhezherya Y, Annett J (2006) Imteractions of nanoscale ferromagnetic granules in a London superconductors. Supercond Sci Technol 19:381–384
6. Dzhezherya Y, Novak IY, Kruchinin S (2010) Orientational phase transitions of lattice of magnetic dots embedded in a London type superconductors. Supercond Sci Technol 23:1–5
7. Kruchinin S, Nagao H (2012) Nanoscale superconductivity. Inter J Mod Phys B 26:1230013

8. Slichter CP (1963) Principles of magnetic resonance. Harper and Row, New York
9. Andrew ER (1955) Nuclear magnetic resonance. Cambridge University Press, London
10. Vonsovskii SV (1973) Magnetism of microparticles [in Russian]. Nauka, Moscow
11. Dorfman YG (1955) Magnetic properties and structure of matter [in Russian]. Gostekhizdat, Moscow
12. Soldatov AV, Bogolyubov NN Jr, Kruchinin SP (2006) Method of intermediate problems in the theory of Gaussian quantum dots placed in a magnetic field. Condes Matter Phys 9(1):1–9
13. Kruchinin S (2016) Energy spectrum and wave function of electron in hybrid superconducting. Inter J Mod Phys B 30(13):1042008
14. Dorfman YG (1961) Diamagnetism and chemical bonding [in Russian]. Fizmatgiz, Moscow
15. Morozova IV, Morozov VI (2010) Improvement of performance properties of FL by means of an electrophysical influence. In: Proceedings of international science technology conference on "Problems of chemmotology" [in Russian]. NAU, Kiev, pp 158–163
16. Morozov VI, Morozova IV (2008) Influence of the electrophysical action on the economic and ecological parameters of the operation of vehicles. Neft' i Gaz 9:90–92
17. Morozov VI, Morozova IV (2009) Results of tests of a unit for the electrophysical action on diesel fuel. In: Proceedings of IX international science technology conference on "AVIA – 2009" [in Russian]. NAU, Kiev, vol III, pp 18.37–18.40
18. Tereshchenko YM, Kulyk MS, Mitrakhovych MM et al (2015) Theory of heat engines. Gas-dynamical calculation of elements of gas-turbine engines [in Ukrainian]. NAU, Kiev
19. Garg R, Agarwal AK (2013) Fuel energizer: the magnetizer (A concept of liquid engineering). IJIRD 2(4):25–35
20. Kunz R. Magnetic treatment of fuel. http://www.probonoscience.org/pennysolutions/recipes/automobile/buyer_beware/magnetic_fuel_treatment.htm
21. Jain S, Deshmukh S (2012) Experimental investigation of magnetic fuel conditioner in I.C. engine. IOSR J Eng (IOSRJEN) 2(7):27–31
22. Super Fuel Max FAQ. http://www.o-fuel.ru/faq.html

Chapter 25
Information Technology for Security System Based on Cross Platform Software

S. Kondratiuk, K. Kruchynin, Iu. Krak, and S. Kruchinin

Abstract The technology, which is implemented with cross platform tools, is proposed for modeling of gesture units of sign language, animation between states of gesture units with a combination of gestures (words). Implemented technology simulates sequence of gestures using virtual spatial hand model and performs recognition of dactyl items from camera input. With the cross platform means technology achieves the ability to run on multiple platforms without re-implementing for each platform.

Keywords Cross platform · Sing language · Dactyl modeling · Gesture recognition

25.1 Introduction

Communication via gestures is one of the three main means of transmission of information between people, among character (text) and voice (speech) communication. Sign language is usually used by people with hearing disabilities to communicate with each other and with their environment, increasing the number of people who need to know sing language. Note that sign language is universal in the sense that you can send text information via gestures and in case if certain words don't have corresponding gestures (names, cities, areas, etc.), they may be shown via letters one-by-one using dactyl alphabet.

Modern hardware is able to collect information fast and almost without restriction, process data both in cloud computing (model, which provides a universal, easy access on demand through the network to the virtual cluster computing resources)

S. Kondratiuk · K. Kruchynin · Iu. Krak (✉)
Department of Theoretical Cybernetics, Taras Shevchenko National University of Kyiv, Kyiv, Ukraine

S. Kruchinin
Bogolyubov Institute for Theoretical Physics, Kyiv, Ukraine

J. Bonča, S. Kruchinin (eds.), *Nanostructured Materials for the Detection of CBRN*, NATO Science for Peace and Security Series A: Chemistry and Biology,
https://doi.org/10.1007/978-94-024-1304-5_25

[1] and locally on the device, and through data channel processing results are returned to the user. All this is also true for sign language. Sings can be stored and reproduced via a variety of devices and platforms, stationary or mobile, high performance or energy efficient. The actual problem is the reproduction of sign language on all these platforms, for further usage by people with hearing disabilities in particular and everyone in general. Deployment of a single unified technology on various platforms (android, ios, windows, linux, web) without need to port it or to implement it under each platform is a major problem.

One way of solving the stated problem of visualization and reproduction of sign language is cross platform software development. Unlike single-platform technologies that operate only on a specific platform under which they were developed, "cross platform software provides the ability to perform on more than one platform with identical (or nearly identical) functionality" [2]. The term "platform" in this context may refer to one of or a combination of several definitions: (1) the type of operating system (such as Microsoft Windows, Mac OS X, Linux, Solaris, Android, iOS); (2) processor type (such as x86, PowerPC, ARM); (3) the type of hardware (e.g., mainframe, workstation, personal computer, mobile device) [2]. Cross-platform technologies are on a par with the platform independent technologies (those that can operate on any platform, such as Web application) [2] and cross-platform virtual machines (technologies that support individual processes or systems, depending on the level of abstraction at which is virtualization) [3].

In this article the proposed solution of the problem is via cross platform development, taking into account characteristics of different classes of devices (such as hardware, CPU power, amount of memory, presence on the Internet) and setting the number of polygons of the three-dimensional hand model and gesture animation step. Gesture modeling and gesture recognition is performed via cross platform means as a part of proposed communication technology.

25.2 Software and Methods

Sign modeling is a problem that is considered both independently and as part of the problem of modeling and recognition of gestures and thus as a technology learning and evaluating sign language. One of the systems to display the sign language is American Sing Language Online Dictionary [4], which consists of a video database of words and phrases displayed via sign language. Control via gestures is an actual problem in the development of platform independent human-computer interaction [5]. These developments were involved in a number of commercial agencies [6, 7], but the systems they propose are configured to pre-determined number of gestures, and therefore do not solve the problem of modeling sign language. Also all of them lack functionality of gestures recognitions, thus not allowing to evaluate quality of sing language performed by a user.

Creating a model hand is the first step in the task of sing language modeling. In their work [8], authors analyze existing approaches of hand modeling, which are divided into two main groups: spatial and temporal. Former consider the characteristics of different positions for the hand gestures, while latter refer to the description of the dynamics of gestures. Modeling hands in the spatial area can be completed in two and three dimensions.

In [9] proposed system by authors is able to simulate sign animations for a given text. As a part of this system a statistical model is used to analyze input text and generative algorithm is used when creating the appropriate simulated kinematics of sign animations. Within the article, the authors have provided ANVIL tools for input text annotation, gesture generator NOVA, and DANCE library developed in [10] is used for gesture animation. The system is built on the Microsoft Windows platform and x86 processor. In [11] authors discuss the modeling of virtual character for spatial reproduction of sign language on the platform of Microsoft Windows. In [12], authors developed a system of signed language training, which consists of two modules – gesture demonstration module via video and gesture recognition module (required gloves), based on Hidden Markov Model. The training system is based on Microsoft Windows platform and x86 processor. Gesture recognition for mobile platforms is developed in [13], but gesture modeling on mobile devices is not performed.

25.3 Formulation of Problem

Zeolite The proposed technology should perform modeling of sign units (morphemes [14]) of sign language, and reproduce animation of gestures structures (words, sentences) via state transitions between shown units using spatial virtual model hand. The proposed technology should perform recognition of sign language based on camera input from the device in order to evaluate sing language performed by user. The technology should be a combined solution for learning sing language via gesture modeling and recognition.

Technology should solve the problem of running on existing platforms using cross platform development without implementing the functionality for each platform separately. The effectiveness of the proposed approach is shown in building cross platform technology for modeling and recognition of Ukrainian dactyl alphabet.

25.4 Proposed Approach

To address the modeling of sign language and perform animation of sign structures using spatial virtual model hand the cross platform technology based on cross platform framework Unity3D [15] is proposed. Cross platform framework Unity3D is also used for the user interface, both libraries and technology are implemented with programming language C#. Proposed tools can solve the problem of running the technology on multiple existing platforms. The novelty of the proposed technology is that it is cross platform and has customizable level of polygons for three dimensional hand model and animation step for gesture transitions. This allows to run proposed technology without changes on multiple platforms (different types of processors, operating systems and hardware).

Advantage of cross platform technology over technologies developed for a single platform is that there is no need to modify or re-implement the functionality already available for other platforms (porting) [2], which speeds up the process of developing and deploying technologies, and increases the number of potential users. The advantage of cross platform technology over cross platform virtual machine emulators is performance speed and absence of necessity to install additional software (software dependencies).

25.5 Infologic Model

The core of the technology is composition of three cross-platform modules as shown in Fig. 25.1: three dimensional hand model (which is implemented with cross platform framework Unity3D), user interface (implemented also with cross platform framework Unity3D) and gesture recognition module (implemented with cross platform framework Tensorflow [16]). Core functionality is implemented with C# and Python and runs on desktop OS (MacOS, Linux, Windows) and on mobile OS (Android, iOS).

Hand model module is cross platform and provides hand model representation for gesture recognition module. Hand renderer receives hand model representation and gesture specifications from gesture storage module, and provides a high-polygon rendered hand model. Gesture learning module and gesture modification

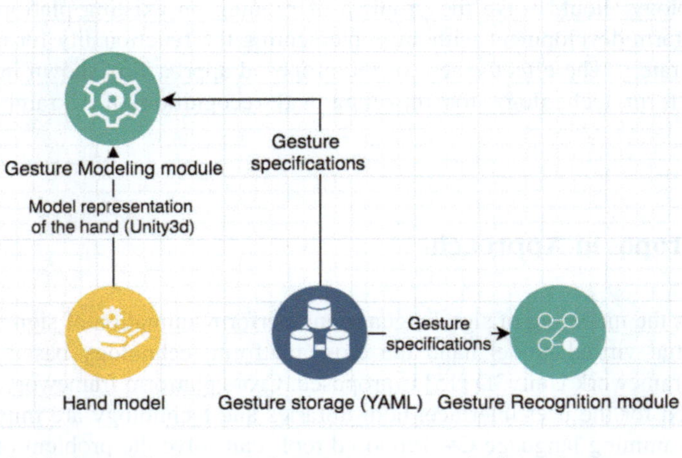

Fig. 25.1 Infologic model of cross platform gesture communication technology

module are implemented with cross platform Unity3D, both taking as input results of hand model renderer. Gesture modification module provides updated gesture specifications and transmits them to gesture storage. Gesture recognition module is proposed to be implemented with Tensorflow framework and receives as input hand model, gesture specifications and input from camera.

25.6 Gesture Modeling

The hand model which is built in gesture modeling module has 27 bones, 8 of the bones are in wrist, 3 are in the thumb (one metacarpal and 2 phalanx) and 4 metacarpus and 12 phalanges are in other fingers [17]. Each bone is connected to the other through different types of joints. Designing your own cross platform engine for simulating the hand is non-trivial task, thus as the core technology for modeling three-dimensional hand model and gesture animations between morphemes cross platform framework Unity3D was selected. Unity3D framework is able to effectively reproduce a realistic hand model which consists of more than 70,000 polygons as shown in Fig. 25.2. Based on the anatomy of the hand within Unity3D hand model was developed with 25° of mobility, four of them located in the metacarpal-carpal joint, to the little finger and thumb to provide movement palm. The thumb has 5° of mobility, middle and index fingers have 4° of mobility (metatarsophalangeal joint with 2° of mobility, and the distal and proximal interphalangeal joints each have one.

To preserve the gesture YAML format was selected [18], in which the gesture is stored as follows:

%YAML:1.0 rotation: [–9.5845758914947510e-02,
8.3791027449819921e-09, –9.9539619684219360e-01,
–7.5690525770187378e-01, –6.4944797754287720e-01,
7.2881683707237244e-02, –6.4645802974700928e-01,
7.6040601730346680e-01, 6.2246840447187424e-02]
hand_joints: finger1joint1: [0., 0., 0.] finger1joint2: [0., 0., 0.]?...

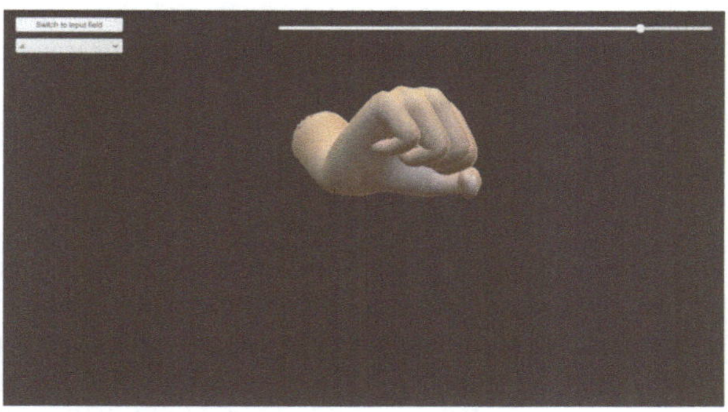

Fig. 25.2 Gesture modeling under iOS platform

25.7 Gesture Recognition

Gesture learning and gesture recognition modules, developed with cross platform tools (frameworks based on Python, C++) can be embedded into information and gesture communication cross platform technology. Multiple approaches were considered as an approach for gesture recognition. Automatic sign language recognition can be approached similarly to speech recognition, with signs being processed similar to phones or words. Most previous work has used approaches based on hidden Markov models (HMMs) [19, 20]. Conventionally, sign language recognition consists of taking an input of video sequences, extracting motion features that reflect sign language linguistic terms, and then using pattern mining techniques or machine learning approaches on the training data. For example, Ong et al. propose a novel method called Sequential Pattern Mining (SPM) that utilizes tree structures to classify signs [21]).

Convolutional Neural Networks (CNNs) have shown robust results in image classification and recognition problems, and have been successfully implemented for gesture recognition in recent years. In particular, deep CNNs have been used in researches done in the field of sign language recognition, with input-recognition that utilizes not only pixels of the images. With the use of depth sense cameras, the process is made much easier via developing characteristic depth and motion profiles for each sign language gesture [22]. Multiple existing researches done over various sing languages show that CNNs achieve state-of-the-art accuracy for gesture recognition [23–27].

Convolutional neural networks have such advantages: no need in hand crafted features of gestures on images; predictive model is able to generalize on users and surrounding not occurring during training; robustness to different scales, lightning conditions and occlusions. Although, selected approach has couple of disadvantages, which may be overcome with a relatively big dataset (1,000 images for each gesture, among more than 10 people of different age, sex, nationality and images taken under different environment conditions and scales): need to collect a rather big and labeled gesture images dataset; black-box approach which is harder to interpret.

Usage of cross platform neural network framework such as Tensorflow allows to implement gesture recognition as a cross platform module of proposed technology and serve trained recognition model on server or transfer it to the device.

25.8 Application of Cross Platform Tools

Due to selected cross platform implementation tools, the proposed technology solves the problem of executing on multiple platforms without the implementation under each platform separately. Software offered and used in the implementation of information technology is cross-platform and operates unchanged regardless of operating system (Windows, Linux, Android, iOS), CPU type (x86, arm), and the type of hardware (mobile or stationary device).

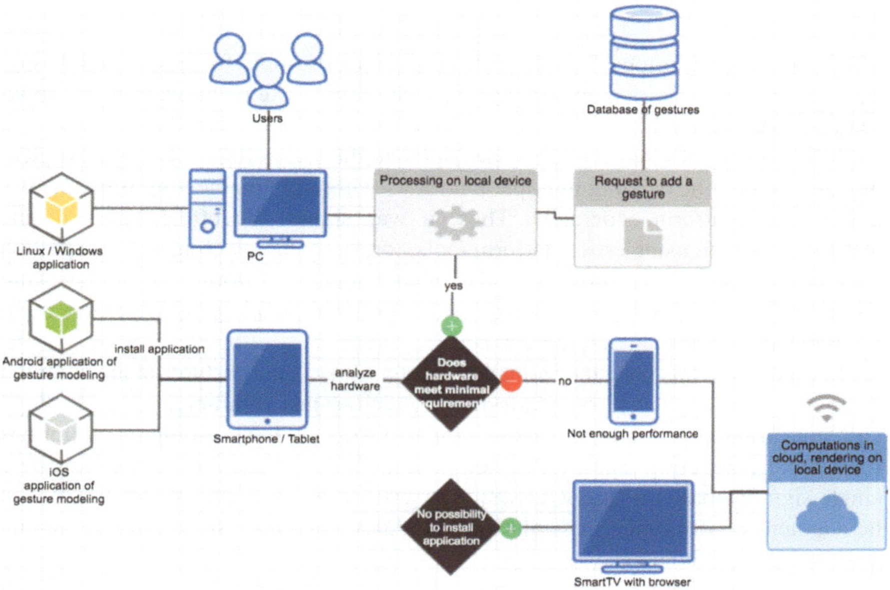

Fig. 25.3 General scheme of cross-platform and adaptive execution of information technology

With its cross platform build system Unity3D it is possible to create applications for each platform without porting or changing the original code.

As there are no specific hardware requirements for information technology for modeling sign language, there are objective obstacles for performance speed of older generations devices. To overcome this problem, the following adaptive approach to information technology was proposed as shown in Fig. 25.3.

Further modules implementation will leverage from existing cross-platform technology. Gesture learning and gesture recognition modules, developed with cross platform technologies (Python, Tensorflow) will be embedded into information and gesture communication cross-platform technology. In case of the mobile app (iOS, Android) or application on the device with a stationary operating system (Windows, Linux), during installation on the device, information technology analyzes the existing hardware and, depending on its capacity, conducts a series of adjustments: (1) number of polygons of the hand model changes to priority for performance speed; (2) during rotation hand model changes pitch angle at which it rotates, with priority for speed. If the available hardware does not meet the minimum requirements of information technology, the user is given the recommendation to choose "online" mode, in which the calculation is not performed on hardware.

25.9 Conclusion

The proposed technology is built with cross platform tools for gesture modeling, gesture transitions animation and gesture recognition. The technology uses virtual spatial model of hand.

With the help of cross platform development, the technology solves the problem of execution on the existing multiple platforms without implementing functionality under each platform separately. Thus, it was shown the effectiveness of the technologies built using cross platform tools, for example modeling and recognition elements of dactyl Ukrainian alphabet sign language. Information and gesture communication technology was developed with further scaling capabilities in mind for gestures of other languages alphabets.

To implement this idea, the validation mechanism of new gestures to the common database can be applied. Cross platform information and communication technology and standardized protocol and data format (YAML) allows a range of solutions for remote computing using cloud computing, Web servers, local servers using a single sign database PostgreSQL [28]. The gesture communication technology can be augmented with other cross platform modules, such as gesture recognition and gesture learning modules.

References

1. Mell P, Grance T (2011) The NIST Definition of Cloud Computing (Technical report), National Institute of Standards and Technology: U.S. Department of Commerce. https://doi.org/10.6028/NIST.SP.800-145. Special publication 800-145
2. The Linux Information Project, Cross-platform Definition
3. Smith J, Nair R (2005) The architecture of virtual machines. Comput IEEE Comput Soc 38(5):32–38
4. ASL Sing language dictionary. http://www.signasl.org/sign/model
5. Graschenko LA, Fisun i dr AP (2004) Teoreticheskie i prakticheskie osnovyi cheloveko-kompyuternogo vzaimodeystviya: bazovyie ponyatiya cheloveko-kompyuternyih sistem v informatike i informatsionnoy bezopasnosti: Monografiya, red. A.P. Fisun. OGU, Orel, p 169
6. Samsung TV Gesture book. http://www.samsung.com/ph/smarttv/common/guide_book_3p_si/waving.html
7. Apple Touchless Gesture System for iDevices. http://www.patentlyapple.com/patently-apple/2014/12/apple-invents-a-highly-advanced-air-gesturing-system-for-future-idevices-and-beyond.html
8. Khan RZ, Ibraheem NA, Meghanathan N et al (eds) (2012) Comparative study of hand gesture recognition system. SIPM, FCST, ITCA, WSE, ACSIT, CS & IT 06:203–213
9. Neff M, Kipp M, Albrecht I, Seidel H-P (2008) Gesture modeling and animation based on a probabilistic re-creation of speaker style. ACM Trans Graph 27(1):24 pages. Article 5. https://doi.org/10.1145/1330511.1330516
10. Shapiro A, Chu D, Allen B, Faloutsos P (2005) Dynamic controller toolkit. http://www.arishapiro.com/Sandbox07_DynamicToolkit.pdf
11. Kruvonos IG, Krak YV, Barchukova YV, Trocenko BA (2011) Human hand motion parametrization for dactilemes modeling. J Automat Inform Sci 43(12):1–11

12. Aran O (2006) Sign language tutorial tool. In: eNTERFACE'06, July 17th–August 11th 2006 Dubrovnik, Croatia. Final Project Report
13. Raheja JL (2015) Android based portable hand sign recognition system, Mar 2015. https://doi.org/10.15579/gcsr.vol3.ch1. Source: arXiv
14. Stokoe WC (2005) Sign language structure: an outline of the visual communication systems of the American deaf. J Deaf Stud Deaf Educ 10(1):61–67
15. Unity3D framework. http://www.unity3d.com
16. Tensor flow framework documentation. http://www.tensorflow.org/api/
17. Tubiana R, Thomine J, Mackin E (1996) Examination of the handed wrist, 2nd edn. Martin Dunitz, London. ISBN:1853175447/1-85317-544-7
18. YAML – The Official YAML Web Site. http://www.yaml.org
19. Koller O, Forster J, Ney H (2015) Continuous sign language recognition: towards large vocabulary statistical recognition systems handling multiple signers. Comput Vis Image Underst 141:108–125
20. Dreuw P, Rybach D, Deselaers T, Zahedi M, Ney H (2007) Speech recognition techniques for a sign language recognition system. Interspeech, Antwerp, pp 2513–2516
21. Ong E-J et al (2012) Sign language recognition using sequential pattern trees. In: IEEE conference on computer vision and pattern recognition (CVPR) 2012. IEEE, pp 2200–2207
22. Agarwal A, Manish TK (2013) Sign language recognition using Microsoft Kinect. In: IEEE international conference on contemporary computing, 8–10 Aug 2013
23. Dieleman S, Kindermans P-J, Schrauwen B (2015) Sign language recognition using convolutional neural networks. In: ECCV 2014 workshops, part I, LNCS, vol 8925, pp 572–578
24. Garcia B (2015) Real-time American sign language recognition with convolutional neural networks. Stanford University, Stanford
25. Bobic V (2016) Hand gesture recognition using neural network based techniques. School of Electrical Engineering, University of Belgrade
26. Sugahara M, Kruchinin SP (2001) Controlled not gate based on a two-layer system of the fractional quantum hall effect. Mod Phys Lett B 15:473–477
27. Kruchinin S, Nagao H, Aono S (2010) Modern aspect of superconductivity: theory of superconductivity. World Scientific, p 232. ISBN: 9814261602
28. PostgreSQL official web site. http://www.postgresql.org

Chapter 26
Active Remote Sensing of Chemical and Biological Agents: Detection, Remote Sensing, and Risk Analysis

V. Yatsenko

Abstract This paper focuses on a new methodology and an experimental optical device for detection and identification of biological and chemical agents including various toxins and viruses. The proposed approach is based on pattern recognition optimization techniques, a new optical device, an information technology and most informative parameters. Another objective of this paper is to develop and evaluate the performance of strategies for cooperative control of autonomous air vehicles that seek to gather information about biological and chemical agents.

Keywords Detection · Chemical · Biological · Agents · Risk analysis · Unmanned vehicle system · Quantum computing · Quantum optimization

26.1 Introduction

The ecological interest is in response to a proliferation of agents development and threats for human health. Additionally, there is a pervasive interest across diverse application areas such as medicine, environmental protection, and vegetation processing [1, 2] to achieve a rapid detection and identification capability of various agents. In many cases, a standoff capability is desired. Technologies such as optical spectroscopy measurements, laser induced fluorescence, pattern recognition and optimization methods will be used in our device [1, 3]. This device is intended to be used for ground measurements or can be installed on small airplanes for remote sensing.

V. Yatsenko (✉)
Space Research Institute of NASU-NSAU, Kyiv, Ukraine

J. Bonča, S. Kruchinin (eds.), *Nanostructured Materials for the Detection of CBRN*, NATO Science for Peace and Security Series A: Chemistry and Biology,
https://doi.org/10.1007/978-94-024-1304-5_26

26.2 Problems

Accurate estimation of chemical and biological agents by remote sensing is a challenging problem. Such estimation is especially needed in an ecologically dangerous environment. Our goal is to develop new methods that allow estimating danger agents concentration using remote sensing data for multiple kinds of soil and vegetation. We also concentrates on the following problems: theoretical investigations: methods, models, and algorithms for remote sensing of chemical and biological agents, development of models for on-board detection and identification of biological and chemical agents, an analysis of laser emission impact on absorption and reflectance spectra, vegetation indexes estimation, modeling with dynamics crop models, data assimilation and precision agriculture, optimization problem with constraints on risk, and robust recursive Bayesian estimation and quantum minimax strategies. The main objective of this paper is to develop and estimate the performance of strategies for cooperative control of autonomous air vehicles.

26.3 Methods

A variety of research methods and approaches have been used to investigate the main problem:

- synergetic approach;
- dynamical-information methodology;
- non-linear dynamic classification;
- SVM-classification and SVM-regression;
- Independent Component Analysis (ICA);
- modified PCA;
- K-Means Clustering with SAM Distance;
- global optimization methods;
- reflectance properties of vegetation are dominated by biochemical components in the visible-near-infrared region covered by a sensor instrument;
- dynamical-information approach to a proliferation of agents;
- optical devices for detection and identification of biological and chemical agents.

26.4 Airborne Hyperspectral System and Software for Solving Remote Sensing Problems

This system based on new conception of the optical device for detection of biological and chemical agents. The system includes the following subsystems and models: intelligent subsystem for robust detection of biological and chemical

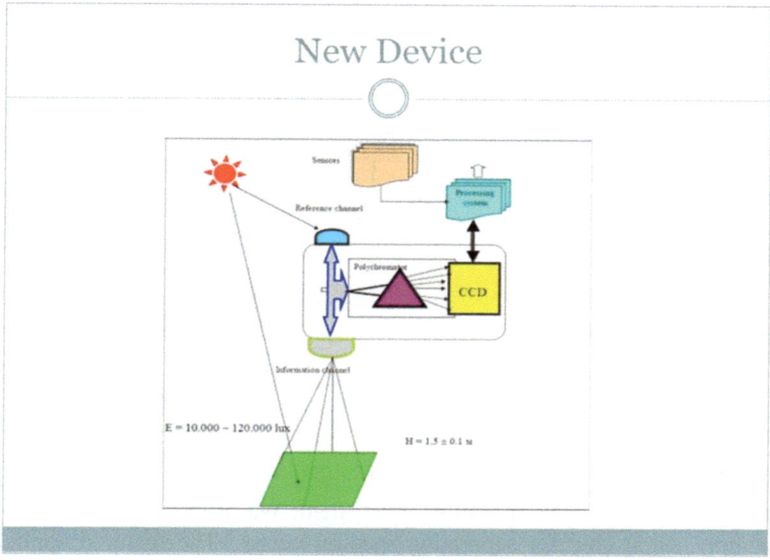

Fig. 26.1 Optical device for detection of biological and chemical agents

agents, subsystem for classification of hyperspectral remote sensing image data, neural chip, optical devices for detection and identification of biological and chemical agents, and cell models of danger object (Fig. 26.1).

Technical characteristics of the optical device:

- type for hyperspektrometer – grating;
- spectral range – 0.45–2.5 microns;
- spectral sampling interval – 1–2 nm;
- sensor (detectors) – CMOS or CCD camera;
- data encoding – 10–14 bits;
- sources of radiation – laser diodes.

26.5 Separating Spectral Curves by a Biological or Quantum Neural Chip

Over the past decade, deep artificial neural networks (DNNs) have become the state-of-the-art algorithms in Machine Learning (ML), speech recognition, computer vision, natural language processing and many other tasks. This was made possible by the advancement in Big Data, Deep Learning (DL) and drastically increased chip processing abilities, especially general-purpose graphical processing units (GPGPUs). All this has created a growing interest in making the most of the

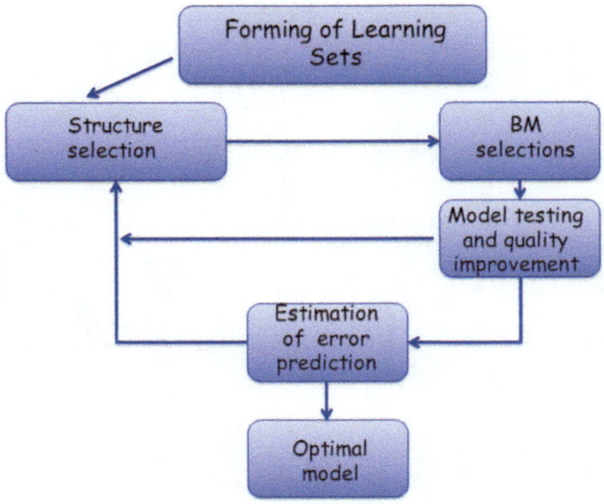

Fig. 26.2 Software

potential offered by DNNs in almost every field. We propose to use a biological or quantum neural chip for separating spectral curves. The featured applications are: detection of biological and chemical agents, virtual screening (VS), and other data mining (Fig. 26.2). The Deep Artificial Neuron–Astrocyte Networks (DANAN) could overcome the difficulties in architecture design, learning process and scalability of the current ML methods. We also propose to use Hopfield Networks, Probabilistic Neural Network, and Boltzmann Machine [3].

26.6 Risk Analysis

Let $z = f(\mathbf{v}, \mathbf{u})$ be a loss function of a bio-object depending upon the control vector $\mathbf{v}$ and a random vector $\mathbf{u}$ (Fig. 26.3). We assume that the random vector $\mathbf{u}$ has a probability density $p(\mathbf{u})$. We can define a function

$$\Phi_\beta(\mathbf{v}, \beta) = (\alpha - \beta)^{-1} \int\limits_{f(\mathbf{v},\mathbf{u})>\alpha} (f(\mathbf{v}, \mathbf{u}) - \alpha) p(\mathbf{u}) d\mathbf{u}.$$

In this case we have the following optimization problem:

$$\min \mu(v)$$
$$v \in V, \ \Phi_\beta(x) \leq C_\beta, \ \Phi_\gamma(x) \leq C_\gamma.$$

The solution to this optimization problem can be found using the Matlab Optimization Toolbox.

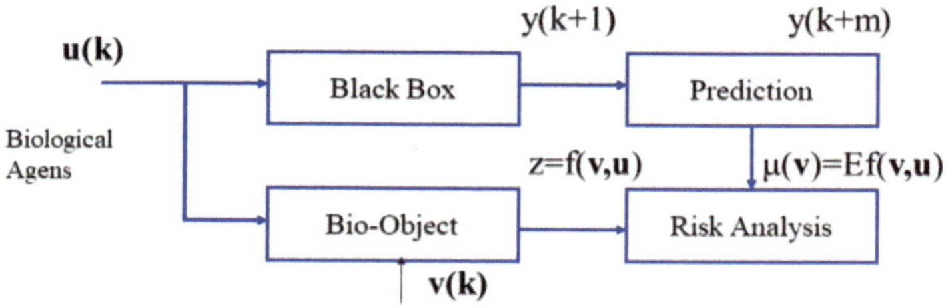

Fig. 26.3 Risk analysis

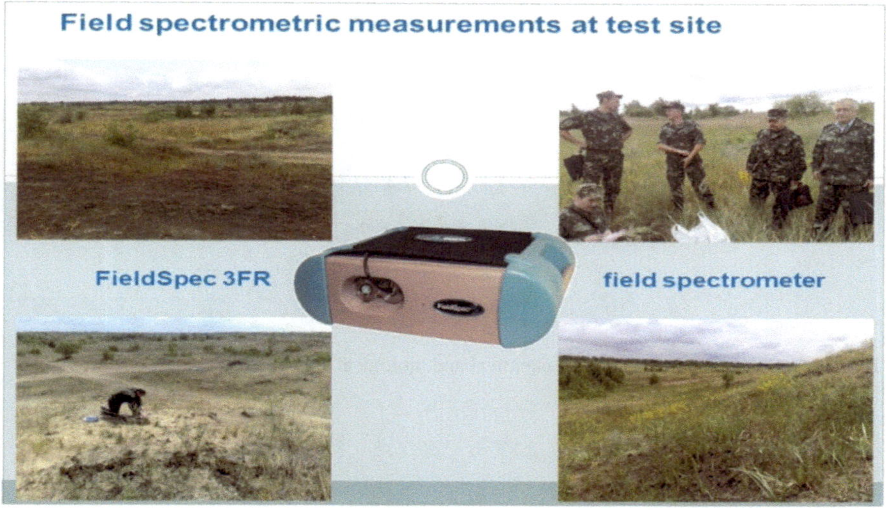

Fig. 26.4 Field spectrometric analysis

26.7 Experimental Results

Experimental results were obtained at the military test site (Fig. 26.4). Figures 26.5, 26.6, 26.7 and 26.8 show results of spectral measurements.

26.8 Conclusions

The system analysis has been conducted to creation of new methods for detection of the biological and chemical agents concentration across several different sensor stages. New modelling approaches to remote sensing of biological and chemical agents using component analysis and SVM regression are proposed. They allow estimating biochemical components based on the information about the type of plant, soil, and projective covering. It is shown that the quantitative system analysis

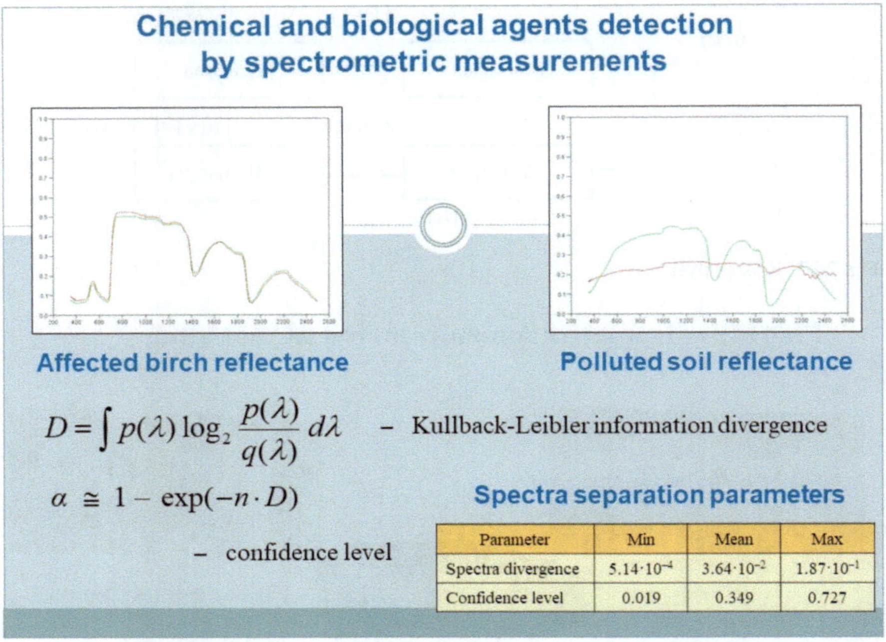

Fig. 26.5 Diagrams of detection of chemical and biological agents

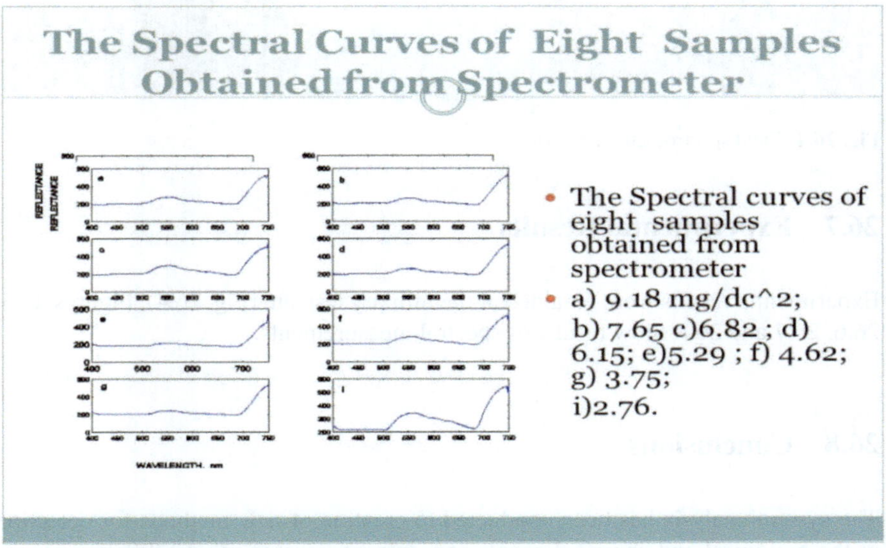

Fig. 26.6 Spectral function obtained from spectrometer

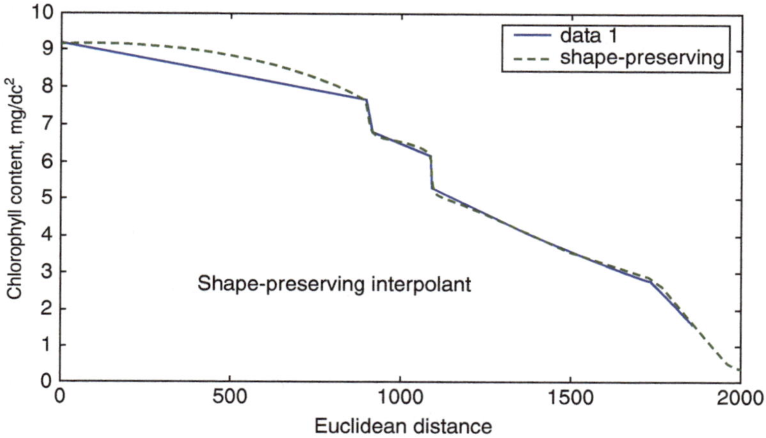

Fig. 26.7 Function of chlorophyll component

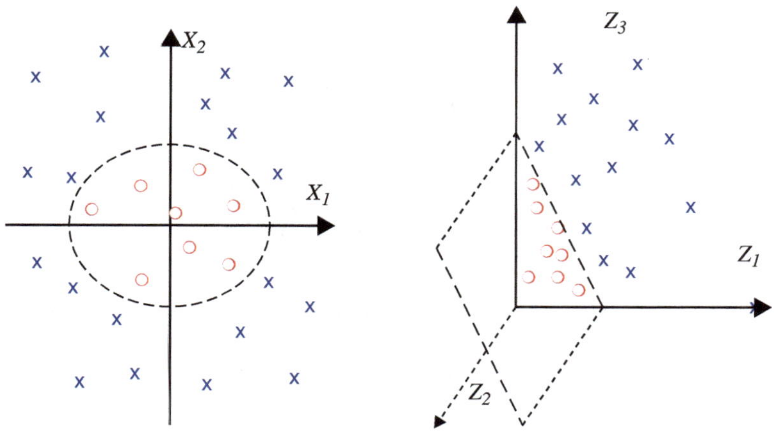

Fig. 26.8 Vector of principal components

of spectral reflectance curves is a useful for developing new adaptive algorithms for extraction of most information characteristics. The use of remotely sensed imagery is becoming a valuable tool in many agricultural and other plant applications.

The conception of the hypespectrometer has been proposed. It is based on the novel principles of information processing. The hypespectrometer is elaborated on the basis of: (1) optical and information technologies; (2) most informative characteristics of spectral curves.

It has been shown that the principal components of the spectral curve contain most essential Shannon information. These components can be used as independent variables of nonlinear regression models. It is demonstrated that there is a nonlinear dependence between the vectors of principal components and biochemical components concentration.

Advances in distributed computing and wireless communications have enabled the design of distributed agent systems. One of the key issues for a successful and wide deployment of such systems is the design of cooperative decision making and control strategies. New algorithms of rapid search and earlier detection of biological/chemical agents using an unmanned aerial vehicle (UAV) have been proposed. New algorithms and software for early warning of agents have been developed.

An idea about separating spectral curves by a biological or quantum neural chip has been proposed [3–7]. We also planning to use deep artificial neural networks (DNN) for signal processing. DNNs are used every day unknowingly, since in our smartphones there are numerous applications based on Deep Learning. For example, some cameras use a DNN to perform face recognition, while others employ a voice recognition piece of software, which is also based on DL. There are many other applications with DNNs that perform state-of-the-art results.

References

1. Yatsenko V, Cifarelli C, Boyko N, Pardalos PM (2008) Advanced in modeling agricultural systems. Springer, New York, pp 421–431
2. Ermakov V, Kruchinin S, Fujiwara A (2008) Electronic nanosensors based on nanotransistor with bistability behaviour. In: Bonca J, Kruchinin S (eds) Proceedings of the NATO ARW "Electron transport in nanosystems". Springer, New York, p 341–349
3. Yatsenko V (1996) Determining the characteristics of water pollutants by neural sensors and pattern recognition methods. J Chromatogr A 722(1–2):233–243
4. Yatsenko V et al (2010) Space weather influence on power systems: prediction, risk analysis, and modeling. Energy Syst 1(2):197–207
5. Pardalos P, Yatsenko V, Butenko S (2002) Robust recursive estimation and quantum minimax strategies. In: Pardalos P, Murphey R (eds) Cooperative control and optimization. Kluwer Academic, Dordrecht/Boston/London, pp 213–230
6. Rodionov VE, Shnidko IN, Zolotovsky A, Kruchinin SP (2013) Electroluminescence of Y_2O_3:Eu and Y_2O_3:Sm films. Mater Sci 31:232–239
7. Vlaskina S, Kruchinin S, Kuznetsova E, Rodionov V et al (2016) Nanostructures in silicon carbide crystals and films. Inter J Mod Phys B 30(13):1042015

Author Index

J. Bonča, S. Kruchinin (eds.), *Nanostructured Materials for the Detection of CBRN*,
NATO Science for Peace and Security Series A: Chemistry and Biology,
https://doi.org/10.1007/978-94-024-1304-5

Subject Index

J. Bonča, S. Kruchinin (eds.), *Nanostructured Materials for the Detection of CBRN*,
NATO Science for Peace and Security Series A: Chemistry and Biology,
https://doi.org/10.1007/978-94-024-1304-5

MIX
Papier aus verantwortungsvollen Quellen
Paper from responsible sources
FSC® C105338

If you have any concerns about our products,
you can contact us on
ProductSafety@springernature.com

In case Publisher is established outside the EU,
the EU authorized representative is:
Springer Nature Customer Service Center GmbH
Europaplatz 3, 69115 Heidelberg, Germany

Printed by Libri Plureos GmbH
in Hamburg, Germany